# Aviation

# The Complete Story of Man's Conquest of the Air

Consultant Editor Bill Gunston

# Contents

First published in 1978 by
Octopus Books Limited
59 Grosvenor Street London W1



Hardback ISBN 0 7064 0879 9
Paperback ISBN 0 7064 0899 3

Printed in England by Jarrold & Sons Limited

# From Myth to Reality

# 1 Early Ideas

Man's desire to fly is one of his oldest and strongest aspirations. Poetry and the visual arts have always been full of aeronautical imagery expressing hope in the possibility of rising to the heavens and attaining a state of bliss. While until recent times technology could not provide a means of realising this aim, historical annals include many accounts of fruitless attempts by experimenters whose longing to fly overcame the dictates of caution and common sense.

Although it is among the most deeply rooted of human urges, the desire to move freely in the three dimensions of the air has always been associated with awe and fear. Even today, when sophisticated modern aircraft provide one of the safest forms of transport, some people are reluctant to move away from their natural element. To men of past ages there appeared to be two related dangers associated with flying, one physical, the other spiritual. Since to rise above our earthly station might be looked upon as evidence of a desire to emulate the gods, attempts to fly might incur divine wrath – an idea which finds powerful expression in the legend of Daedalus and Icarus.

In order to escape from imprisonment on the island of Crete, Daedalus, who was credited with many ingenious inventions, made wings for himself and his son Icarus. The feathers from which the wings were mainly constructed were held on to their arms by wax. Before embarking on the flight Daedalus gave careful thought to the matter of navigation. In early times many thinkers believed that in the moist regions close to the sea the atmosphere was very dense, sharing some of the buoyant qualities of water and making it potentially possible for a man to fly with a motion analogous to that of a swimmer.

Accordingly, Daedalus advised his son that they should take advantage of these conditions by flying low over the water. After a successful take-off, Icarus grew so intoxicated with the delights of flying that he ignored his father's advice, rising higher and higher until he was close to the Sun. There the additional warmth melted the wax, causing the wings to collapse and Icarus to fall into the sea, where he was drowned. The gods, and in particular the Sun-god Apollo, had duly punished the young man for his lack of humility and excessive daring. Although Daedalus himself had also exceeded the normal limits laid down for human activities, he had stayed sufficiently close to the Earth to be spared.

**Previous two pages** William Henson's 'Aerial Steam Carriage', patented in 1842–3, became one of the most widely used subjects for aeronautical illustrations in the 19th century. Although a full-scale version was never built, several generations of designers drew inspiration from the features Henson proposed to include in it. It was to have a wing span of 45.75 m (150 ft) and two six-bladed pusher propellers driven by a 25 hp engine.

**Below** During the 1870s Alphonse Pénaud built a rubber-powered model which was one of the most significant ancestors of the modern aeroplane. Called the *Planophore*, it had main planes with wing tips set at a dihedral angle, a pusher propeller, and a vertical rudder (not shown in this illustration).

## Fixed-wing Models

While the legend of Daedalus was often invoked as a cautionary tale to discourage would-be bird-men, it must have seemed less of a deterrent to a small number of inventors who wisely confined their experiments to the use of models. The first European attempt at artificial flight of which we have any knowledge was in fact made with a model. Archytas of Tarentum, a Greek mathematician and friend of Archimedes – the discoverer of the displacement principle which keeps *aerostats* (lighter-than-air craft) aloft – is reputed to have built a wooden dove which was capable of free flight. How this was achieved is by no means clear, the earliest accounts of the dove being rather obscure: 'Archytas made a wooden model of a dove with such mechanical ingenuity and art that it flew; so nicely balanced was it, you see, with weights and moved by a current of air enclosed and hidden within it.' The reference to a current of air has led some historians to believe that the dove may have been operated either by some system of compressed air or by rocket power. While the use of compressed air would not have been beyond the skill of the Greeks, who delighted in small, ingenious contrivances of this sort, no such model could have been made to fly for more than a few seconds. As gunpowder rockets did not reach Europe until well over 1,000 years later, any form of rocket propulsion seems even less likely.

As accounts such as those of Archytas's dove have usually been recorded by historians with little or no mechanical sense they are notoriously unreliable and often exaggerated. It may well be that the story, if it contains any element of truth, concerns nothing more than a rigid glider. With hindsight one might indeed suppose that the easiest way to gain knowledge of the principles of aerodynamics would be to build fixed-wing gliders which could be made to imitate the readily observable flight of soaring birds. But we must be cautious about using

According to Greek legend Daedalus and his son Icarus attempted to fly from captivity in Crete, using wings made of feathers held on by wax. When Icarus flew too near the sun, the heat melted the wax, causing him to crash into the sea, where he was drowned.

present-day knowledge to interpret earlier attitudes and assumptions. For most men of the remote past, to fly meant to fly as a bird usually does, not by soaring but by flapping. Doubtless owing to the dominance of this idea, very few gliders or other fixed-wing models appear to have been built. In comparison with the 20 or more stories of experiments with full-sized flapping wings, the period up to the end of the Renaissance contains only two or three accounts of fixed-wing models. One of these concerns an ingenious rocket-bird built in the 15th century by Giovanni da Fontana, while another tells of wooden sparrows (probably hand-launched gliders) built early in the 16th century to amuse Emperor Charles V of Spain after his retirement from the throne.

R. O. Cambridge's satirical poem *The Scribleriad* (1751) contains a vivid account of this aerial contest between a German and an Englishman. While the Englishman's wings are simple rectangular flappers based on a design tried by Besnier in the 1670s, the German is using hinged vanes which spread open on the downstroke and fold together on the upstroke.

## Flapping Models

Ever since classical times inventors have been fascinated by the idea of making self-powered working models of moving objects. In many medieval manuscripts one finds detailed instructions for the use of steam, clockwork, or weight-and-pulley systems to animate dancing figures, grimacing devils, and the like. Among the most popular of such automata were models of singing birds with small reed pipes in their beaks and realistically flapping wings hinged to their bodies. These static models inspired some imaginative men to speculate on the possibility of building flapping models which would actually fly. In the 15th century Regiomontanus of Nuremberg was credited with having built an eagle which flew over the city, to the astonishment of a visiting dignitary; some commentators were even more impressed by a second creation of Regiomontanus's: an iron fly which could make a circuit of his dining room. A 16th-century English writer, William Bourne, had no doubt that such things were possible:

> And for to make a bird or fowl made of wood and metal, with other things made by art, to fly, it is to be done to go with springs, and so to beat the air with the wings as other birds or fowls do, being of a reasonable lightness, it may fly.

To allay suspicions that black magic might be involved, Bourne insisted that such flights could be achieved 'by good arts and lawful'.

Although some such models were probably built, it seems highly unlikely, in view of the general ignorance of the true nature of flapping flight, that many of them were even briefly successful. While most of the stories are no doubt exaggerated, at least one claim, made by the great scientist Robert Hooke (1635–1703), deserves to be treated with respect. According to his own notes, Hooke built, in 1655, a clockwork-powered model bird which 'raised and sustained itself in the air'. As Hooke was not given to making untrue statements it seems probable that, by a mixture of luck and insight, he managed to produce sufficient lift to achieve a short flight.

Towards the end of the 19th century aerodynamicists again turned their attention to flapping models in order to gain some insight into the principles of flight. A rubber-powered model was flown in 1874 by the Frenchman Alphonse Pénaud, whose later design for a full-sized monoplane was far in advance of its time. Among other notable experimenters with *ornithopters* (flapping-wing aircraft) was the

Australian Lawrence Hargrave, who in the late 1880s built a series of finely engineered models before undertaking his classic work on the design of kites (*see* Chapter 3).

## Full-sized Flappers

Owing to the relationship between power required, power available, and weight, there is a definite upper limit to the size of a flying animal, the theoretical maximum weight having been calculated at something like 12–5 kg (26–33 lb). Small birds, such as swallows, can fly for days, weeks, or even months without alighting, while large birds of prey must conserve energy as much as possible by soaring on rising thermals or on the up-currents which they seek out near rising ground. As is well known, birds such as the emu and ostrich, which have evolved to sizes well beyond the theoretical maximum, have lost all ability to fly. While this relationship of weight to power was independently understood by Robert Hooke, it was not fully formulated until 1680, when the first part of G. A. Borelli's *De motu animalium* was published.

The limiting weight of 12–5 kg is a maximum for a 'practical' animal which must be able to fly more or less whenever it wishes, with sufficient power in reserve for take-off, climb, and fast manoeuvre. As the growing successes of man-powered machines have shown in recent years, flight of a more restricted kind can nevertheless be achieved, in very good conditions, by much heavier 'animals'. Fixed wings and highly efficient propellers are necessary, however, there being no possibility of a man's sustaining himself in the air by the inherently inefficient to-and-fro flapping motion of wings. Despite the physical limitations, there was doubtless a number of occasions when dedicated ornithopterists achieved 'powered hops', leading them to hope that they might be within reach of attaining true, sustained flight. In the 17th century one such tireless worker is mentioned by Bishop Wilkins, who emphasises the degree of patience needed:

> He that would effect anything in this kind, must be brought up to the constant practice of it from his youth. Trying first only to use his wings running on the ground, as an ostrich or tame geese will do, touching the earth with his toes, and so by degrees learn to rise higher, till he shall attain unto skill and confidence. I have heard it from credible testimony that one of our own nation hath proceeded so far in this experiment that he was able by the help of wings in such a running pace to step constantly ten yards at a time.

Leonardo da Vinci made countless sketches for flying machines such as this arm-and-leg-powered ornithopter. There is no evidence that he built any of them, and none of his designs would have been capable of flight.

## Beliefs About the Nature of Flight

No doubt many of those who attempted to fly with the aid of flimsy wings were as much out of touch with the science of their day as are the bird-men who from time to time still throw themselves recklessly from bridges and rooftops. While some of the speculations about flight set down in the Middle Ages by learned men may also be absurd, much of what was written on the subject appears to be a good deal more rational when it is placed in the context of contemporary beliefs and assumptions about the nature of the physical world.

Most of what one may call the 'prehistory' of European aviation falls in the post-classical period, during which science and technology were based on suppositions and theories current since the time of Aristotle and before. In general these ideas, which might be termed applied common sense, were in fairly direct accord with observable experience. What follows here is an attempt to summarize the theories and beliefs which were immediately relevant to aeronautical speculation and experiment.

All things below the sphere of heaven were held to be composed of four 'elements' – earth, water, air, and fire – which, in their free form, had their naturally allotted places in the universe. At the centre of the universe lay the Earth itself, surrounded by a sphere or ring of the next heaviest element, water. Above the water lay the sphere of air, while between the air and the heavens, in which the stars and planets were embedded, there was supposed to lie the region of the lightest element, 'fire'. As this did not consist of flames but was the *principle* of heat in purified, refined form, there

**Above** Leonardo, possibly inspired by string-pull toys (see next page), sketched this design for a small powered helicopter. The full-screw helical blades, shown in the clockwork-powered reconstruction (**left**), were inefficient but could be made to work after a fashion.

was disagreement as to whether a human being penetrating that sphere would come to harm.

The air itself was often held to contain three main layers or subdivisions. The lowest level, warmed by the sun's rays, was the familiar region of which man had direct everyday experience. The second, extending to the tops of the highest mountains, was a turbulent and watery band in which all the normal weather phenomena were generated. Above that, forming the uppermost subdivision in contact with the region of fire, was a layer of great beauty and serenity which, despite its attractions, might nevertheless prove dangerous to any human venturing to rise too high.

While only one or two medieval thinkers speculated about flight in the strife-torn middle region of the air, many argued about what might happen if one reached the upper atmosphere. On the one hand, since any attempt to fly was potentially an act of spiritual arrogance, experiments with flying machines ought perhaps to be confined to the lowest regions, where the dangers were relatively familiar. On the other hand, the clarity and serenity of the upper region offered a prospect of peace and fulfilment beyond what was normally attainable on Earth. There, in what represented an enticing, if hazardous, goal, one might experience something of the bliss of the celestial spheres.

As it seemed clear that imitating the birds offered the best hope of success, the principles of bird-flight, as then understood, were frequently discussed in passing. Until comparatively recent times it was believed by most people that birds propelled themselves by pushing their wings downwards and backwards through the air, using a stroke like that of a swimmer. A bird moved forward, so it was thought, by pushing against the resistance of the air which lay behind it. (After the coming of Christianity, the relationship of swimming and flying seemed all the closer because of a passage in Genesis which was sometimes interpreted as meaning that God created the birds out of water.) This belief went unchallenged for centuries, in spite of the fact that careful observation would have shown that a bird does *not* move its wings backwards during the downstroke.

In order to be able to fly, a bird needs to be light enough to sustain itself in the air with comparatively little effort. For most medieval thinkers, lightness was not merely the absence of weight, but a positive, exploitable quality of certain materials. They assumed, for instance, that a stone would fall and smoke would rise because the former had a *natural tendency* to move downwards towards the centre of the Earth while the latter had a natural tendency to move up away from the centre. The four elements had varying degrees of 'lightness' and 'heaviness', earth being purely heavy, fire purely light, and water and air containing mixtures of the two. As each element tended towards its natural place in the universe, air would move up through water, while smoke, being a mixture of air and fire, would tend to rise through ordinary air.

Since differing combinations of the basic elements were to be found in natural objects, some of these contained more of the lightness of air and fire than did others. It therefore seemed clear to many thinkers that, although they were predominantly 'watery' and 'earthy', birds contained a sufficiently high proportion of the light elements to be able to sustain themselves in the air with little difficulty. The 'lightness' in their constitution was continually trying to rise, carrying the 'earthier' parts of the bird upwards with it. If man were ever to be able to fly he would need, among other things, to 'gather' as much lightness as possible and make it serve his purpose. One way of doing this would be to make his wings of feathers, which contained a comparatively high proportion of the lighter elements and which, because of their many small interstices, were able to trap large quantities of air within themselves.

An attempt to gather 'lightness' underlies the tactics of a Turk who, in the 12th century, jumped from a tower with wings of cloth attached to his arms. An account written not long after his exploit describes him standing with his arms apart for a long while, 'the better to have gathered the wind, as birds do with their wings'. The more the Turk could fill the 'pleats and foldings' of his wings with air, the more lightness he would, in the common estimation, have harnessed for his intended flight.

The apparently self-evident need to imitate nature by using light-weight flapping apparatus in order to row oneself forward through the air led to the almost total neglect of two aerodynamic objects which lay close to hand and which might well have stimulated useful experiment: the kite and the airscrew. While medieval European writers once or twice commented on the possibility of building kites large enough to lift men, no one seems to have conceived of them as the basis of any kind of free-flying machine. Again, although it was introduced into Europe as early as the 12th century, in the form of the windmill, the airscrew seems to have caught the imagination of no one except Leonardo da Vinci (1452–1519).

Soon after the introduction of the windmill, passive airscrews rotating freely on the ends of sticks became familiar as children's toys. The

Early in the 14th century a new toy appeared in Europe: the string-pull helicopter. The string was wound around the shaft, passing through a hole in the side of the 'nut'. A sharp pull caused the horizontal windmill blades to spin, lifting them and the shaft out of the nut. Toys based on this principle are still made.

active (driven) airscrew quickly followed, with the invention of another and more sophisticated toy, the string-pull helicopter. (A small airscrew is rapidly rotated by pulling on a string wound round the stick, to which the screw is in this case firmly attached. Lift is then created, carrying the airscrew and the stick upwards.) Possibly inspired by these toys, which were frequently illustrated from as early as the 14th century, Leonardo once sketched a design for a helicopter using a full helical (spiral) screw. Not until the 18th century, however, did anyone think of using the airscrew for propulsion in the horizontal direction. What may seem like a strange blindness to exciting possibilities is readily explained by contemporary attitudes. As a kite's sails do not flap, and as rotary motion does not occur in animate nature, neither of these objects would have struck early speculators as obvious starting points. Ironically, the aeroplane may with only slight exaggeration be said to be a combination of the two.

## Aerial Ships

Although it was the most direct expression of man's desire to fly, the use of wings attached to the body was only one of many techniques considered by medieval scientists. Almost equally attractive was the notion of an aerial vessel which would sail the sky just as a ship sails the sea. It was thought that a ship which could somehow be made to stay aloft could be manoeuvred by means of the usual sails in combination with large oars and rudders. The inadequacy of contemporary concepts of relative motion did not allow thinkers to perceive that such airborne ships would always move freely with the winds, making the sails useless.

Fantasy and reality appeared to meet when people saw, or thought they saw, strange and spectacular phenomena in the sky. For many centuries it was believed that in stormy weather flying dragons, generated in the turbulent middle region of the air, were sometimes to be seen among the clouds. These, while frightening enough, were by no means so awe-inspiring as the flying ships which plied back and forth on behalf of the celestial masters of the tempests. As late as Renaissance times, European annals include accounts of flying ships, often described in great detail. Although scornful of these reports, Bishop Agobard of Lyons (*c.*769–840) testifies to the power of such imaginings over the minds of ordinary people:

> We have seen and heard many, so struck with madness, so crazed with stupidity, that they believe and say there is a place called Magonia, from which ships sail into the clouds and in these the crops knocked down by hail and ruined by storms are carried to that region, the aeronauts paying the gods of the tempests, and receiving in exchange the corn or other crops. We have seen several, so blinded by profound stupidity that they believe such things to be possible, exhibiting in an assembly four people, three men and a woman, who had been bound because they were thought to have fallen from such ships. These, having been kept prisoner for some days, were at last presented to the whole assembly of men, in our presence, as I have said, as people fit to be stoned.

While flying ships and flying chariots remained terrifying possibilities, some people were prepared to contemplate ways of constructing them, the simplest machinery being a pair of wings operated by hand. In the 13th century, before Leonardo had designed his complex apparatus designed to make optimum use of human muscle power, Friar Roger Bacon asserted confidently: 'It's possible to make engines for flying, a man sitting in the midst whereof, by turning only about an instrument, which moves artificial wings made to beat the air, much after the fashion of a bird's flight.'

If only it had been as easy as that! More ingenious was the idea of Albert of Saxony, who put forward a suggestion based on the principle of displacement. Conceiving the boundary between the region of air and the region of fire to be analogous to the boundary between the sea and the atmosphere, he announced that 'the upper air, where it is contiguous with fire, is navigable, just as the water is where it is contiguous with the air. Hence if a ship is placed on the upper surface of the air, filled not with air but with fire, it will

Many early inventors failed to understand that balloons are carried along at the same speed and direction as the wind. This late-18th-century design illustrates the popular misconception that a free-floating balloon could be manoeuvred, like a ship at sea, by the use of sails.

not sink through the air.' In making this interesting proposal Albert had, of course, forgotten the need to build the aerial ship of comparably light materials; nor is it clear from his account how the pilot of the ship, immersed in elemental fire, would have been able to breathe.

Although with the coming of the 'New Science' in the 17th century older ideas about the nature of the world began to change, an understanding of the principles of flight was slow to develop. Even after the first successful balloon ascents in 1783, most experimental work was little more than trial and error. Although the theory of *aerostation* (lighter-than-air flight) was well understood throughout the 19th century, it is sobering to reflect that *aerodynamic lift* (the principle of heavier-than-air flight) was not satisfactorily explained until the 1890s. Despite the rapid growth of our knowledge since then, the old dream of emulating Daedalus is hard to forget, and even to this day some ardent ornithopterists can still be found energetically flapping feather-covered wings in an effort to beat the laws of physics.

# 2 Balloons and Airships

**Above** In recent years there has been a remarkable revival of interest in hot-air balloons. Typical sizes range from 1,000 to 4,000 m³ (about 35,000 to 140,000 cu ft). These balloons participated in a mass ascent at a meeting in Albuquerque (New Mexico).

**Above** In 1966, at Dunstable (Bedfordshire) enthusiasts flew a carefully decorated replica of the first full-scale Montgolfier balloon.

Man's first success in his long series of attempts to rise into the air and stay there for more than a few seconds was achieved in 1783 with a hot-air balloon. From the earliest times it had of course been possible to observe that smoke and warm air tend to rise – a phenomenon which, however little understood, led a number of imaginative thinkers to contemplate ways of harnessing the forces involved. Several times, in fact, it seems as if medieval scientists, puzzling over the question of inherent lightness and heaviness, might have been on the verge of inventing the hot-air balloon. In a tantalizing passage written in the 13th century, the German philosopher Albertus Magnus observed that a bladder one blows up with air warmed by the lungs will be measurably lighter than when it is empty. As his argument was, however, entirely concerned with the theory of relative weights, he gave no consideration to practical applications.

Two hundred years after Albertus, the Venetian inventor Giovanni da Fontana took the idea further, describing not only the principle but also the practice of hot-air ballooning. Despite the exciting new possibilities which were offered, Fontana was far from enthusiastic, attributing the invention to an unnamed predecessor whose ideas he condemned as foolhardy and dangerous. This man, he said,

> conceived of making, from thick cloth and rings of wood, a pyramid of very great size whose point would be uppermost, and of firmly tying across the diameter of the circle at the base a bar of wood, on which a man might sit or ride, holding in his hands burning brands made of pitch and tallow or other material producing intense fire which is long-lasting and creates a great deal of thick smoke. He suggested that because of the fire the air enclosed within the pyramid would be made lighter and rarer, and consequently that as it would move upwards, and could not, of course, escape, the pyramid and the man sitting in it would be raised.

It seems that the attempt was in fact made. Fontana goes on to say that the experiment did not succeed, 'either because the pyramid was too small, or too heavy, or because it had some leak in its cover, or because there had been too little fire, or because too little of the vapour had been enclosed'. Fontana points to many real dangers, especially the likelihood that the whole thing may catch fire, and the difficulties of making a safe descent when the fuel has been consumed.

Not all speculations about ballooning were as close to reality as Fontana's, many writers being misled by false theories of relative weight which remained current well into post-medieval times. Even as late as the 17th century a Jesuit called Lauretus Laurus saw nothing unrealistic in the notion that one might build an artificial flying bird by enclosing within its body an eggshell filled with dew. If the model were left in sunlight, the warmth would evaporate the dew, which would then rise from the Earth carrying the model with it. A more daringly imaginative application of the idea was proposed at about the same time by the French poet and soldier Cyrano de Bergerac (1619–55), who thought of harnessing solar energy for manned flight by strapping flasks of dew to his belt. When the Sun had warmed the flasks sufficiently, they would carry him gently upwards.

Although theories based on the concepts of inherent lightness and heaviness lingered on, the revolutionary ideas being developed by the best scientific minds of the 17th century began to put aeronautics on a new footing. For future balloonists the most significant change lay in the discovery that air has weight.

Taking this as his starting point, an Italian priest named Francesco de Lana de Terzi reasoned that by evacuating the air from a vessel one could make it lighter and give it buoyancy. If emptied of air, lightweight globes would then tend to rise through the atmosphere as a bubble rises through water. In 1670 Lana de Terzi published a design for an airborne ship which would rise owing to the buoyancy of four evacuated copper globes. Up to a point his theory was sound, although in practice no globes could have been made sufficiently rigid to withstand the enormous forces of the air pressure tending to crush them.

## The First Balloons

Most older books on the history of flight credit the French brothers Joseph and Étienne Montgolfier with having built the first practicable hot-air balloon, but recent research shows that their success may have been at least partially anticipated by as much as three-quarters of a century. An illustration of a strange aircraft called the *Passarola* (Great Bird), the invention of a Brazilian priest, Bartholomeu de Gusmão, has long been familiar. The fantastic mixture of flying boat, parachute, and ornithopter, commonly dismissed as the product of an undisciplined imagination, has sometimes been supposed to represent, in highly exaggerated form, some kind of model glider.

It now seems probable, however, that the *Passarola* is a thoroughly distorted representation of a small hot-air balloon which Gusmão is said to have built and demonstrated indoors in the presence of the king and queen of Portugal. According to the contemporary reports, fire placed in an earthenware bowl suspended

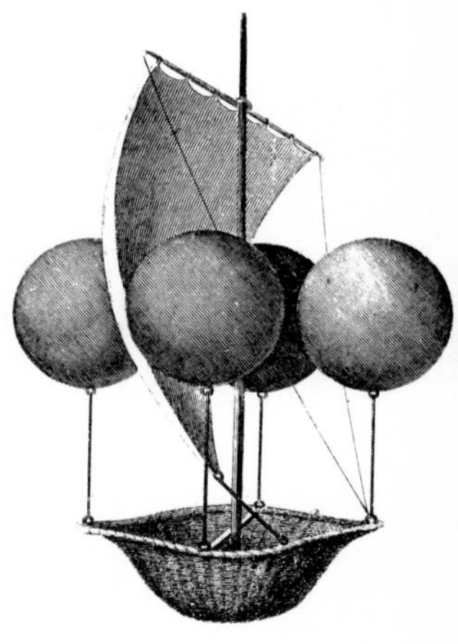

**Above** In 1670 the Jesuit Francesco de Lana de Terzi published a book in which he proposed a flying ship that would be lifted by four evacuated copper spheres. Although theoretically sound, his idea was impractical owing to the impossibility in his day of making large, lightweight spheres of sufficient rigidity.

**Below** Early in the 18th century the Brazilian priest Bartholomeu de Gusmão experimented with hot-air balloons. This imaginary flying machine called the *Passarola*, published in 1709, misrepresented Gusmão's experiments.

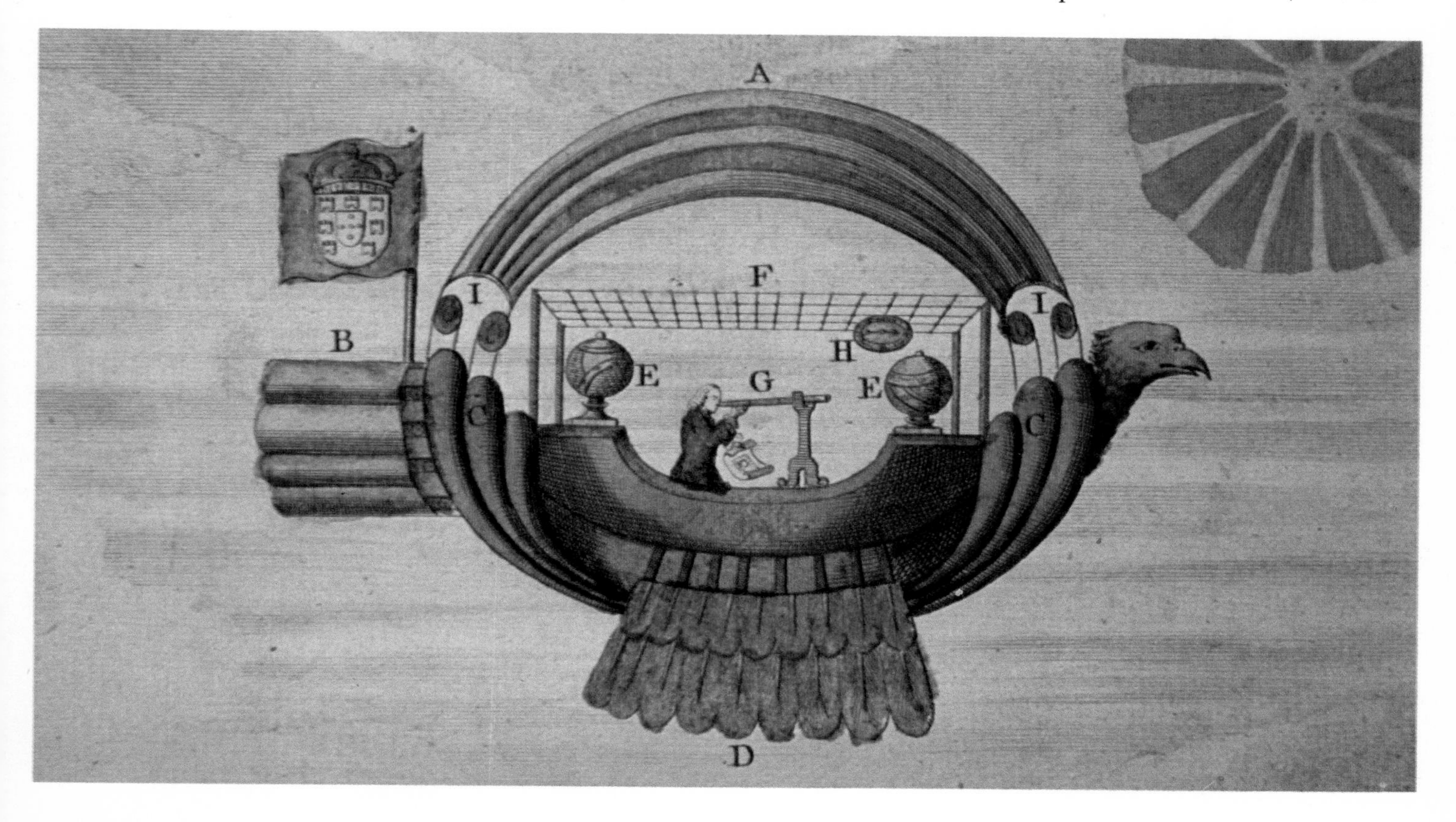

Above Vincenzo Lunardi, an employee of the Italian embassy in London, made the first manned balloon ascent in England on 15 September 1784. The car of this larger balloon, completed in 1785, could accommodate up to 10 passengers.

Right Modern hot-air balloonists use propane burners to inflate their balloons, which are made of non-inflammable materials.

beneath a balloon made of thick paper was sufficient to raise the model several feet into the air. As there seemed to be a danger of its setting fire to the curtains in the room, servants soon knocked it to the floor. Other reports, less convincing, tell of the flight of a larger, man-carrying version of the balloon. While it is not impossible that Gusmão may have flown a hot-air balloon, some details of the accounts are open to question. In order to be sufficiently light, a small balloon needs to be made of thin rather than thick paper, while any model small enough to be conveniently flown indoors would certainly be incapable of lifting an earthenware bowl.

There is no such doubt, however, about the Montgolfiers' use of models. In 1782, having begun to take an interest in scientific works on gases, especially Joseph Priestley's *Experiments and Observations on Different Kinds of Air* (1774–86), Joseph Montgolfier, a professional paper-maker, built a balloon which he filled with the newly discovered hydrogen gas. Having found that hydrogen very rapidly penetrated envelopes made of paper and silk, Joseph, soon to be joined by his brother Étienne, turned his attention to the possibilities of using the 'rarefied air' which appeared to be generated by fire. Despite the advances of science, many people still believed, as in the days of Giovanni da Fontana, that some kinds of smoke were lighter than others and therefore potentially more satisfactory for use in balloons. Not realizing that the relative lightness of the smoke is almost entirely dependent on its temperature, the Montgolfier brothers experimented with a variety of smoke-producing fuels, including old shoes and rotting meat – the smell of which proved highly offensive to interested bystanders, who included the fastidious Queen Marie Antoinette.

In June 1783 the Montgolfiers felt confident enough to embark on a public demonstration of their invention. For this purpose they built a spherical balloon 33 m (110 ft) in circumference which, when filled with hot air, needed eight men to hold it down. After the signal had been given for its release, the balloon rose to a height of about 1,800 m (about 6,000 ft) before landing two or three kilometres (about $1\frac{1}{2}$ miles) away after a flight of 10 minutes.

This demonstration attracted the attention of King Louis XVI, who commanded a further balloon to be flown for him. In September 1783, after a false start during which a large and beautifully decorated balloon was destroyed by a storm, the Montgolfiers successfully released a cloth and paper balloon about 13 m (41 ft) in diameter. Although there had been tentative plans to allow a man to fly, the king vetoed the attempt as too hazardous. Instead, the first passengers in the history of modern flight were a sheep, a cock, and a duck. On 19 September this strange company flew to about 550 m (1,800 ft) before descending some 3 km (2 miles) from their starting point. After the flight, which lasted about eight minutes, the cock was found to be a little worse for wear, having probably been sat on by the sheep.

## The First Men to Fly

After strenuous representations had been made, the king was persuaded to lift his embargo so that a manned flight might be undertaken. Even then, however, Louis at first insisted that only a pair of criminals should be permitted to undertake so dangerous a journey, a proposal which was abandoned only after further protests from a young scientist, Pilâtre de Rozier, who did not wish to grant to such unworthy men the honour of being the first to fly.

A balloon somewhat larger than the previous one was built and flown on 15 October 1783, with Pilâtre de Rozier and his friend the Marquis d'Arlandes as passengers. Caution still prevailed, however, the balloon being kept tethered to a rope about 26 m (84 ft) in length. The first manned ascent, lasting about 4½ minutes, was followed later that day by a number of other tethered flights during which Rozier gained experience in controlling the airborne fire, which he fed with straw.

On 21 November that year the first free flight was made, with the same passengers aboard. Rozier and his friend were aloft for 25 minutes, during which time they covered a distance of about 8 km (5 miles).

Although the Montgolfiers had turned to hot air as easier to contain within a paper or cloth envelope, the French scientist Jacques Charles persisted with hydrogen, which he generated by the action of sulphuric acid on iron filings. After overcoming many technical difficulties he managed, on 27 August 1783, to launch a small, unmanned balloon made of rubber-coated silk which flew for about three-quarters of an hour before landing in a village some 25 km (15 miles) away, where the terrified inhabitants, thinking it was some kind of monster, tore it to shreds.

A few months later Charles was ready to attempt a manned flight, which he successfully accomplished on 1 December 1783 in the company of one of his assistants, Marie-Noël Robert. An enormous public interest was generated by these exploits, the ascent of the manned *charlière* from the Tuileries gardens in Paris being watched by nearly half a million people. For decades thereafter ballooning was among the most talked of and widely illustrated subjects, and, although few people had the resources to undertake a flight themselves, the making and flying of miniature balloons became something of a craze.

The art of ballooning developed rapidly. By 1785 the English Channel had been crossed, and in 1794, only 11 years after the first manned flight, the French military authorities formally established a balloon corps whose main function was to act as aerial observers.

During the American Civil War (1861–5) the

Using this hydrogen balloon Jean-Pierre Blanchard and Dr John Jeffries set off from Dover on 7 January 1785 to make the first flight across the English Channel. Despite its useless encumbrance of wings and tail, their balloon landed safely in a wood about 20 km (12 miles) beyond the French coast.

Union army also made use of balloons for reconnaissance, and during the Franco-Prussian War of 1870–1 a series of flights was undertaken by the French to circumvent the siege of Paris. Before the end of January 1871 a total of 66 balloons had been flown out of the city, carrying passengers, dogs, mail, and pigeons, the last being used to despatch return correspondence from friendly territory.

Meanwhile there had been serious accidents, including the death of Pilâtre de Rozier in June 1785 during a second attempt to cross the English Channel. Despite the dangers balloons, some of enormous size, were used by many 19th-century showmen, and scientists turned to them for first-hand experience of the atmosphere. In 1862 James Glaisher FRS and the professional balloonist Henry Coxwell undertook one of the most remarkable ascents of all time when they climbed to an estimated 9,000 m (30,000 ft). Suffering from the bitter cold and lack of oxygen, Glaisher lost consciousness, and it was only with the greatest difficulty that Coxwell managed to free the balloon's release valve so that a descent could be made.

Scientists, especially meteorologists, have continued to make effective use of balloons, both manned and unmanned. In 1932 the Swiss physicist Auguste Picard, in an enclosed gondola beneath a balloon, climbed to what was then the record altitude of over 16,000 m (53,000 ft). By the 1960s US aeronauts had more than doubled this figure, a notable flight being that of Commander Malcolm D. Ross, who on 4 May 1961 reached 34,668 m (113,740 ft).

## Steerable Balloons

A free balloon drifts at the speed of the wind, allowing the passengers to enjoy the peaceful surroundings of air which appears to be totally motionless. Not fully appreciating the significance of this fact, some early designers, including Lana de Terzi, suggested the addition of sails to enable their aerial ships to be flown in any direction, as one manoeuvres a yacht. Since, apart from the effects of occasional gusts, the relative motion of the wind and the balloon is zero, all such sails would have hung limply in the apparently dead calm air along with which they were being carried.

Although alternative suggestions for propelling balloons by the use of flappers or manually operated airscrews were more soundly based in theory, the development of *dirigible* (steerable) balloons became a practical possibility only with the emergence of a suitable power unit. Steps towards the creation of a dirigible balloon, or airship, were tentatively taken

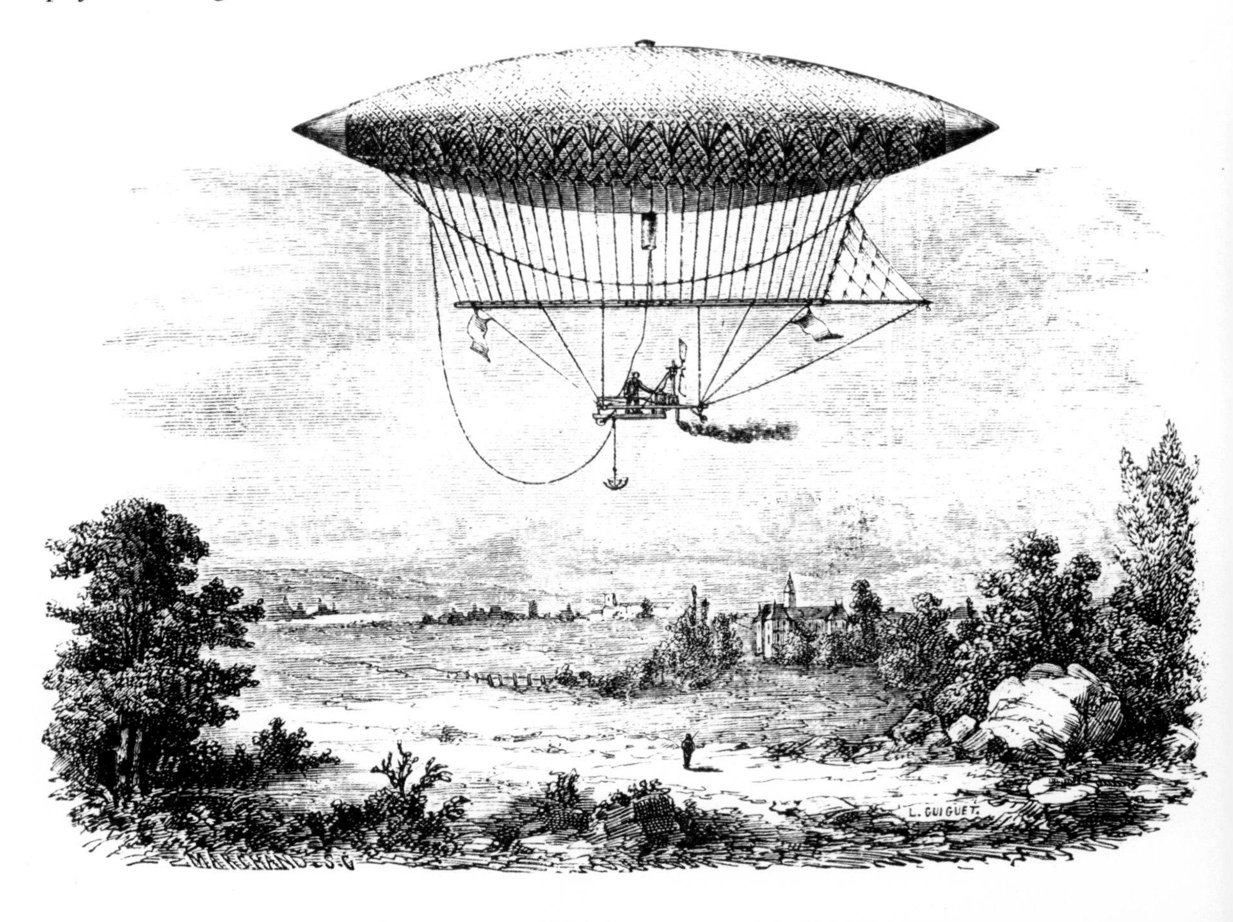


On 24 September 1852 Henri Giffard made the first powered flight when his steam-driven airship flew from the Paris Hippodrome to Trappe, covering a distance of about 27 km (17 miles).

during the first half of the 19th century, including in particular the development of the elongated, cigar-shaped envelope. Nevertheless it was not until the 1850s that significant practical advances could be made.

In 1852, having familiarized himself during the previous year with the control of free balloons, the Frenchman Henri Giffard built and flew the first powered aircraft in the history of flight. Beneath an elongated envelope, sharply pointed at both ends, he suspended horizontally a long pole, from which he hung on rigging lines an open car containing a steam engine driving a three-bladed propeller; directional control was provided by a triangular rudder attached to one end of the pole. On 24 September 1852 Giffard made a flight of 27 km (17 miles) ending with a safe landing. As his airspeed was only about 10 km/h (6 mph) he needed almost windless conditions in order to be able to exercise true navigational control.

Twenty years later unsuccessful but nevertheless significant steps were taken when an Austrian, Paul Hänlein, built an airship powered by the first airborne coal-gas-powered engine. Twice as powerful as Giffard's, Hänlein's engine was too heavy to be satisfactorily carried by his airship, which never achieved free flight. An encouraging speed of 14 km/h (9 mph) was reached in tethered trials, and but for lack of money Hänlein might well have been able to develop a successful dirigible.

Giffard's coke-fired steam engine was hung some 12 m (40 ft) below the airship envelope. Delivering about 3 hp, it drove a three-bladed propeller which provided an airspeed of about 10 km/h (6 mph).

In the 1880s the emergence of electric power encouraged two brothers, Gaston and Albert Tissandier, to try an airship driven by an electric motor. In 1883, having successfully demonstrated an electric model, they built a full-sized dirigible using a 1.5 hp electric motor powered by 24 heavy batteries. Although the Tissandiers just managed to fly, under marginal control, it was not until the following year (1884) that a fully controllable airship was built. *La France*, designed by Charles Renard and Arthur Krebs, was 50 m (165 ft) long and had a capacity of about 1,800 m$^3$ (64,000 cu ft). A 7.5 hp electric motor was suspended beneath it in a long slim car. At an airspeed of 24 km/h (14.5 mph) *La France* was able to manoeuvre against moderate breezes and several times made flights during which she was able to return to her starting point.

Despite these qualified successes it was not until compact internal-combustion engines were developed that a really satisfactory power unit became available. Following some brief flights in 1888 a German, Dr Karl Wölfert, began a series of experiments which he continued sporadically into the 1890s. Together with an assistant, Wölfert attempted in 1897 to make an ambitious round trip in a hydrogen-filled airship powered by a petrol motor rigged dangerously close to the envelope. When they had reached an altitude of about 1,000 m (about 3,300 ft), a jet of flame from the engine's exhaust set fire to the gas, resulting in a crash which killed both men.

## Rigid Airships

Until the end of the 19th century most airships used either non-rigid or semi-rigid envelopes. Late in the century the introduction of aluminium made possible the construction of truly rigid craft, the first of which, wrecked on its maiden flight at Berlin in November 1897, was a tubular structure designed by the Austrian David Schwartz. Not only did he use a rigid aluminium framework, but the envelope itself was constructed from very thin aluminium sheeting. Although the airship was a failure, Schwartz's work proved the practicability of using aluminium in the construction of dirigible envelopes.

Meanwhile the Paris-domiciled Brazilian Alberto Santos-Dumont, later to become the first man in Europe to make a sustained and controlled flight in a heavier-than-air machine, had begun to study the problems of aeronautical design. Using a small motorcycle engine as a power plant, he built his first airship in 1898. Two years later, after making several successful flights and having had some lucky escapes, he piloted his sixth airship to win the Deutsch prize of 100,000 francs offered to the first man to take off from the flying base at St Cloud (Paris), circumnavigate the Eiffel Tower, and return to base – a round trip of about 11 km (7 miles). Following up this success, Santos-Dumont built several more airships in rapid succession until, during the early years of the 20th century, his trips around Paris, at quite low level, had become a familiar sight. In *My Airships* (1904), Santos-Dumont amusingly described such incidents as his circumnavigation of the Arc de Triomphe, his landing at a favourite café, and parking the airship at his own front door on the Champs-Elysées.

More significant development of dirigibles was undertaken by the German, Count Ferdinand von Zeppelin. In 1899, after much serious thought had led him to believe that airships might be of military value to Germany, he began the construction of an enormous envelope 128 m (420 ft) long, with a capacity of 11,300 $m^3$ (400,000 cu ft). The two Daimler engines which he used during the test flights of 1900 proved far too small and the airship's control surfaces were ineffective. In 1901 the *LZ1*, as it was designated, was scrapped, to be succeeded four years later by the *LZ2*. While this also was a failure, it provided much useful information which was applied, with the help of Zeppelin's new assistant, Hugo Eckener, to the construction of the successful *LZ3* in 1906.

Between that time and the outbreak of World War I, rigid-airship technology advanced very rapidly in Germany and elsewhere: propulsion was provided by engines developing horse power in the hundreds, streamlining techniques considerably improved, control systems were made more efficient, and constructional methods were greatly refined. In addition to the work carried out at the Zeppelin factory, important developments occurred in France, where Paul and Pierre Lebaudy built several semi-rigid airships with characteristically pointed ends, the first of which is credited with having made, in November 1903, the first fully controlled aerial journey over a significant distance. Flying from Moisson to Paris, where it moored at the Champ-de-Mars, the Lebaudy airship covered 62 km (38 miles) in 1 hr 41 mins. Pioneering work was also carried out in America and England. Ambitious American attempts to fly an airship over the North Pole failed in 1907 and 1909, while a flight across the Atlantic from New Jersey in 1910 ended in the crew's having to be rescued from the sea by a British ship. At Farnborough (Hampshire) the

**Left** The first truly practicable airship was built by the Lebaudy brothers, Paul and Pierre. A Daimler engine gave it a speed of about 40 km/h (25 mph), allowing it to fly in conditions of moderate wind. Many successful flights were made in 1902 and 1903.

**Right** During the early years of the 20th century Alberto Santos-Dumont, a Paris-domiciled Brazilian, built and flew several successful airships. In 1901 he delighted the Paris crowds by flying around the Eiffel Tower.

**Right** Zeppelin LZ.129 *Hindenburg*, completed in 1936, was 245 m (804 ft) long, had a maximum diameter of 41 m (134.5 ft), and a volume of about 200,000 m³ (7,036,000 cu ft). Four diesel engines gave it a maximum speed of 131 km/h (81 mph). The accommodation for its 75 passengers was of an opulence comparable with that of an ocean liner.

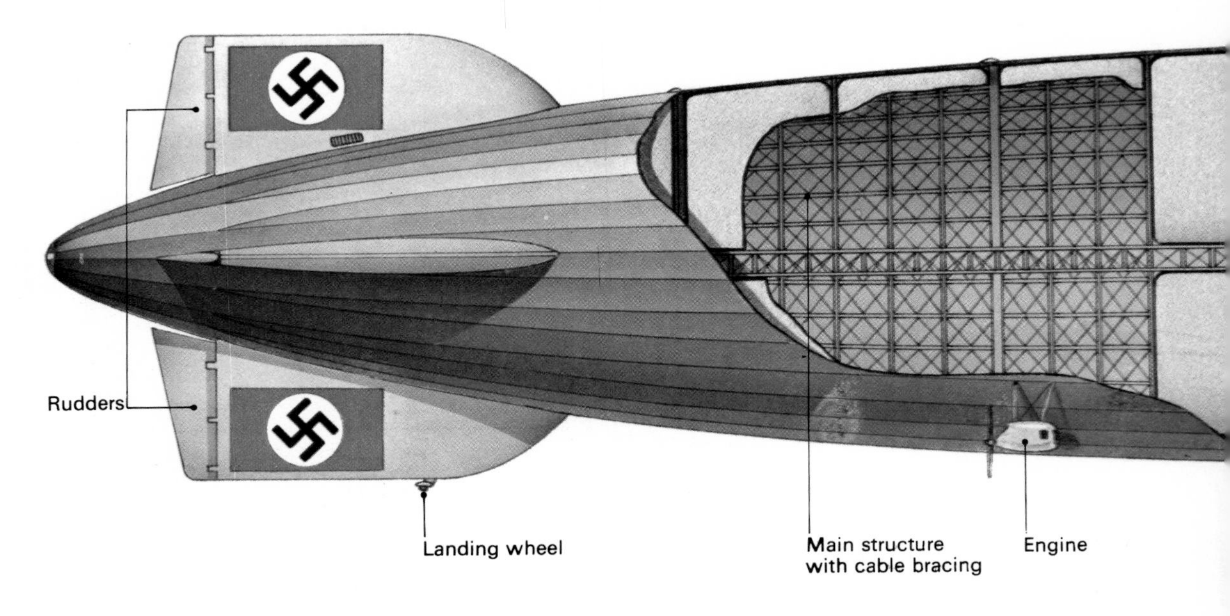


British authorities experimented with a number of airships, including the *Dirigible No. 1* (usually known as the *Nulli Secundus*) with which an American expatriate, Samuel F. Cody, a designer of man-lifting kite systems and a pioneer heavier-than-air pilot, was closely associated.

After World War I, during which the Zeppelins of the German Navy played a significant role, development of airships continued both in Germany and in the Allied countries. In 1919 the British *R.34* made the first aerial crossing of the Atlantic, covering the outward journey in 108 hours and returning in a little over 75.

## Hydrogen v. Helium

All the earliest airships had derived their lift from the exceedingly dangerous hydrogen gas. The hazards, evident since the earliest days of ballooning, were demonstrated with dreadful clarity when, in October 1930, the poorly built and inadequately tested British airship *R.101* crashed and was consumed in a fierce fire on its maiden operational flight, killing 48 people and bringing British airship development to an abrupt halt. The safe alternative gas, helium, which had been first isolated only in 1895, was not only somewhat heavier but was also comparatively rare and therefore very much more expensive to produce. Natural helium, most of which occurs as a constituent of natural gases in the southern United States, began to be extracted in sufficient quantities for airship use only towards the end of World War I. Envisaged for the inflation of British and American airships during the war, it was not in fact used until 1923, when the US (Zeppelin) airship *Shenandoah* was launched. Although the Zeppelin company achieved great success with its hydrogen-filled *Graf Zeppelin*, launched in 1928, the disaster of the *R.101* led it to plan the use of helium for the 245 m (804 ft) *Hindenburg*, which first flew in 1936. Since, in the event, the American authorities refused to supply helium, the *Hindenburg* was slightly modified to allow hydrogen to be used instead, a substitution which led to a further terrible fire. On 6 May 1937, when mooring at Lakehurst, New Jersey, the *Hindenburg* burst into flames, killing 35 of the 97 people on board.

## Modern Airships

Although one more rigid airship was built shortly after the *Hindenburg* – the *Graf Zeppelin II*, dismantled at Frankfurt in 1940 – the fire at

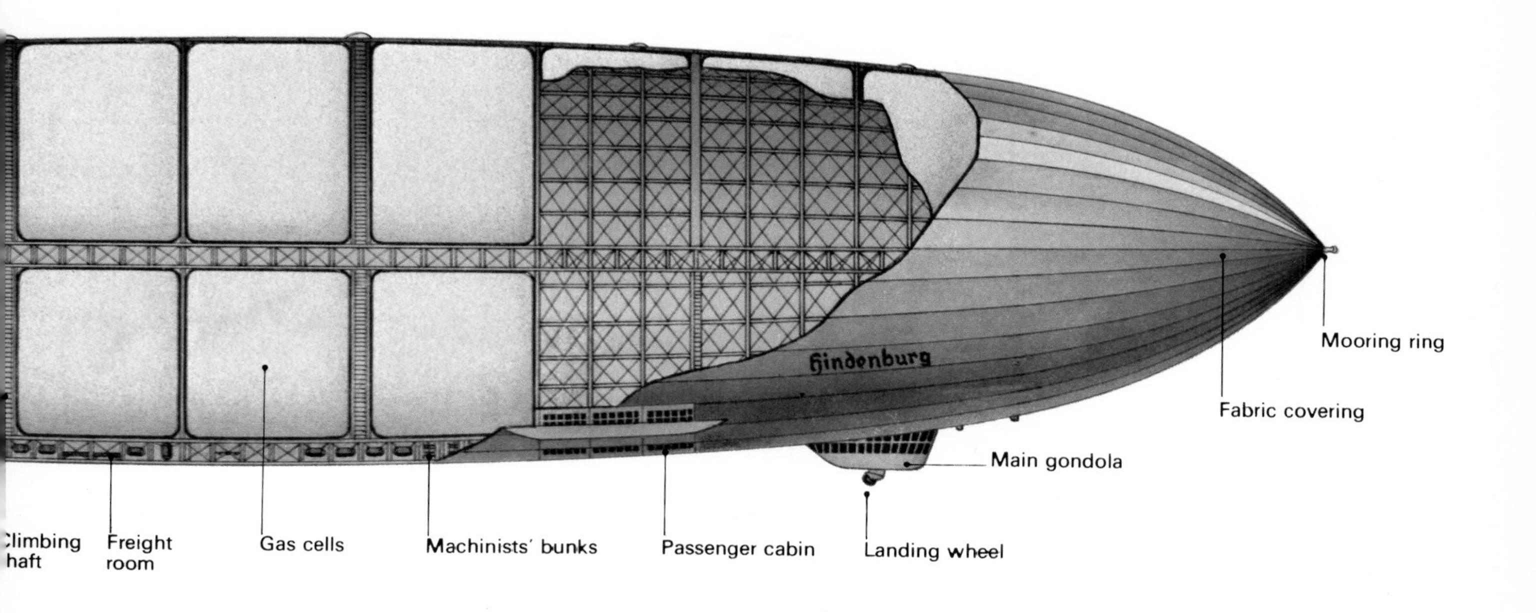


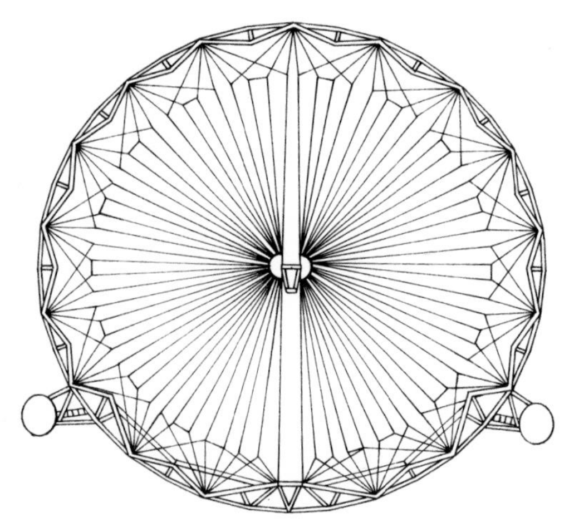

**Left** Section through the *Hindenburg* and its two forward engines (not shown in main drawing).

Lakehurst effectively brought development of this type of dirigible to an end. During the 30 years of progress since the first controlled flights, their lifting capacities had increased twenty-fold; speeds, which before World War I were of the order of 70 km/h (45 mph), had risen to over 125 km/h (80 mph); passenger comfort had been vastly improved; and much had been learned about control techniques.

While for a time it seemed as if airships were things of the past, a modest revival of interest has been seen in recent years. New airships have been built, some of which, like the small fleet flown by the Goodyear Company, are used partly for advertising purposes, while others have been developed for practical and conservationist reasons. As energy is not needed to maintain lift, the airship can be, in applications such as continuous reconnaissance and patrol, much more economical to fly than aeroplanes or rotorcraft. Since it can be manoeuvred at very low speed, it is also useful for reaching inaccessible areas. Being potentially capable of carrying quite heavy loads at low running cost, it offers a quiet, efficient form of transport if high speed is not essential.

Plans have been drawn up for building large freight-carrying airships as much as 400 m ($\frac{1}{4}$ mile) in length and capable of lifting loads of up to 500 tons. Very large craft such as this could well make use of nuclear energy, a source of power conferring important advantages on the airship designer. For long-distance flights not only would a well-engineered nuclear plant be lighter than an equivalent internal-combustion engine with its associated fuel, but there would be no ballast problems such as arise when fuel is consumed in conventional power plants. For passenger flights accommodation would be much more luxurious than is possible in aeroplanes, and although the initial costs would certainly be high, the use of large modern airships for both cargo- and passenger-carrying purposes could well prove economical and attractive. As Max Pruss, last surviving captain of the *Hindenburg*, used repeatedly to insist: 'If you want to travel quickly, take an aeroplane; if you want to travel in comfort, take an airship.'

# 3 Kites

Chinese and Japanese kites are often intricately built and painted to represent a great variety of animals and objects. Most spectacular of all are the multi-disc dragons, which may be hundreds of feet long. The example here is flying on Parliament Hill, London.

No one knows when kites were invented. Although it seems likely that they were known in parts of Asia as much as 3,000 years ago, the earliest firm evidence of their use dates from the 3rd century BC. One of the most famous stories of Chinese warfare tells how General Han Hsin, who died in 196 BC, flew a kite over the walls of an enemy palace in order to measure the length of the tunnel which his troops would have to dig in order to gain an entrance. Since this tale, and a number of others which are similarly connected with warfare, treat kites as familiar objects, it is reasonable to suppose that they were invented a good deal earlier.

How the first kite was made, and why, is a matter for speculation. Suggestions have included the idea that sails or banners might have been allowed to fly freely from the ends of cords, at first perhaps by accident, and then later by design in order to produce a pleasing effect. One historian thought that kites might have developed from the tethered arrows used by some hunters to enable them to pull in their prey, while another saw a possible origin in the large hats of Chinese coolies, which were sometimes tied to the body to prevent their blowing away in the wind.

Two basically different types of kite spread from China to Japan, Korea, and parts of South-East Asia. The first to be widely used was a simple rectangular structure which could be built either very small or scaled up to impressive proportions many times bigger than a man. A little later there began to emerge a family of smaller and more intricate kites, built to represent birds, animals, insects, flowers, and even artificial objects such as fans and houses. In addition to kites of Chinese origin, other types may have been independently invented in other parts of the Orient. These include the leaf and wickerwork kites used by a number of Pacific island peoples for ceremonial purposes and for kite-fishing, and the various diamond-shaped kites flown in parts of South-East Asia mainly for recreation and kite-fighting.

Whatever their true origins, kites rapidly developed mystical, religious, and ceremonial importance. They were variously used as demon-quellers, as aids to fortune-telling, and as a means of establishing communication with the heavenly regions. In parts of Polynesia they were sometimes thought of as personifications or incarnations of divine powers, and there are legends of encounters between gods who have transformed themselves into kites. One of these explains the origin of kites by saying that they (or at least the spirits which inhabit them) are all descended from a pair of primordial divine kites which mated and produced offspring.

Eastern kites were not, in the early days, looked upon as children's playthings, but were flown by adults who put them to a great variety of practical as well as spiritual uses. In warfare they were used to carry lights over enemy lines at night, to drop leaflets encouraging prisoners to rebel, to carry incendiary flares, and to lift men to inaccessible places. At least 1,000 years ago man-carrying kites were commonplace in China and Japan, there being many stories of warriors and adventurers using them to achieve apparently impossible feats. As Marco Polo (*c.*1254–1324) points out, the aeronauts were not always entirely willing: in the account of his travels Marco describes the flight of a Chinese prisoner sent aloft by sailors who wanted to judge the strength and direction of the wind.

## Early European Kites

In spite of the long history of kites in the East, some parts of the world remained in almost total ignorance of them until recent times. They were unknown in Australia, pre-Columbian America, and southern Africa. By early medieval times they may have reached Muslim North Africa, but they certainly were unknown in classical Greece and Rome. The ordinary diamond-shaped children's kite which is now familiar all over Europe was derived from kites imported as curiosities by sailors returning from the East Indies after the sea routes from Europe to the Orient were established in the 15th and 16th centuries; but Chinese rectangular and figure-kites remained little known.

The history of kites in Europe is complicated, however, by the existence of a totally different and perhaps independent type of uncertain origin. This had a large and often dragon-shaped head below which a long wide cloth tail was allowed to blow freely in the wind. The first clear evidence of the use of such a kite appears in an English book of 1327 by Walter de Milemete, showing three soldiers flying a winged dragon over the walls of an enemy castle. That it is large enough to exert a powerful pull is indicated both by the efforts the soldiers are making to hold it steady and by the use of the kite to drop a bomb into the enemy territory. Since Milemete's text includes no comment on the scene, the bombing raid might be dismissed as fantasy if we did not have, in a German manuscript written about 100 years later, a description of a very similar kite whose reality is beyond question. The latter, a long and detailed account, reveals that the writer was an experienced maker and flyer of kites essentially identical to Milemete's.

Although these flying dragons bear some resemblance to snake-kites found in parts of Asia, they were probably independent inventions rather than importations. Early in the

**Above** This large, dragon-shaped kite appears in an English manuscript of 1327 by Walter de Milemete. The soldiers are using it to drop a bomb over the wall of an enemy castle.

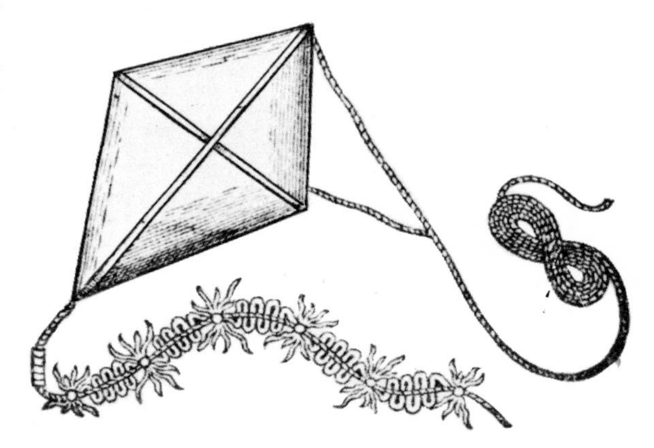

**Left** The first printed illustration of an English kite occurs in a book published in 1634. The tail of this one consists of a string of 'jumping jacks' used at firework displays.

**Below** Some medieval meteorologists thought that dragons appeared among the clouds during violent storms. Their shape influenced the design of the dragon-kites during the 14–16th centuries.

Early in 1893 Lawrence Hargrave built the first cellular or 'box' kite. Combining rigidity, stability, and great lifting power, the type was used widely in meteorological research and influenced the design of many early gliders and powered aircraft.

Christian era the Roman army borrowed from Dacia (modern Romania) and Persia a special kind of banner which consisted of a carved wooden head set on a pole. Behind the head a cloth tube was allowed to billow out like a windsock, while a lighted torch placed in the mouth made the whole thing very effective in frightening superstitious enemy soldiers. Although they gradually died out, the dragon standards were occasionally seen in Europe until the 16th century and may have inspired the simpler but similarly shaped dragon kites.

By the 17th century the large, heavy dragon kites, which often needed all of a man's strength to hold, were almost wholly supplanted by the newer, lighter kites from the Orient. Although for a time they continued to be flown mainly by European adults, especially to raise fireworks at pageants and festivals, kites rapidly became associated with children's play. The earliest printed illustrations come from the period of change-over: the earliest of all (1618) shows one flown by a Dutch boy, while in the first English woodcut (1634) a man is launching a kite whose tail consists of a line of 'jumping jacks' tied together for use as an aerial firework.

Perhaps because they were regarded mainly as toys, kites were rarely given serious attention by European scientists until late in the 19th century. The only significant exceptions were the early meteorologists, a number of whom used them to explore the atmosphere. In 1752 the American philosopher-statesman Benjamin Franklin flew a kite on a wire line in order to demonstrate that lightning is caused by electricity – a very dangerous experiment, repetitions of which have led to several deaths. At about the same time two Scottish meteorologists undertook safer experiments when they raised thermometers which were released from kites by slow-burning fuses. In spite of their interesting findings it was not until over 100 years later that kites were systematically applied to similar ends. Beginning in the 1880s, meteorologists devoted half a century to the refinement of kite-flying techniques, developing special lightweight weather-recording instruments and building rotatable winch-houses from which to handle the bulky reels of high-tensile 'piano wire' used as flying lines.

**Right, above and below**
Samuel F. Cody further developed the Hargrave box kite, adding wings to increase lift. A system of man-lifting kite trains which he patented in 1901 was officially adopted by the British War Office. Modifications and variants of the Cody 'bats' (as he sometimes called them) continue to appear from time to time. The upper picture shows a recent example (only the lowest kite in the train is visible), which uses a sophisticated system to keep the kites the required distance from each other on the line.

## Kites and Early Aerodynamics

From time to time during the 19th century kites were used or discussed by Europeans and Americans who were engaged in experiments related to the problems of flight. Surprisingly, however, few of them paid much attention to this ready source of information about lift, stability, and control. Not until the 1880s and 1890s, when Lawrence Hargrave (1850–1915) began his aeronautical work in Australia, were kites treated with full seriousness.

Hargrave began by examining the behaviour of small flying models which he constructed with great care. The earliest of these were powered by twisted rubber; later he used beautifully made compressed-air motors of his own design. After some experiments with models propelled both by screw propellers and by flapping wings, Hargrave turned his attention to kites, trying many different kinds until early in 1893 he invented the now-familiar box kite (which he preferred to call a 'cellular kite'). Many configurations of the box kite were tried before, in the mid-1890s, Hargrave settled on the simple two-box shape. Once the virtues of the box kite – its simplicity, rigidity, and stability in flight – had become widely known, it was adopted as the structural basis for a number of early gliders and aeroplanes. With minor modifications it also became the standard configuration used by meteorologists, who were soon raising trains of box kites to heights of several miles.

## Man-lifting Kites

As a result of the rapidly accelerating interest in flying machines during the 1890s, a number of people turned their attention to the man-lifting capabilities of kites. In 1894 Captain B. F. S. Baden-Powell devised for the British army a system of man-lifters to be used as a means of extending an observer's range. His first trials were made with a huge plane-surface kite 11 m (36 ft) high, but as this proved far too difficult to handle he soon turned to trains of 4 m (13 ft) plane-surface hexagonal kites, which he patented under the name of *levitors*.

Although Baden-Powell's man-lifting system worked after a fashion, it was too unstable for regular use. More successful man-lifters were evolved from box kites by other experimenters, among whom Samuel F. Cody deserves special mention. (He was a friend, but not a relative, of his better-known namesake 'Buffalo Bill' Cody). After tentative beginnings with large standard box kites, Cody modified the basic design, adding wings for extra lift and stability, and finding, like Baden-Powell, that trains of smaller kites were more manageable than a single large one. His patented man-lifting system was so well engineered and so useful that the British army officially adopted it for a time in the early 1900s. The rapid development of the aeroplane, however, made man-lifting kites redundant almost as soon as they had been perfected, and after experimenting with gliders Cody himself turned his attention to powered flight, using modified box kites for his first experiments (*see* Chapter 5).

## Kites in the 20th Century

Even after the aeroplane had become firmly established as the vehicle of the future, a few people continued to work with kites. Among the most impressive of the early 20th-century designs were the huge compound kites of the Scottish-born American Alexander Graham Bell, whose more enduring claim to fame is his invention of the telephone. Like Cody, Bell began with rectangular box kites, after which he tried many other shapes, including rings, double spools, and triangular boxes, the results of his experiments being carefully written up in a series of interesting and detailed articles. Bell finally settled on a structure which is ultimately derived from the Hargrave box, but which differs markedly from it: the tetrahedron. The basic shape of a tetrahedral kite is a four-sided pyramid, each side of which is an equilateral triangle. If two sides are covered, the result is a kite with two triangular wings folded upwards at a sharp dihedral angle (*see* Chapter 7). Bell made kites which contained several thousand such tetrahedral cells and were large enough to lift a man. Although experiments with powered versions were made, they led nowhere: the machines were fragile, produced enormous amounts of drag (*see* Chapter 7), and were almost impossible to manoeuvre or control.

For some years after Cody and Bell, kites faded from the scene as powered aircraft increasingly caught the attention of the public. They were nevertheless far from wholly neglected, playing in particular an important minor role during both World Wars. During World War I the Germans used man-lifting box kites to carry an observer over a surfaced submarine, a technique which they exploited again during World War II using a neat and effective autogyro-kite. During World War I the Allies used kites to carry barrage wires above convoys, and in the United States Paul E. Garber developed a fine, manoeuvrable gunnery target consisting of a large, plane-surface diamond kite equipped with a controllable rudder. Two flying lines were used, manipulation of which could cause the kite to perform

vigorous aerobatic manoeuvres. With an aeroplane painted on the surface, these kites provided useful target practice for trainee gunners. Most of the aerobatic control-line kites which have become popular as toys in recent years dispense with the rudder, relying instead on a technique of banking the kite from side to side. As a result, they are much slower to manoeuvre than the more complex Garber type, versions of which have also become available as toys in recent years.

Soon after World War II Francis Rogallo, then with the US National Advisory Committee for Aeronautics – now the National Aeronautics and Space Administration (NASA) – began to experiment with 'stickless' kites. Entirely flexible kites had been known for some time: in the late 19th century Baden-Powell had flown one, while in subsequent decades simple 'parachute kites' were occasionally to be seen. Rogallo, who was concerned to develop something more sophisticated, evolved in 1948 the stable, non-rigid delta-wing design which has now become generally familiar as the 'Rogallo flex-wing'. The true flex-wing is entirely stickless, the delta shape being sustained by air pressure alone. Rigid and semi-rigid versions of the Rogallo kite have become still better known as the basic configurations of ski-kites, hang-gliders, and a number of para-gliders and short-haul vehicles with which several American aviation companies, notably North American and Ryan, experimented in the 1950s and 1960s.

An equally revolutionary advance in design was made by another American, Domina C. Jalbert. It is to Jalbert that we owe the more complex non-rigid kite known as the parafoil. This is essentially a high-lift wing of conventional aerofoil shape, the upper and lower surfaces of which are made of cloth or sheet plastic.

Early in the 20th century Alexander Graham Bell perfected the tetrahedral kite. Although his designs for powered man-lifters were failures, modern versions based on large numbers of tetrahedral cells are widely used for sporting purposes.

**Right** Kites of almost any shape can be made to fly as long as they have sufficient stable lifting surfaces. In recent years western designs, such as this ship, have vied with oriental kites for elegance and ingenuity.

**Left** After World War II Francis Rogallo perfected the first fully practicable non-rigid (stickless) kite, which was later widely used as the basis of para-gliders and short-haul vehicles. Rigid-frame versions of Rogallo's delta-wing design are used in ski-kites and, as here, in hang-gliders.

The surfaces are separated by vertical ribs of fabric, and the leading edge is open to allow the airflow to enter and inflate the wing. Ventral fins and shroud lines are used to stabilize the shape and to maintain the correct angle of attack (*see* Chapter 7). Unlike many other kites, parafoils fly at angles below the stall, so they are highly efficient and have found many applications for military, meteorological, and sporting purposes. One of their commonest uses is as the basis of a steerable, rectangular parachute which is in effect a kind of hang-glider and similarly capable of long glides.

Development of the kite is continuing, partly for sporting purposes in connection with the many new kite-flying clubs and associations which have been founded in the 1970s, and partly as a potential means of harnessing the energy of the winds. Among the most exciting possibilities are tentative plans to use trains of kites for raising wind-operated generators into the fast-flowing jetstreams (*see* Chapter 6), where they would be able to run for long periods without having to be brought down.

**Above and right** One of the most successful modern kite designs is Domina C. Jalbert's parafoil, a non-rigid, high-lift aerofoil which is kept inflated by wind entering through the open leading edge. Versions range from the small kites seen here to large, manoeuvrable models used as sporting parachutes.

# 4 Gliders

Whereas it is impossible for a man to rise and sustain himself in the air with the use of flapping wings powered by his own unaided muscles, modern achievements with hang-gliders have shown that soaring flight with fixed wings is by no means out of the question. Although ignorance of the principles of flight must have condemned most early attempts at flying by winged men to total failure, it is possible – even likely – that in some cases their wings may have been capable of supporting them in some semblance of a controlled glide over a short distance.

Outside the realms of myth and legend, the first successful hang-glider flight may have been made in England by a monk known as Eilmer of Malmesbury. The date of Eilmer's birth is unknown, but he was an old man at the time of the Norman Conquest of Britain. As he was said to have undertaken the flight 'in his earliest youth', the event probably occurred in about AD 1000. Eilmer 'fastened wings to his hands and feet . . . and, collecting the breeze on the summit of a tower, he flew for more than the distance of a furlong' (about 200 m or 600 ft). The landing resulted, we are told, in his breaking both his legs – a crash which Eilmer intelligently attributed to his having omitted to equip his glider with a tail.

In about 1498 a young Italian mathematician and inventor, Giovanni Battista Danti of Perugia, made a similar attempt. Like many experimenters in later centuries, he first practised over water, using wings attached to his body by means of iron bars. Having learned to control his glides to his own satisfaction, he then chose an important public spectacle – the marriage of a duke – at which to demonstrate his skill in the centre of the city. During this flight one of the iron bars broke, the wings collapsed, and he crashed onto the roof of a church, where he broke a leg.

Broken bones turn up again in the story of John Damian, an Italian charlatan who lived at the court of James IV of Scotland. In 1507 Damian tried to impress the king by flying, possibly with a fixed-wing glider, from Stirling Castle to France. Instead of flying, however, Damian fell by the castle walls, where he broke his thigh. In common with Eilmer, Danti, and many others, Damian used feathers as the basis of his wings. After the crash he attributed his failure to an unwise choice of materials: instead of the feathers of the high-flying eagle, he had used those of the common earthbound hen.

Many similar attempts at man-powered flight were made in the years preceding the invention of the aeroplane. In France a locksmith called Besnier and the Marquis de Bacqueville were both credited with partial success. In the 17th century the painter Guidotti may have managed some kind of glide. In Germany an aeronautically minded cantor broke an arm, after which he presumably returned to the less hazardous business of church music.

As far as one can judge, Danti was the only one among these courageous experimenters who gave any serious thought to the mechanics and aerodynamics of what was being attempted, and it may be that he intuitively grasped the advantages of gliding over flapping. As he committed nothing to paper, however, it fell to Leonardo da Vinci, a dedicated 'flapper', to write the first extended analyses of the theory and practice of human flight. A professional engineer as well as a great painter, Leonardo was fascinated by aeronautical theory and filled scores of notebook pages with sketches and speculations. After several years of observation and thought, he devoted a small manuscript book to what he took to be the fundamental basis of the whole subject: bird flight.

Elsewhere in Leonardo's manuscripts there is draft after draft of potential flying machines of many different configurations. Nearly all of these were to have had wings flapped by systems of levers and wheels driven by the flyer's arms or legs, or both. The suggested construction is often highly ingenious, and the details are usually drawn with sufficient accuracy to enable the whole machine to be built. Sometimes the aviator was to lie prone, as in a hang-glider, and sometimes he was to sit or stand in a boat-shaped nacelle. While Leonardo may have built and even tested some of these machines, there is no evidence that he ever did so: attempts to fly in any of them would have resulted in immediate failure. Great thinker though he was, Leonardo shared many contemporary misconceptions about aerodynamic principles, and in particular about the movements of a bird's wings. This led him to design machines which, apart from the inadequacy of human muscle-power to operate them, would have been totally unstable in flight.

On at least one occasion, however, Leonardo managed a partial break away from his usual concern with flappers. At some time about the year 1500 he made intriguing sketches for a large flying machine with fixed monoplane wings, from the centre of which the pilot was to be suspended in a harness. Although he included flappable hinged wingtips, operated by a system of cords, most of the lift would have been provided by the rigid inboard parts of the wing. This powered hang-glider, the first of its kind ever to be designed, is a remarkable precursor of machines with which Otto Lilienthal was to gain so much valuable experience of the air nearly 400 years later.

Among the most interesting of Leonardo da Vinci's ideas for flying machines is this series of sketches for a semi-ornithopter. Only the wing tips were to be flapped, the inboard parts of the wing remaining rigid. As with the hang-gliders used by Otto Lilienthal, the pilot was to have been suspended in a harness.

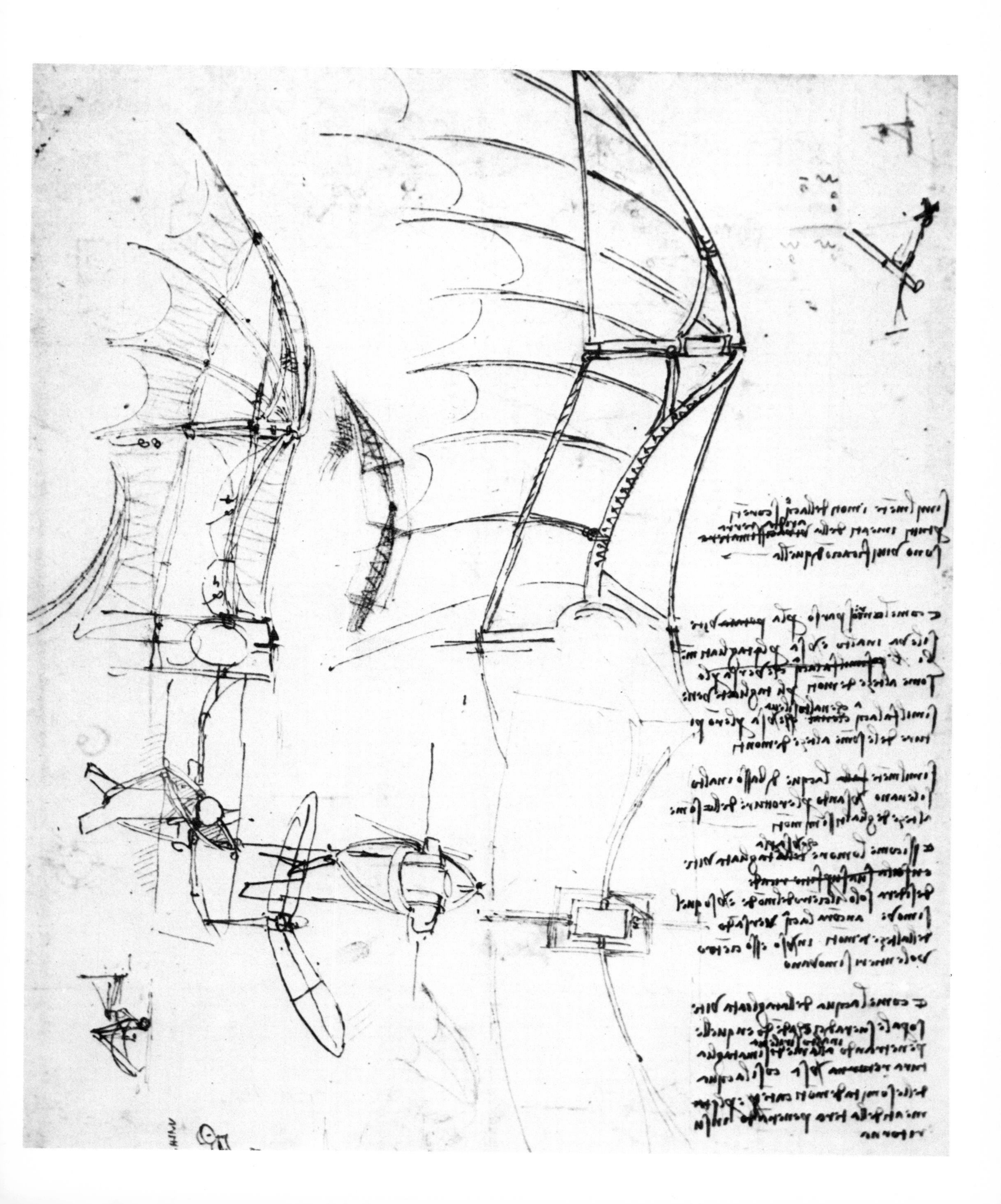

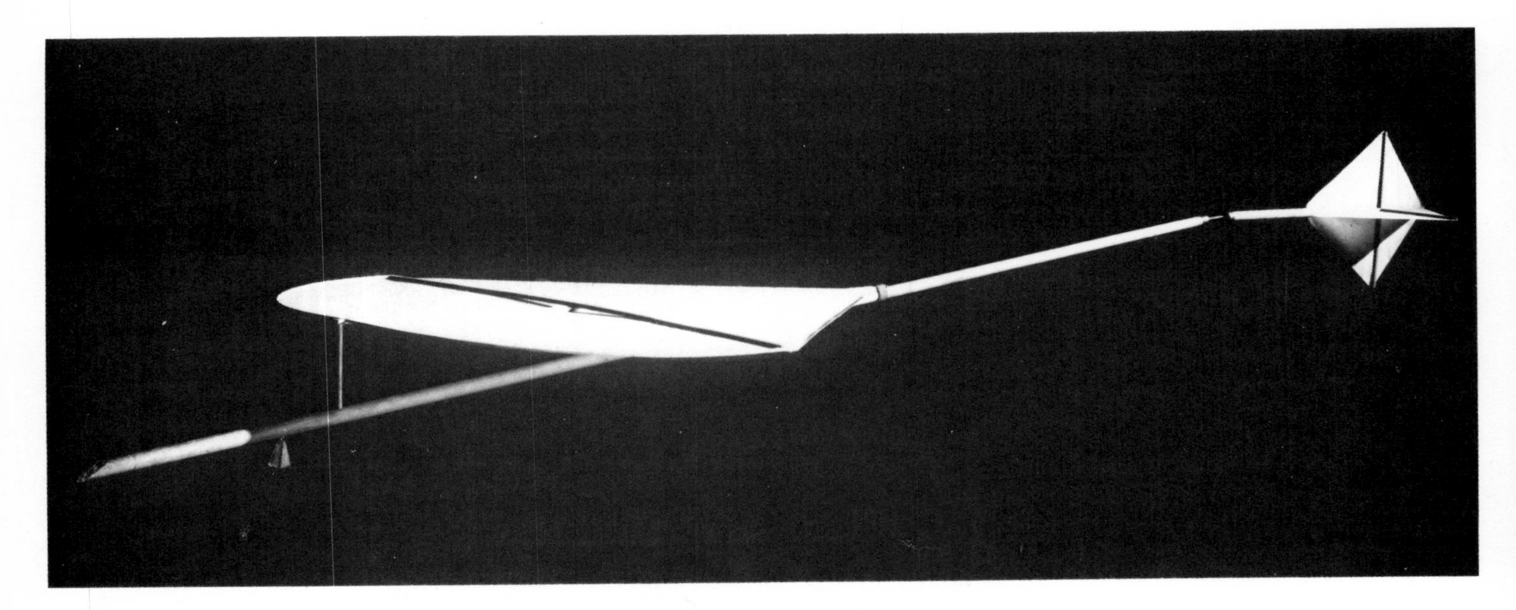

Above Sir George Cayley made this model glider in 1804. The wing, set at a six-degree angle of incidence, consisted of a kite with a semi-circular leading edge. A movable, cruciform tail unit provided directional control, and a sliding weight at the nose allowed Cayley to alter the position of the centre of gravity.

## Gliding in the 19th Century

After Leonardo almost no serious work was carried out in heavier-than-air flight before the investigations of Sir George Cayley (1773–1857). In 1804 this brilliant and careful workman, whom William Henson rightly called 'the father of aerial navigation', made a now-famous model glider 1.5 m (5 ft) long, from which he derived much useful information. The glider, a replica of which is in the Science Museum, London, consisted of a small round-topped kite attached to a pole. An adjustable cruciform tail unit was added at the rear, and a sliding weight was fitted to the nose for altering the position of the centre of gravity. Describing the glider's performance, Cayley wrote:

> It was very pretty to see it sail down a steep hill, and it gave the idea that a larger instrument would be a better and safer conveyance down the Alps than even the surefooted mule, let him meditate his track ever so intensely. The least inclination of the tail towards the right and left made it shape its course like a ship by the rudder.

Rightly sensing that the problems of stability and control must be solved before he considered possible power units, Cayley made other gliders, one of which, a full-sized machine built in 1809, he tested for a time in short hops. After three decades of further experiment and theoretical investigation, Cayley returned when quite an old man to the construction of full-sized gliders. Having since 1843 been convinced that a multiplane structure was likely to be better than the less-rigid braced monoplane, he built, in 1849, a triplane glider which once or twice carried a boy for a few yards. In 1852 he returned to the monoplane, designing (but not building) a graceful glider embodying many refinements of construction and control, and in 1853 he built the famous 'coachman carrier', a glider in which Cayley's coachman was flown across a small valley.

Public enthusiasm for aeronautics was kept alive during the 19th century by a number of well-publicized gliding exploits, notably a series of descents from a balloon made by the Frenchman Louis-Charles Letur in 1853–4. Letur's large and fragile machine was a curious combination of parachute and glider. After a number of successful glides, he was killed when trying to rectify a fault that had occurred before he released himself from his balloon. Also in France, an enterprising sea captain, Jean-Marie Le Bris, built two graceful but fragile bird-shaped gliders, which he tested in 1857 and 1868.

## Otto Lilienthal

Neither these experiments, nor the tests carried out in 1858–9 by the Englishman F. H. Wenham on a five-winged glider, advanced the practice of gliding to any important degree. By the 1880s, however, gliding had been put on a solid theoretical and practical footing through the brilliant work of Otto Lilienthal (1848–96). Although in his book *Der Vogelflug als Grundlage der Fliegekunst* (Bird Flight as the Basis of Aviation) he expressed the belief that the way to success lay in imitating the flapping flight of birds, he also believed that before studying methods of propulsion it was essential for the aspiring aviator to learn to control a glider in the air. Accordingly in the 1890s he built a series of 18 different types of fixed-wing glider

Right The first really serious experiments with full-sized gliders were carried out in the 1890s by Otto Lilienthal. Although he preferred braced monoplane gliders, his biplane types (as here, with Lilienthal aboard, in 1894) were probably easier to control. Both types were manoeuvred by swinging the body to shift the centre of gravity – a method akin to that used on modern hang-gliders.

which he flew himself. For his early work he launched his gliders from natural heights, but in 1893, after he had built a large hangar, he began flying from its roof; finally he made an artificial mound about 15 m (50 ft) high, from which he could fly in any direction, according to the dictates of the wind.

Lilienthal built his machines, all of which were hang-gliders, with great engineering skill. Although in the last years of his life he was beginning to consider the use of control surfaces, his basic form of control, as with most modern hang-gliders, was the swinging of the pilot's body to bring about a shift of the centre of gravity. Lilienthal kept meticulous records of his work, publishing specifications and tables which were useful to many later experimenters. He might well have advanced aviation a great deal further had he not been tragically killed in August 1896, when a sudden gust caused his glider to stall. A wing dropped sharply when Lilienthal was about 15 m (50 ft) from the ground, giving him insufficient time to recover.

## Pilcher and Chanute

Work of a similar nature was undertaken in Britain by the Scot Percy Pilcher (1866–99), who observed Lilienthal's progress very closely and visited him in Germany. Although by no means as original a thinker as Lilienthal, Pilcher worked with enormous enthusiasm and towards the end of his life was making interesting plans for a powered machine which might have proved successful. Regrettably he also was killed in an accident, in September 1899, when he was flying in bad weather. A bamboo rod snapped, causing his glider to plunge out of control.

The French-born American Octave Chanute (1832–1910) was over 60 when he began to take an active interest in manned flight. Having published his classic book *Progress in Flying Machines* in 1894, he soon became one of the most influential aeronautical experimenters of his time. In 1896 Chanute began building triplane gliders closely related to the Lilienthal designs but incorporating structural refinements to make them still more rigid and stable. At the age of 64 he did not, of course, attempt to do his own test flying, employing instead an enthusiastic young engineer, A. M. Herring, who was later to experiment with a powered biplane of his own design.

Chanute's main importance lies in his encouragement of the Wright brothers, Wilbur (1867–1912) and Orville (1871–1948), who in 1900 wrote to him for advice. A close relationship developed from which the Wrights benefited in many ways, particularly in their adoption of Chanute's excellent braced biplane wings. There were nevertheless many fundamental differences in their respective approaches to the problems of aviation, notably in their attitudes to the problem of stability and control. Chanute had always been among those who gave absolute priority to stability, seeking, as he put it, 'automatic equilibrium exclusively'. He nevertheless responded without prejudice to the advice of his pilot, Herring, and always stressed the need to work from direct experience of the air.

## The Wright Gliding Experiments

In 1899, after carefully considering the available literature and after closely studying the flight of birds, the Wrights began the rationally ordered sequence of experiments which led to their ultimate and much-deserved success. The essential advance in the Wrights' thinking lay in their pursuit not of inherent stability but of controllability. Having observed that buzzards appeared to retain their equilibrium in gusty conditions of flight by (as Wilbur put it) 'a torsion of the tips of the wings', the brothers first built a biplane kite whose wings could be warped (twisted) at the tips. A fixed tailplane was used to create fore-and-aft stability, and the wings were so arranged that they could be moved backwards and forwards in relation to each other to alter the 'stagger' and so shift the relative position of the centre of gravity.

Following their correspondence with Chanute the Wrights used the experience of their kite to help them experiment with a full-scale biplane glider. This they flew over the sand dunes of Kitty Hawk, North Carolina, during the autumn of 1900, making a few short manned glides, but mainly testing it as a kite. The wing-warping system was incorporated, as was the forward elevator which was to remain characteristic of all Wright machines for a decade, but as yet there were no vertical surfaces (rudder or fin).

They built a second glider in 1901 and a third in 1902. During experiments with the second glider the Wrights found that, although the wing-warping system produced the required banking action, it also caused the upgoing wing to be retarded, thus turning the glider against the direction of bank. The Wrights' most important single insight, which led to what was probably their greatest contribution to the development of powered flight, was their diagnosis of the trouble. The upgoing wing, they realized, was retarded by the now familiar phenomenon of warp drag, resulting in adverse 'yaw' (movement about the glider's vertical axis). Simultaneously with the increase in lift produced by the warp, the upgoing wing contributed more drag than the downgoing wing, resulting in a yaw in the unwanted direction.

In their third glider the Wrights set about trying to cure this problem. At first they fitted a double fixed fin at the rear. As this was only partly successful, they took the vastly important step of converting the double fin into a rudder linked by cables to the wing-warping system. The *No. 3* glider could now be fully controlled about all three axes, the linked rudder making it possible to keep the machine more or less balanced in a turn. This glider formed the basis for the Wrights' first successful powered machine, which they built and flew in 1903.

With their *No. 3* glider, flown in 1902, the Wright brothers finally solved the basic control problems they had been studying since 1899. A movable rudder, coupled to the wing-warping system, enabled them to fly balanced turns. Here Orville Wright begins one of more than 900 controlled glides the brothers made at Kitty Hawk, North Carolina, in preparation for their epoch-making powered flights in 1903.

## Gliding in Europe

While the Wrights were engaged in their painstaking work, a few men in Europe were independently experimenting with gliders. Among these was Captain Ferdinand Ferber, a French artilleryman who, in 1901, was attempting to develop a hang-glider built on Lilienthal lines. After corresponding with Chanute, and after reading accounts of the Wrights' experiments with their first and second gliders, Ferber began experimenting with a simple Wright-type glider which had a forward elevator but no vertical surfaces. By 1904 he had modified this primitive machine to create a new configuration which was to have an important influence on the history of aviation: while retaining the forward elevator Ferber took the significant step of adding at the rear a fixed horizontal tailplane. Departing, as did virtually all the Europeans, from the Wrights' idea of inherent instability, Ferber created a degree of stability in the rolling plane by angling his wings upwards, creating dihedral. It was a machine broadly along these lines that Ferber used in 1905 as the basis of a moderately successful powered aircraft.

In April 1903, during a visit to Europe, Chanute gave a lecture which had considerable influence on glider design. As well as Ferber, his audience included Ernest Archdeacon, a wealthy French enthusiast who, in 1904, built and tested a modified version of the Wright *No. 3* glider. After Archdeacon had been joined by a young colleague, Gabriel Voisin, the two men decided to try yet another configuration. Seeking greater inherent stability they built, in 1905, a large float-glider using the Hargrave box-kite principle. Both the biplane wings and the tailplane were fitted with vertical dividing surfaces ('side curtains'), while a forward elevator, in the Wright style, was used for control in pitch. In the same year Voisin collaborated with Louis Blériot to build a second float-glider, which may most readily be distinguished from its predecessor by the sharply angled outer side-curtains on the wings. Machines of this kind were to form the basis of many powered flying machines in Europe before the basic configuration of the modern aircraft had been firmly established.

## Later Bird-men

Once powered flight had been achieved, gliding was given little further attention until after World War I, when the Germans, prohibited by the Treaty of Versailles from developing powered aircraft, applied their skills for a time to perfecting the construction and handling of gliders (*see* Chapter 22). In the case of would-be bird-men, however, the invention of the aeroplane seems to have stimulated rather than dampened their enthusiasm. In the 20th century, more than ever before, men have vigorously pursued the goal of flying by means of wings attached to the body. The carefully engineered hang-glider of today was preceded by the energetic experiments of stunt-flyers. Perhaps the most celebrated stunt-flyer of all time was the American Clem Sohn, who allowed himself to be dropped from aeroplanes in order to learn the skill of manoeuvring with wings attached to his arms and legs. So as to land safely he used a parachute for the last part of the descent. Although Sohn did much exhibition flying in order to raise money, he was seriously interested in perfecting the art of manipulating his wings. He might have contributed significantly to free-fall techniques had he not died in 1937, after his parachute had failed to open at 550 m (1,800 ft).

## The Invention of the Parachute

Although it is likely that many unrecorded attempts to build parachutes were made in the remote past, the earliest-known drawing of one is in a Siennese manuscript now in the British Library. This probably antedates, by a few years, the better-known parachute drawn by Leonardo in about 1485. Both designs have rigid frame members at the base, presumably on the assumption that they would not otherwise remain open. Although both parachutes look impossibly small, the true scale, at least in the case of Leonardo's, is much greater: his was intended to be well over 7 m (23 ft) wide.

About 100 years later another square-rigged parachute appeared in a book on machines by the Venetian Fausto Veranzio. It was not until after the first hot-air balloons had flown, however, that anyone attempted a jump. On 22 October 1797 Jacques Garnerin released himself from a balloon over Paris at an altitude of about 1,000 m (3,200 ft). He descended safely, and started something of a fashion which was followed by stunt balloonists throughout the 19th century. Inevitably there was some loss of life, including that of Robert Cocking, who died in 1837 at the age of 60 when he jumped over Kent in an untested parachute of new design. Sir George Cayley had suggested that a shallow conical structure, with the apex pointing down, would give a parachute much greater stability and avoid the often violent swinging of some of the earlier types. Although the theory was sound, Cocking's parachute was insufficiently strong and broke up in flight.

The first parachute drop from an aeroplane using a modern canopy was on 1 March 1912 by the American Captain Albert Berry.

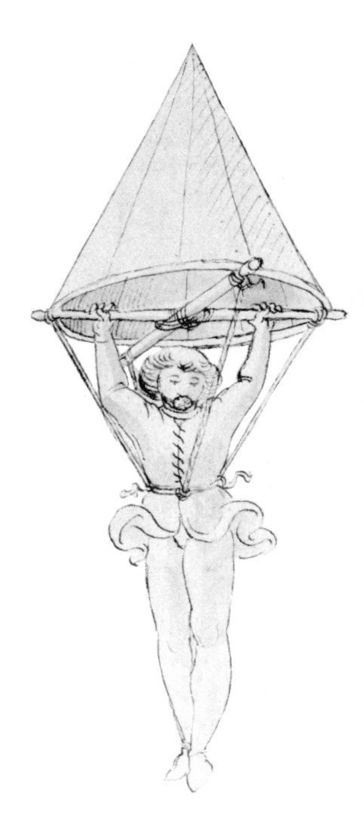

**Above** The first western parachute was designed in about 1470 by a Siennese engineer. The conical cloth canopy, held open by a wooden ring at the base, was probably intended to be much larger than the scale of the drawing indicates.

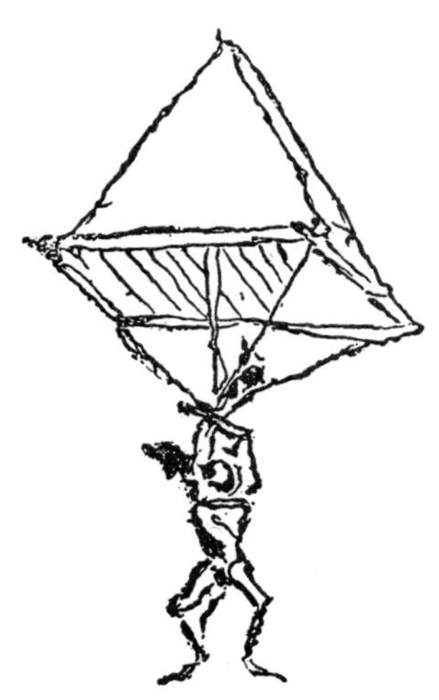

**Above** Leonardo's drawing of a pyramid-shaped parachute dates from about 1485. The canopy, over 6 m (20 ft) wide, was to have been of linen.

# 5 Powered Flight

**Right** In 1843 Sir George Cayley drew up detailed plans for his 'Aerial Carriage' convertiplane. The eight segments of each circular rotor were to be angled upward for vertical lift, and then made to lie flush to form smooth wings for forward flight, which was to be powered by pusher propellers.

The first serious aerodynamic experiments, as distinct from speculations, were undertaken by Sir George Cayley, who in 1804 used a whirling arm to examine the properties of lifting surfaces. During the next five years he investigated the crucial questions of angle of attack, movement of the centre of pressure (the point on the surface of a wing through which the sum of the lifting forces may be said to act), the effects of camber (curvature), the correct shape for streamlining, and the value of dihedral for creating lateral stability. Cayley can also be credited with having had the first mature insights into the true movements of a bird's wing in flight. While looking into the problems of propulsion by flapping, he came to understand that the primary feathers of the wing (the 'flight' feathers at the tip) twist under the pressure of the air to form small propeller blades driving the bird along by the production of 'lift' acting in the direction of flight.

Before he began systematic work on aerodynamic problems, Cayley's imagination had been caught by an 18th-century development of the medieval string-pull helicopter. In 1784, when Cayley was still a child, two Frenchmen, Launoy and Bienvenu, had made a toy helicopter consisting of two contra-rotating airscrews powered by a small bow-drill motor. Using rotors made from feathers stuck into corks, Cayley built his own version of this toy in 1796. Having given the possibilities of the helicopter a good deal of further thought, he published a modified design, with a commentary, in 1809. The little flying machine quickly became well known, and may be looked upon as the true origin of all subsequent helicopter development.

Prompted by the ideas of a younger man, Robert Taylor, Cayley returned to the idea of the helicopter in 1843. This time he designed something much more ambitious: a *convertiplane* using twin full-circle rotors placed on either side of a boat-shaped fuselage. For forward flight the inclined blades of the rotors could be made to lie flat, forming smooth, circular, biplane wings. Although the convertiplane was never built, it was highly imaginative and anticipated many later rotorcraft ideas.

After Cayley's early experiments on fixed-wing aircraft, very little important further work was done until 1843, when William Henson published his famous design for an 'Aerial Steam Carriage'. Although this remarkably modern-looking aeroplane was never built, the drawing was so strikingly original and prophetic that it was repeatedly reproduced. Together with his colleague John Stringfellow, Henson built a small-scale model of the machine with a wingspan of 6 m (20 ft). In the years 1845–7 this was several times tested, without success, after which Henson abandoned aeronautical work. In 1848 Stringfellow took up the project again, building an even smaller model which he tested after suspending it from an overhead wire. To what extent this new model managed to sustain itself under its own power has always been in doubt, but it seems that Stringfellow can be credited with only limited success. Another model, built in 1857–8 by the Frenchman Félix du Temple, was more satisfactory, but neither his work nor that of Henson and Stringfellow added significantly to our understanding of aerodynamics.

Apart from the work of Cayley the most important contribution to the subject in the mid-19th century was a paper entitled 'Aerial Locomotion' delivered to the Aeronautical Society in 1866 (the year of its foundation) by Francis Wenham. Wenham had made important findings about camber, centre of pressure, and aspect ratio (the relationship of a wing's length to its width). In common with Cayley, who had always feared long, braced monoplanes, Wenham advocated superposed planes of the kind with which he himself had experimented in the late 1850s. In 1871 Wenham collaborated with John Browning to test aerofoil (wing) surfaces in a wind-tunnel, the first ever to be built.

A further significant step in the development of aerodynamics was taken when Horatio Phillips (1845–1924) began systematic experiments with aerofoils of widely differing section. In 1884 he patented a series of shapes of varying thickness and camber, having demonstrated conclusively that the greater part of the lift is produced by the low-pressure region on the upper surface. Phillips's work became widely known and exerted considerable influence. In 1893, having further developed his ideas, he tested a large powered multiplane which travelled around a circular track and successfully demonstrated his theories of lift.

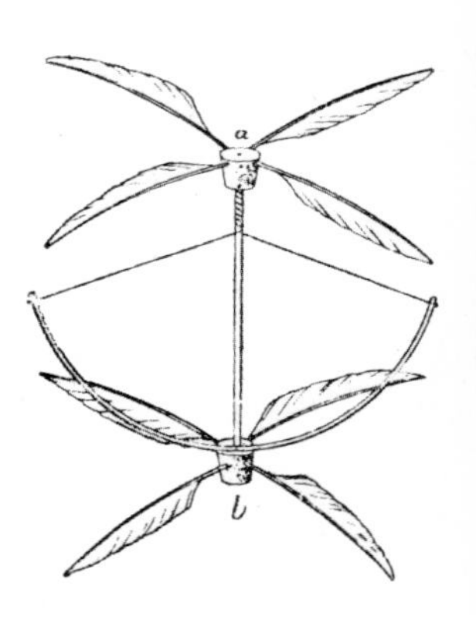


**Above** As a young man Cayley constructed this toy helicopter, basing it on an earlier French design. Contra-rotating propeller blades made of feathers were driven by a bow-string motor.

Cayley and Phillips had shown that lift results from low pressure on the top surface of an aerofoil, assisted by a region of higher pressure below. But it was not until F. W. Lanchester formulated the so-called 'circulation theory' – presented in 1894 to the Birmingham Natural History and Philosophical Society in a paper entitled 'The Soaring of Birds and Possibilities of Mechanical Flight' – that the reasons why lift is created were properly understood. Broadly speaking, circulation theory shows that lift is generated because the velocity of the airstream over the upper surface of an aerofoil is greater than that over its lower surface, and so exerts less pressure. Unfortun-

ately, Lanchester did not publish his findings until 1907, in his book *Aerodynamics*, and even then he presented them in a rather indigestible form. The circulation theory of lift was developed almost simultaneously in Germany by Ludwig Prandtl, to whom must go the credit of formulating the ideas in readily comprehensible mathematical language. His work, together with that of Wilhelm Kutta and the Russian Nikolai Joukowski, placed aerodynamic theory on a firm foundation and profoundly influenced the development of 20th-century aerodynamics.

## Early Attempts at Powered Flight

Nearly all the experimental work which ultimately led to the building of practical aeroplanes was nevertheless undertaken by a combination of trial-and-error and intuitive insight. The second half of the 19th century saw the creation of many powered models and full-sized aeroplanes, some of which achieved brief flights or hops. In about 1874 Félix du Temple built and tested an aeroplane powered by a hot-air engine. It became airborne after travelling down a ramp with a young sailor as pilot. Ten years later a large steam-powered machine, built by the Russian Nikolai Mozhaiski, took off after a run down a ski-jump and remained briefly airborne. It is sometimes stated that these hops, and one or two others like them, should be recorded as the first true powered flights, antedating the Wrights. What the Wrights achieved, however, was not merely to get off the ground, which is not so very difficult, but to build a machine which could be fully controlled in all three axes of movement while sustaining itself aloft. In this respect they have unrivalled priority.

Before the Wrights others had, nevertheless, developed some insights into crucial matters of stability and control. As long ago as 1868 M. P. W. Boulton had envisaged the use of ailerons for the control of roll, although he did not fully understand their potential. Wing-warping was suggested by Lilienthal (1895), while a remarkably detailed and prophetic design for an aeroplane controllable in all directions was patented in 1876 by Alphonse Pénaud. This machine, which lacked only a positive system of control in roll, was to have had rear elevators, rear fin and rudder, steering airbrakes on the wings, a retractable under-carriage, an enclosed cockpit, and a single control column for movement of the elevators and rudder.

Most of the practical attempts to fly full-scale aircraft in the 1880s and 1890s were undertaken with comparatively crude machines. Large claims have been made for some of them, especially for Clément Ader's 20 hp, steam-driven *Eole*, which managed to take off under its own power in 1890. The *Eole*, a curious bat-like monster with enclosed cockpit, four-blade propeller, and the most rudimentary of control systems, managed a hop of some 50 m (165 ft). In the mid-1890s, with the help of a subsidy from the French military authorities, he continued his work, first building a machine called *Avion II*, which he abandoned, and later completing *Avion III*, which was unsuccessfully tested in 1897.

Better work was done in the United States by Samuel Pierpont Langley, an astronomer and secretary of the Smithsonian Institution in Washington, D.C. After successful trials with small-scale models, he progressed in 1903 to a full-sized machine which he called the *Aerodrome*. This he attempted to test over water, launching it from the top of a house-boat. In two launching attempts the *Aerodrome*, with Charles M. Manley on board, fouled the launching mechanism and plunged into the Potomac river. The machine, poorly constructed, underpowered, and insufficiently controllable, was eventually flown in 1914, after secret modifications had been made. Although the configuration which Langley adopted – a tandem-wing monoplane – was unsatisfactory, and although like so many others he devoted far too little attention to control, he was a careful and intelligent worker who gave due consideration to the theory of lift and drag. Had he been prepared to learn from experience in the air, as did the Wrights, he might well have preceded them in achieving powered, controlled flight.

During the very early years of the 20th century, before the progress achieved by the Wright brothers was widely understood and accepted, many other inventors were attempting to profit from the many comparatively lightweight internal-combustion engines that were being made. In 1903, in Germany, Karl Jatho achieved a short hop in a curious little biplane powered by a 9 hp petrol engine. The next year Horatio Phillips built a 20-wing aircraft based on his 'venetian blind' test rig of 1893. Although he was able to get this machine to lift from the ground, it was virtually uncontrollable. Only slightly better results were obtained in 1907, when Phillips built a still more remarkable and totally impractical machine using four banks of slat-wings arranged in a tandem configuration.

## Successful European Aeroplanes

Although by 1905 the Wrights had created, in their *Flyer 3*, a fully practical aeroplane, no one in Europe had yet achieved anything like sustained and controlled flight. The first truly free powered flight in Europe was no more than a powered glide from an overhead cable, achieved in May 1905 by Captain Ferdinand Ferber, who had equipped his Wright-type glider with a 12 hp motor. Although by late 1905 and early 1906 the control system invented by the Wrights had been fully publicized, European workers were slow to accept their claims and slower to understand them. Progress was further slowed by the Wrights' decision to stop flying until they had sold their invention, patents for which had now been granted. Not until 1908 did they return to flying, giving public demonstrations in France (in August) and in the United States (September). Although at that time European flying was still much less advanced, some of the lost ground had been regained, and progress had been achieved with configurations in many cases radically different from that adopted by the Wrights.

In addition to the Wright-type biplanes used by some Europeans (Ferber from 1902, Robert Esnault-Pelterie in 1904), box-kite structures of the Voisin-Archdeacon-Blériot type began to emerge. Louis Blériot (1872–1936), later to fly

**Right** On 12 November 1906 Alberto Santos-Dumont's *14-bis* became the first European aeroplane to carry a man in sustained flight. The barely controllable aircraft was of canard configuration, with forward elevators; its two steeply angled wings had 'side-curtains' inspired by Hargrave's box kites.

**Right** Between 1895 and 1903 Samuel Pierpont Langley experimented with tandem-wing models which he called *Aerodromes*. Tests with a full-sized machine, in October and December 1903, were unsuccessful.

monoplanes with such courage, used a Voisin powered biplane variation of the float-glider of 1905. In its initial (1906) form this machine used, instead of the rectangular boxes, two elliptical wings in tandem. Several variants of this were tried, including a landplane version with the more usual rectangular box-kite structure at the front but retaining the elliptical tailplane. Although Blériot learned much from his experiments, he could not make the aircraft fly in any of its versions.

In the same year, however, success of a qualified kind finally crowned European efforts to fly. In 1906 Alberto Santos-Dumont, the highly enterprizing designer of airships, tried his hand at heavier-than-air flight, designing and building an aeroplane with double forward box-kite elevator, large box-kite biplane wings with a pronounced dihedral, and multiple side-curtains. After trials with the aircraft suspended from his airship *No. 14*, Santos-Dumont flew the machine, which he called the *14-bis*, from the ground. In October and November, after the addition of large octagonal ailerons, he made a series of semi-controlled flights, the last covering 220 m (722 ft).

Although Santos-Dumont's grotesque ma-

On 17 December 1903 the first sustained and controlled flight in a powered aircraft over level ground was made by Orville Wright in the *Flyer* (this painting distorts the gentle undulations of Kitty Hawk sands). The flight lasted 12 seconds, and it was followed by three more on the same day. The Wrights' reputation rests less on their priority than on their painstaking research (which included the use of a wind tunnel) into the principles and practical problems of flight over many years. The control system they built into the *Flyer* included warping wings, biplane elevators at the front, and twin rudders at the rear.

chine had a negligible effect on the development of aviation technology, it was of great psychological value in encouraging others in Europe to continue their efforts. At about the same time, another expatriate living in Paris, Trajan Vuia from Romania, experimented with machines which, although unsuccessful, were profoundly to influence the progress of aviation. In August 1906 he managed to achieve several hops in a small monoplane whose bat-like wings were equipped with a primitive warping system. Later in 1906 and in 1907 Vuia continued monoplane experiments, improving his machines but never achieving sustained and controlled flight. The monoplane had nevertheless been born: it was to flourish until the early part of World War I, when it went into a temporary decline.

Also in 1906 the Dane Jacob Ellehammer, for whom large claims have sometimes been made, tried a primitive biplane which ran on a circular track while tethered to a central post. Ellehammer managed to get his machine to lift off without, however, achieving free flight. Although in 1907 and 1908 he had more success with other machines, he did not contribute significantly to the history of aviation.

Inspired by the partial success of Vuia, Blériot now turned his attention to monoplanes. During 1907 he tried three machines of his own design, the last of which, although only a qualified success, firmly established the basic configuration of the modern monoplane.

Adequate engines had now become readily available, especially Léon Levavasseur's justly famous Antoinette, which powered a great many machines. The Europeans had also learned a good deal about inherent stability. In this respect they were in advance of the Wrights, who continued to adhere to their policy of building unstable machines needing continuous control by the pilot. The Wrights' approach had both advantages and disadvantages. Provided it was not pushed too far, a degree of inherent instability made the aircraft quick and light to manoeuvre. It also, however, required constant vigilance from the pilot to correct the aircraft's twitchy tendency to depart from the desired flight path. The choice of a forward elevator, which the Wrights originally chose in the mistaken belief that it would prevent a sudden dive, was especially unsatisfactory because it created undesirable instability in pitch.

By the time Wilbur Wright came to France to demonstrate the *Flyer* at Le Mans in 1908, the Europeans had created aeroplanes endowed with something more nearly approaching an optimum degree of stability. They were nevertheless still vastly inferior in their understanding and application of control, a fact which became abundantly clear from the moment that Wilbur made his first European flight of 1 min 45 sec on 8 August 1908. By the end of that year Wilbur had made more than 100 flights, one of them lasting 2 hr 20 min. The ease with which he could turn, bank, fly figures of eight, and land gracefully showed that, whatever fundamental weaknesses there might be in the configuration of the *Flyer*, the Wrights had learned how to make a reliable and practical machine, and, above all, they had learned how to fly it with conspicuous skill.

In 1907 Gabriel Voisin and his brother had begun to develop powered machines based on the float-gliders of 1905. These were stable pusher biplanes with large side-curtained tailplanes and forward elevators. One of them, made for Léon Delagrange, flew several times in 1907 but eventually crashed. Another, built in the same year for Henry Farman (a Briton living in France), had a happier career. It was many times modified, the changes including in particular the addition, in the autumn of 1908, of large ailerons for control in roll. In making this modification Farman showed himself to be the first European fully to understand the essential nature of the Wrights' control system. Although before the modifications he had been able to execute yawing turns, his machine was incapable of turning in balance – a manoeuvre possible only by the addition of ailerons or by the alternative system of wing-warping.

Other Voisin machines were being flown at the time, notably by Léon Delagrange (who was killed in an accident on 4 January 1910, flying a Blériot). Until as late as 1910, however, many Voisin machines still lacked any means of control in roll, a failing which was rapidly to make them obsolete.

## Early British Aviation

In spite of its technological skills and the existence of the enthusiastic Aeronautical Society (now the Royal Aeronautical Society), Britain contributed very little to the progress of aviation in the early years of the 20th century. Powered flying began in England with the flamboyant American expatriate, Samuel F. Cody, whose elegant man-lifting kites had already gained official War Office approval. In 1905 Cody built and flew a biplane glider based on a kite and equipped with ailerons. Following these trials he embarked on the design of a powered aircraft, which was completed and tested in 1908. This was a biplane of the Wright type, with forward elevator, rear rudder, and small ailerons. For the next five years Cody put much energy and enthusiasm into promoting the cause of aviation in Britain. He was a skilled mechanic and became a first-rate pilot; but he was not an especially original thinker and he made no significant contribution to the development of the aeroplane. In 1913 British aviation lost one of its most colourful figures when Cody was killed in an air accident near Aldershot.

Although Britain was very late to enter the world of powered flight, Cody was by no means the only one in Britain building and flying aeroplanes in his day. Alliot Verdon Roe, founder of the Avro company, built a Wright-type machine which was tested in 1908; in the same year J.T.C. Moore-Brabazon learned to fly Voisins in France, and he brought one back with him to become the first Briton to fly in England. Roe went on to develop his own line in biplanes, which led in 1913 to the creation of one of the most famous trainers of all time, the Avro 504.

## Before World War I

Although in 1909 aeroplanes were still in the process of fundamental development, that year may be said to mark their coming of age. On 25 July Blériot crossed the English Channel, to the surprise and delight of many and the consternation of some. In August of that year the first and possibly the most celebrated of all aviation meetings was held near Rheims (France), where biplanes and monoplanes competed for a variety of handsome prizes. From that time until World War I aeroplanes developed rapidly in many different ways. In 1910–1 float-planes became a practical possibility; night-flying was tried; airmail services were begun; aircraft were armed for war; and manoeuvring was developed to establish the art of aerobatics, which was pioneered in 1913 by Adolphe Pégoud, flying a Blériot. Typical speeds were in the range 80–160 km/h (50–100 mph), with some planes flying at more than 160 km/h (100 mph), while pre-war altitudes as great as 6,000 m (20,000 ft) were achieved.

Although it was not until shortly before World War II that the monoplane became the standard configuration, many excellent monoplanes were developed before 1914. Had the war not intervened, it is very probable that machines by Deperdussin, Nieuport, Morane-Saulnier, Blériot, and others would have been more rapidly developed. The difficulty of ensuring strength and rigidity in monoplane structures cast doubts on their ability to withstand the strains of war, however, and as a consequence the structural advantages of the biplane prevailed in most types of combat aircraft until the early 1930s.

On 25 July 1909 Louis Blériot won the *Daily Mail*'s prize of £1,000 offered to the first man to fly across the English Channel in a heavier-than-air aeroplane. The Blériot XI monoplane had a wing span of 7.8 m (25.6 ft) and was powered by a 25 hp Anzani engine giving a maximum speed of 75 km/h (46.5 mph). The flight lasted about 37 minutes, Blériot landing on a grassy hillock beside Dover Castle.

# From Principles to Practice

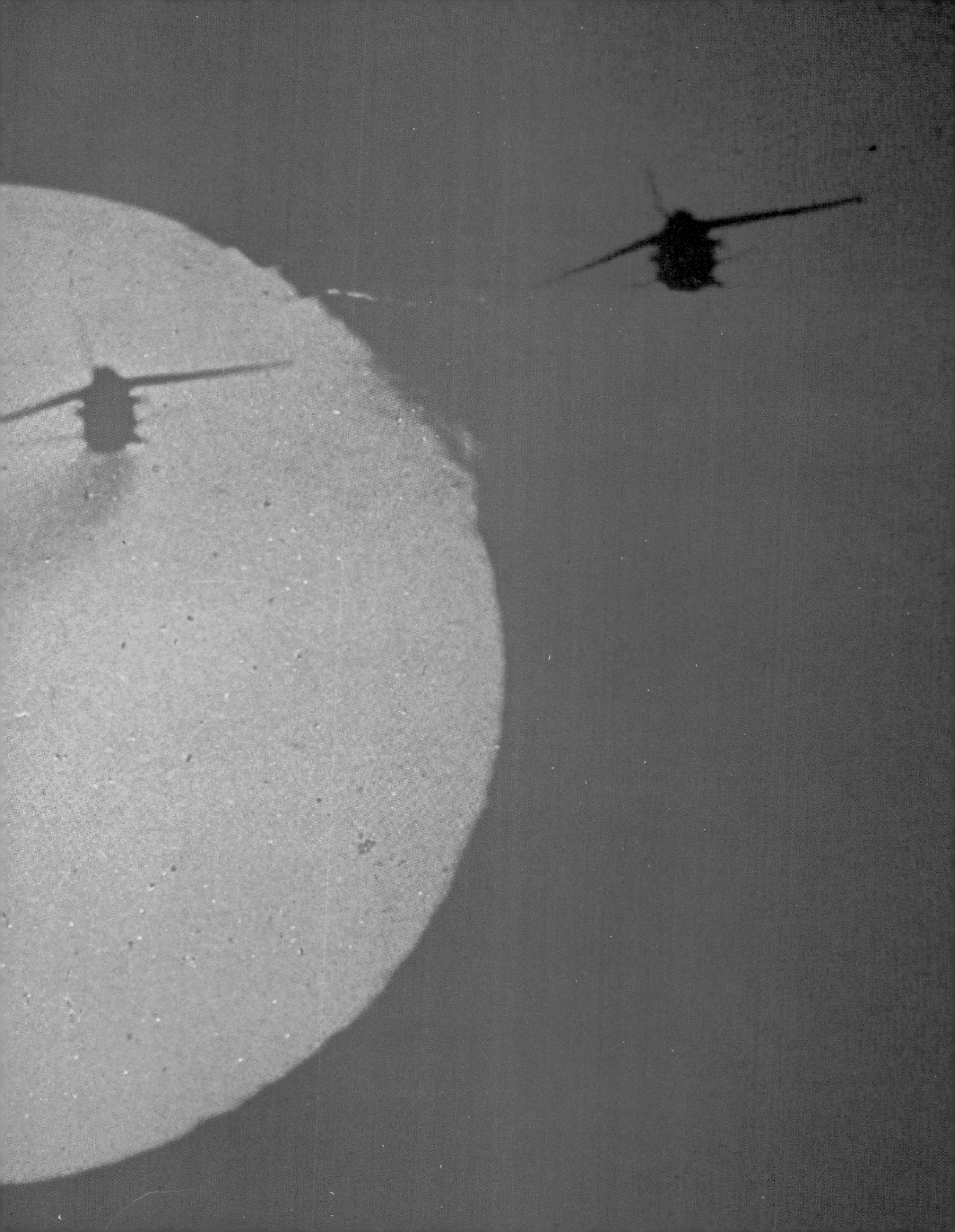

# 6 The Atmosphere

Life on planet Earth is protected from the extremes of space by a thin layer of gas called the atmosphere. This mixture of gases, commonly called air, extends from the Earth's surface out to a distance of about 800 km (500 miles). Aeroplanes, rotorcraft, gliders, balloons, and airships all depend on the atmosphere to make them fly. Air passing over the wing of an aeroplane gives it lift, while balloons and airships are held aloft by differences between the densities of the gases inside and outside their envelopes. All aero-engines except rockets (*see* Chapter 27) make use of air, not only for combustion but also to transmit the power they generate into thrust.

Heat from the Sun and the effects of the rotation of the Earth ensure that the atmosphere is constantly moving and mixing. Its temperature, pressure, density, and humidity vary from place to place on the surface of the Earth and change with distance above the surface. These changes and the phenomena called weather directly affect aircraft operation. Changes in atmospheric conditions, such as temperature and pressure, alter the performance of aircraft and their engines. Winds speed up or slow down progress over the ground and complicate navigation. Storms, turbulence, hail, lightning, and icing may be a physical hazard, while fog, cloud, rain, snow, or dust storms can obscure visibility. The pilot and designer cannot assume the sky will always be friendly to flying.

At sea level, the atmosphere is made up of 75.5 per cent nitrogen by weight, 23.1 per cent oxygen, 1.3 per cent argon, and rather less than 0.05 per cent carbon dioxide, together with some water vapour. There are also small quantities of neon, helium, krypton, hydrogen, xenon, ozone, and radon. The proportions of gases in the mixture change with altitude, and above 80 km (50 miles) the mixture is predominantly hydrogen and helium. Because conventional jet engines use the oxygen in the air for combustion, they cannot fly at altitudes above about 32 km (20 miles); sustained flight at greater altitudes is possible only with rocket engines. Until the launching of man-made satellites little was known about the atmosphere above 32 km. It now appears that it has no distinct upper limit, although for all practical purposes it is about 800 km (500 miles) deep. Density continues to lessen above this altitude, however, with a layer of hydrogen extending outward from 2,400 km (about 1,500 miles). Above 8,000 to 10,000 km (about 5,000 to 6,200 miles) its composition is indistinguishable from that of interplanetary space.

About half the total weight of the atmosphere is contained in the lowest 5.5 km (3.4 miles, or about 18,000 ft). Above this occupants of aircraft need a pressurized air or oxygen supply. Most jet transports except Concorde and the Boeing 747SP operate below 11 km (about 36,000 ft), and only a few military aircraft regularly fly above Concorde's ceiling of 18 km (about 60,000 feet).

The atmosphere can be divided into a series of distinct layers, each having its own characteristics. In temperate latitudes the region up to about 11 km is called the *troposphere*. In this region the temperature falls steadily with increasing height at a rate of about 2°C (3.6°F) per 300 m (1,000 ft) until it reaches a low of about

**Previous two pages** Two US Navy jets flying over Guantanamo Bay, Cuba, are outlined against an atmosphere red-tinged by the setting Sun.

**Below** Cirrus clouds, seen here over the island of Skye, form at high altitude and are composed mainly of ice crystals. They are often the first sign of an approaching depression.

$-56°C$ ($-71°F$). The next region, extending to 56 km (about 35 miles), is the *stratosphere*, which is separated from the troposphere by a thin layer called the *tropopause*. Within the stratosphere the temperature steadily rises to $-20°C$ ($-9°F$) in its upper regions. Layers above this, known as the *mesosphere*, *thermosphere*, and *exosphere*, take the atmosphere to the borders of space.

The temperature profile of the atmosphere above the poles is different from that above the equator, and it is in a state of constant change, hour by hour, season by season. It is not possible, therefore, to lay down strict rules covering the temperatures and conditions to be expected at specific altitudes. General rules devised by scientists and meteorologists do, however, help engineers predict how aircraft will behave and give pilots the data they need to fly. The most widely accepted guide to the variation of conditions with altitude is the International Standard Atmosphere devised by a team of specialists from the International Civil Aviation Organization (ICAO), an agency of the United Nations.

## Properties of Air

The performance of an aircraft or an engine is directly related to the physical properties of the air in which it operates. Because air is a gas, its temperature, pressure, and density are directly related by a series of mathematical formulae. If any two of these characteristics are known, the third can be calculated. Density is a measure of the amount of matter in a given volume. At sea level air has a density of about 1,226 g per cubic metre (0.077 lb per cubic foot). At sea level, the weight of all the air in the layers above produces a pressure of 1 kg/sq cm (14.7 lb/sq in). At altitudes above sea level there is less air above, so the pressure is lower. Because of this, high altitude air is free to expand, and so its density lessens. At 9 km (29,500 feet), about the height of Mount Everest, the pressure has fallen to 0.35 kg/sq cm (5 lb/sq in), one-third of its sea-level value, and by 18 km it is down to 0.07 kg/sq cm (1 lb/sq in). Meteorologists usually measure pressure on the surface of the Earth in terms of *millibars*. Normal atmospheric pressure is taken as one bar, or 1,000 millibars (mb); in the United Kingdom changes in the weather bring variations in pressure from a 'low' of about 970 mb to a 'high' of 1,030 mb. The ICAO Standard Atmosphere suggests that normal sea-level conditions in temperate latitudes correspond to a pressure of 1,013 mb and a temperature of 15°C (60°F).

The properties of a medium determine the speed at which sound moves through it. It can be shown that the speed of sound in air depends not on pressure or density but on temperature. At sea level under normal conditions sound travels at about 340 m (1,116 ft) per second, or 1,224 km/h (760 mph). As air temperature falls with increasing altitude, so does the speed of sound. In aeronautics, high speeds are frequently expressed in Mach numbers, a unit of measurement named in honour of the Austrian physicist Ernst Mach (1838–1916), who made a number of important discoveries about the properties of sound. Mach 1 is 340 m/sec, the speed of sound in air at sea level.

Sound moves in air in much the same way as waves do through water: although waves at sea

**Left** As their name implies stratus clouds develop as horizontal layers – a form evident in this photograph taken from an aircraft over Lake Geneva.

**Right** Cumulus clouds are formed by the upward movement of convection currents. They occur in many different sizes and shapes, and may appear as fluffy, white fine-weather clouds and as dark and threatening thunder clouds.

appear to carry water along with them, in fact the water merely oscillates as the waves pass through it. In the atmosphere sound is transmitted by small movements of the air molecules: the speed at which air molecules vibrate determines its frequency, and the strength of the signal determines its loudness.

The performance of an aircraft is critically dependent on the density of the air through which it is moving. Other things being equal, higher density means higher lift and also higher drag (resistance of the air to an object passing through it). An increase in temperature causes air to expand and, in doing so, its density is reduced. Temperature has a direct effect on the stability of the atmosphere because warm, less-dense air tends to rise, and cold, denser air tends to sink. This vertical motion is called *convection*. Changes in temperature between night and day and between seasons, and differences in the way parts of the surface of the Earth absorb heat, cause variations in air temperature and those phenomena we call weather.

## Wind, Water, and Clouds

Winds are a flow of air from regions of high pressure to regions of low pressure, and the uneven pattern of heating results in the Earth being subject to a basic pattern of winds. Winds do not, however, blow directly from high- to low-pressure regions, owing to the effects of the rotation of the Earth. Surrounding the equator there is a narrow belt of light and variable winds known as the Doldrums. In this region the pressure tends to be low. North of the equator as far as latitude 30°N is a belt of winds known as the North-East Trades, while south of the equator as far as 30°S are the South-East Trades. Between 30 and 40°N and between 30 and 40°S lie narrow belts of sub-tropical high pressure. The Westerlies, which blow across most of Europe and much of North America, are part of a belt of temperate, variable winds between 40 and 60°N. A similar belt of Westerlies, often called the Roaring Forties, is found on the other side of the equator from 40 to 60°S. The north and south polar regions are covered by systems of easterly winds. In the northern summer the wind belts move farther north and in the southern summer they move farther south. This basic pattern of winds is modified greatly by the influence of the continents (especially natural barriers such as mountains) and by the oceans and seas which absorb and store solar heat more readily than does the land.

When the air covering a large area has

Cold and warm fronts. On weather maps fronts appear as bold lines, a cold front being identified by triangles along the line and a warm front by circular discs. The section through the warm front shows a gently sloping boundary between warm air on the left and cold air on the right. As the front moves (from left to right in this diagram) the warm air is pushed up over the wedge of cold air, giving rise to series of clouds. The first signs of an approaching warm front are often high cirrus clouds. These are followed by lower cirrostratus, altostratus, and rain. The section through the cold front shows a steeper boundary, as the cold air pushes the warm air along. A cold front is often preceded, as here, by squalls, cumulus clouds, and rain from typically anvil-topped cumulonimbus thunder clouds.

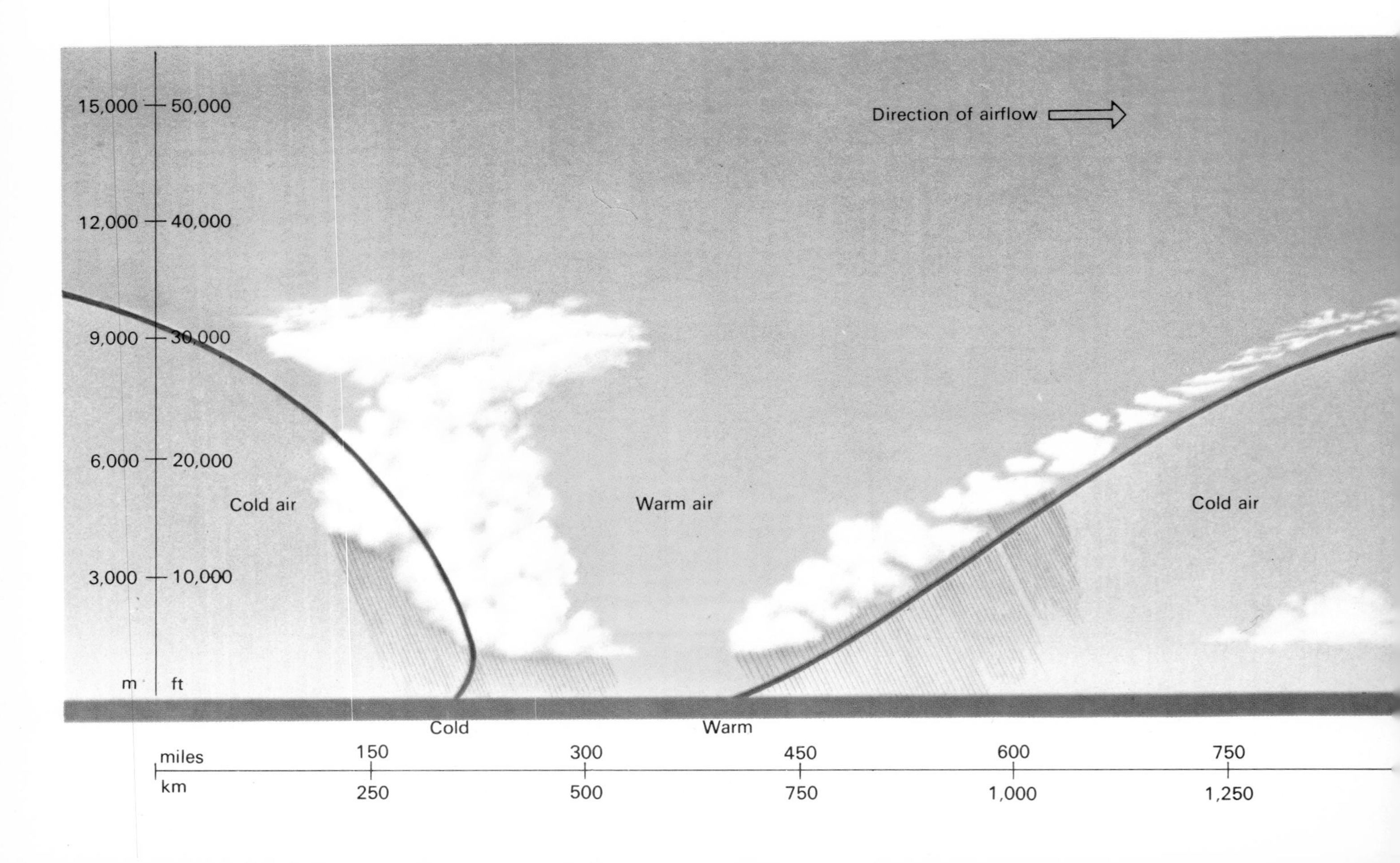

In coastal regions the different rates at which the sea and land absorb heat from the Sun give rise to sea breezes. During the day the land heats up more quickly than the sea. The air above the land gets warm, becomes less dense, and rises, and its place is taken by cooler air which blows in from the sea. After nightfall the land loses heat more rapidly than the sea. Convection currents develop in the reverse direction, creating an off-shore breeze.

broadly the same temperature and humidity it is termed an *air mass*. Air masses may be hundreds or thousands of kilometres wide. The presence of large, stable air masses tends to modify the basic system of winds north and south of the equator and adds to the difficulty of forecasting. The air masses are labelled to indicate their origin – polar, tropical, maritime, or continental. The study and analysis of air masses is called *synoptic meteorology*, or air-mass analysis.

Water on the surface of the Earth and in the atmosphere occurs in three forms – as a solid, as a liquid, and as a vapour. Its presence and its unique properties have a profound effect on the weather and on the operation of aircraft. Like all substances, water absorbs or releases energy, called *latent heat*, when it changes state from one form into another. The latent heats of water are unusually large, so that during changes of state large amounts of heat are released or absorbed. Water provides one of the principal mechanisms for the absorption of heat from the Sun and for the development of weather systems.

Atmospheric air is never dry even when it is apparently clear. When air is cooled, a point is reached when the water, normally invisible as vapour, condenses out as droplets of liquid. This is the way clouds, mist, and dew form. When clouds are cooled still further the result is rain. If the air is then heated again, the condensed water evaporates – that is, it returns to vapour form. These processes of condensation and evaporation are taking place continuously within the atmosphere.

Water is drawn into the atmosphere from seas, rivers, lakes, and puddles, and in general the higher the temperature the greater is the amount of water which will evaporate. Meteorologists use the term *relative humidity* to relate the amount of water vapour in the air at a given time to the amount of water vapour which could theoretically be present under saturated conditions. In the United Kingdom, for instance, relative humidity is rarely less than 60 per cent, and this 'dampness' is one of the reasons why it often feels colder in Britain in winter than in other much colder, but drier, countries.

Condensation makes clouds form when moist air is cooled. This cooling may occur when air is forced over a mountain, when it rises over another air mass, or when it rises owing to convection; it also occurs, of course, at night. Although clouds come in all shapes and sizes, and several different types often appear in the sky at the same time, they are usually classified by their appearance and altitude. There are three basic cloud types: *cirrus* (hair-like), *cumulus* (heaped), and *stratus* (layered); and there are various subtypes.

Cirrus are high clouds at altitudes of about 6,000–12,000 m (20,000–40,000 ft) and are composed mainly of ice crystals; they have a white, feathery appearance and are often the first warning of an approaching depression. Cumulus – the commonest form of clouds – are detached, generally dense, and have sharp outlines. They are formed by the upward movement of air and so often develop as vertically rising mounds or domes resembling cauliflowers; the base is flat, horizontal, and shadowed. Although stratus are cloud-forms in their own right, the term is also applied to any cloud system that is spread out in an extensive horizontal layer. Stratus brings a dull, overcast sky with a fairly uniform base, and may result in drizzle, sleet, or snow; if the base of the stratus is below the level of higher ground it forms hill fog.

Cloud types frequently appear in combinations, of which the following are typical: cirrostratus (high, flat clouds), cirrocumulus (high, patterned clouds), and stratocumulus (low-lying clouds combining layered and heaped forms). Medium-altitude clouds – at

**Above** A condensation halo forms around a Royal Navy Hawker Siddeley Buccaneer. Such formations often develop when aircraft fly in humid atmospheres, and are a result of different air pressures developing on the upper and lower surfaces of wings. The air flowing over the upper surface is reduced in pressure, and so it expands. This causes water vapour in the air to condense, forming a small cloud of tiny water droplets. Such haloes are usually elliptical in shape because the drop in air pressure is greater at the wing roots than at the wing tips.

**Left** Vapour trails formed by wing-tip vortices stream out behind a Panavia Tornado prototype. Such trails are usually shortlived, whereas those formed by the exhausts of turbojet engines may develop into cirrus-like clouds.

between about 2,500 and 6,000 m (8,250 and 20,000 ft) are often given the prefix *alto*. Thus, altostratus are uniform, layered clouds which often allow the Sun or Moon to shine vaguely through them; altocumulus are thin layers, usually with a pronounced small-scale pattern – a 'mackerel sky' being a typical example of this formation.

*Nimbus* (Latin for 'rain storm') is a term used to describe clouds from which rain is falling. Nimbostratus are dark stratus clouds associated with heavy rain; cumulonimbus are large, heaped clouds of rapid vertical development, often with anvil-shaped tops, which bring rain, hail, snow, or thunder and lightning, and strong, gusting winds. *Lenticular* clouds are another distinct, though less common, type. As their name suggests, they are often lens-shaped, like a pointed oval, and form in the lee of air flowing over mountains or ranges of high hills. They appear stationary in the sky as cloud forms on the upward-flowing side and evaporates on the downward-flowing side. Along with cumulus clouds they are of particular interest to glider pilots because they denote areas of exceptional lift.

The passage of an aircraft through damp air may itself cause cloud. Condensation trails are usually associated with high-flying aircraft, but they can form at any altitude provided the conditions are right. Contrails, as they are called, are usually narrow at first, but they sometimes spread into larger bands and may even develop into cirrus clouds.

Contrails develop when the water formed by the combustion of aviation fuel is ejected into the atmosphere as exhaust. Although the exhaust is initially warmer than the surrounding air, it soon cools; and, because of the additional water content, the air may become saturated at the lower temperature. The resulting contrail may be in the form of water droplets or ice crystals. In damp weather a different form of trail may develop immediately behind the wing tips or the propeller tips of aircraft flying at low altitudes. These short-lived trails form because expansion of the air inside a strong vortex (a rotating wake) around the wing or propeller cools damp air to below the dew point (the temperature at which vapour condenses).

## Weather Patterns

The weather maps, or synoptic charts, published in newspapers and shown on television make use of internationally agreed signs and symbols to show the distribution of pressure and to convey information about wind, temperature, cloud cover, and precipitation (rain and snow). The most noticeable feature of a weather map is the pattern of *isobars* – lines which join all points of equal barometric pressure. Isobars are usually drawn at intervals of two, four, or eight millibars. In general, the wind blows parallel to the isobars, whose distance apart is an indication of wind speed. Closely spaced isobars reflect a steep pressure gradient and a strong wind; widely spaced isobars indicate little pressure gradient and light winds. Flight crews use 'actual' charts showing present conditions and 'forecast' charts showing predicted conditions both at sea level and in the upper air at 6,000 or 9,000 m (20,000 or 30,000 ft).

Changing pressure, temperature, and humidity, and variations in the amount of cloud cover all provide a challenge to the aviator. But hostile weather also brings more direct hazards in the form of icing, hail, turbulence, and storms. Ice may form on parts of an aircraft even if there is no rain. The ice not only adds to the weight of the aircraft; it may jam controls and mechanisms, and it also disturbs the airflow, reducing lift and increasing drag. Most commercial and large military aircraft are provided with a de-icing system which prevents any build-up at critical points such as wing and tail leading edges. The carburettors or fuel-injection systems of piston-engined aircraft are also prone to icing and pilots have to prevent the build-up of deposits. Ice may form on aircraft parked in the open, just as it does on parked cars, and they often have to be sprayed with de-icing fluid before take-off.

Hail is associated with cumulonimbus clouds and thunderstorms. Large hailstones are kept aloft by strong upcurrents within the cloud, growing layer upon layer. In regions of violent storms hailstones as large as grapefruit and weighing more than 1 kg (2.2 lb) are not unknown. Hail can cause considerable damage to aircraft and pilots are usually instructed to fly around, or over, intense activity. Weather radar, when fitted, helps to spot the centre of active storms. Surprisingly, perhaps, lightning is rarely dangerous to large aircraft.

In normal flight steady winds are not a hazard to aircraft, although they make navigation difficult and can be a danger during take-off and landing. Aircraft may, however, be endangered by squalls and gusts, and many have been destroyed by rough air. At low altitude turbulence is associated with wind flowing over features on the ground, with fronts, or with convection currents. It is usually possible to forecast when turbulence will occur at low level, either from a knowledge of the terrain or by reading the weather. At high altitude, however, turbulence can occur in clear air. The fact that it is often unexpected and difficult to detect makes clear-air turbulence (CAT) a hazard to passengers and crew unless they are wearing seat belts, and it is the subject of intensive research.

A sudden change in wind speed and direction, or a sudden drop in wind, is usually called *wind shear* and it can be a danger to aircraft flying close to the ground. Aircraft flying into a strong head wind may be moving at high speed relative to the air but slowly relative to the ground. If the wind suddenly drops owing to wind shear, an aircraft takes time to increase speed relative to the ground (necessary in order to make up for the loss of lift), and it is in danger of stalling or losing a great deal of altitude. Wind shear is believed to have caused a number of accidents to airliners during the landing approach.

While many high-altitude air currents make for a rough ride, and occasionally are dangerous, others, known as *jetstreams*, are used by pilots to reduce journey times. Over the North Atlantic jetstreams occur at about 10,000–12,000 m (33,000–40,000 ft); they commonly flow at about 185 km/h (115 mph) but occasionally reach 370 km/h (230 mph), while over South-East Asia speeds approaching 550 km/h (340 mph) have been recorded. Jetstreams may be up to 320 km (200 miles) wide but only 3.25 km (2 miles) deep; as long as they are flowing in the desired direction they effectively add to the cruising speeds of aircraft. Unfortunately they are unpredictable in behaviour, changing their location and direction from day to day.

## Lift and Drag

Balloons and airships are the simplest form of aircraft. They fly because they are buoyed up by the surrounding air much as a ship is kept afloat by the sea. Whereas the pilot of a balloon has no control over the direction his craft flies, an airship has its own source of power, allowing the pilot a considerable measure of directional freedom.

There are basically two forms of balloon, the hot-air balloon and the gas balloon. Each can achieve flight because the gas-filled envelope plus the basket and the pilot are lighter than the atmospheric air the craft displaces until it has climbed to its 'cruising height'. The pilot of the hot-air balloon reduces the density of the air in the gas envelope by heating it; the pilot of a hydrogen- or helium-filled balloon relies on the inherent low-density properties of the gas in the bag. Until recently, the gas envelopes of hot-air balloons were roughly spherical, but advances in the science of balloon design has allowed envelopes in shapes as strange as a pair of trousers to be flown successfully. Hot-air airships have also been built in limited numbers.

Apart from a few special jet-lift designs, aeroplanes use lifting surfaces (wings) to provide the vertical force to balance the effect of gravity. Wings are shaped in such a way that air flowing over them gives rise to a lower than normal air pressure above the wing and an above-normal pressure below. The result is an overall upward force, known as *lift*. The lift force is associated with the less-desirable force known as *drag*, which acts at right angles to the lift. Early experiments aimed at developing heavier-than-air aircraft made use of flat plates inclined to the airflow. While some lift was obtained it soon became clear that more efficient wings could be designed if the section was rounded and shaped. It is now possible to design wings for particular applications which produce more than 40 times as much lift as drag.

By introducing small trails of smoke into a wind tunnel and watching them pass over a model wing, it became clear to early researchers that the pattern could be considered as a series of streamlines showing the direction of flow at any point. Experiments with smoke and other more sophisticated techniques also showed that conventional wings tend to produce a strong vortex (whirlpool effect) in the air at a point close to each wing tip. The strength of such *trailing vortices* is proportional to the lift being generated by the wing.

As aircraft fly faster and the speed of the air flowing over the wing approaches the speed of sound, the flow pattern begins to alter, and as the speed of Mach 1 is reached a shock wave is produced, usually at the fattest point on the wing. At supersonic speeds shock waves form at both the leading and trailing edges and give rise to the characteristic 'boom-boom' of the sonic bang heard on the ground when a faster-than-sound aircraft passes overhead.

Air flowing over the surface of an object causes drag, partly owing to the effects of friction, and if the object is moving quickly the friction produces a rise in temperature. The heat generated at slow flying speeds is negligible, but on a jet-powered airliner such as a Boeing 747 the temperature of the leading edges

**Inset** Schlieren photography is used to take pictures (called shadowgraphs) of the shock waves that aircraft models develop in a supersonic wind tunnel. In actual flight these powerful waves, forming at the nose and leading edge of the wing of an aircraft flying above the speed of sound, cause the characteristic boom-boom of a sonic bang heard on the ground.

may rise by about 15°C (27°F). On Concorde, which cruises at more than 2,000 km/h (1,300 mph), rises in temperature of more than 160°C (288°F) occur in the nose; but when the aircraft slows before its landing approach the nose temperature may fall to that of the outside air – about −15°C (0°F) – within 10 minutes. These great variations in temperature raise problems that had to be dealt with at an early stage by Concorde's designers.

Space vehicles returning to Earth rely on drag to slow them down. Since such vehicles are travelling at speeds of more than 29,000 km/h (18,000 mph) when they enter our atmosphere, frictional heat is a major problem and the special heat-shields used commonly reach temperatures of more than 1,600°C (2,900°F). These shields use *ablative coatings*, made of plastic resins mixed with asbestos, nylon, or glass, which melt and vaporize and so dissipate much of the heat. The ablative material is backed by a heat-sink made of steel or copper which absorbs conducted heat and so offers additional protection to the occupants of the vehicle. Space vehicles with wings, such as the Space Shuttle, can enter the atmosphere at a shallow angle and the heating problem is less severe. Nonetheless, the leading edges of their wings, tail, and fuselage also need to be protected with heat-absorbent materials, which ablate during re-entry and must be repaired or replaced before each flight.

Air flowing over wing surfaces forms into complex patterns of movement, especially at high speeds. Delta wings such as those of Concorde, for instance, produce a series of powerful vortices. Aircraft designers use coloured smoke when testing small-scale and full-size mock-ups of fuselage and wings in wind tunnels to check the predicted lift and drag properties of different configurations.

# 7 Stability and Control

Aeroplanes are heavier-than-air flying machines which rely on wings to provide lift. There are basically four forces acting on any conventional, powered aeroplane: lift, weight, thrust, and drag. In steady, straight, and level flight the four forces are in overall balance, with the lift acting vertically upwards equal to the weight acting downwards, and with the thrust acting forwards equal to drag acting backwards. If lift is greater than weight the aeroplane will rise. Conversely, if lift is less than weight the aeroplane will descend. Unless thrust and drag are in balance the aeroplane will either speed up or slow down. The pilot is free to manoeuvre by increasing lift or changing the throttle setting (thrust). Drag at a given speed can also be changed by deploying flaps and air brakes. Weight, however, is usually regarded as fixed unless the pilot drops the payload (bombs, for example) or dumps fuel. The ratio between lift and drag is a measure of the aerodynamic efficiency of a design; the higher the ratio, the greater the efficiency. Ratios range from more than 40:1 for a glider to about 18:1 for a subsonic jet transport and 14:1 for a twin piston-engined type; for a simple hang-glider of the Rogallo type the ratio may be as low as 1:1.

The aeroplane designer must take into account not only the four forces as such but also the particular points on the structure at which they act. For example, it would probably be impossible to fly if all the lift of an aircraft was generated by a forward-mounted wing while the centre of the weight (the centre of gravity) was in the rear of the body. Similarly, an aircraft would be difficult to handle if the engines were located high up and the drag force acted low down. Balance is the key to the problem and the designer must ensure that the aeroplane is balanced about its centre of gravity. While the general layout and the positions allocated to the payload are of critical importance, slight imbalances in either can usually be compensated for in flight by use of the aircraft's control systems.

Aeroplanes are, of course, free to manoeuvre and rotate in all directions, but as soon as the pilot strays from straight and level flight the relationship between the four forces becomes much more complex. In order to understand how an aeroplane manoeuvres it is necessary first of all to take a closer look at the way wings generate lift.

## Lift

Wings influence the distribution of pressure in the air through which they move (*see* Chapter 6). That variations in pressure on the surfaces of a wing in flight are a key factor in the generation of lift in a heavier-than-air aircraft was realized and formulated scientifically only towards the close of the 19th century (*see* Chapter 5).

As air flows over a conventionally shaped wing, it speeds up over the top surface. This speeding up causes its pressure to fall. The air flowing below the wing slows down, leading to

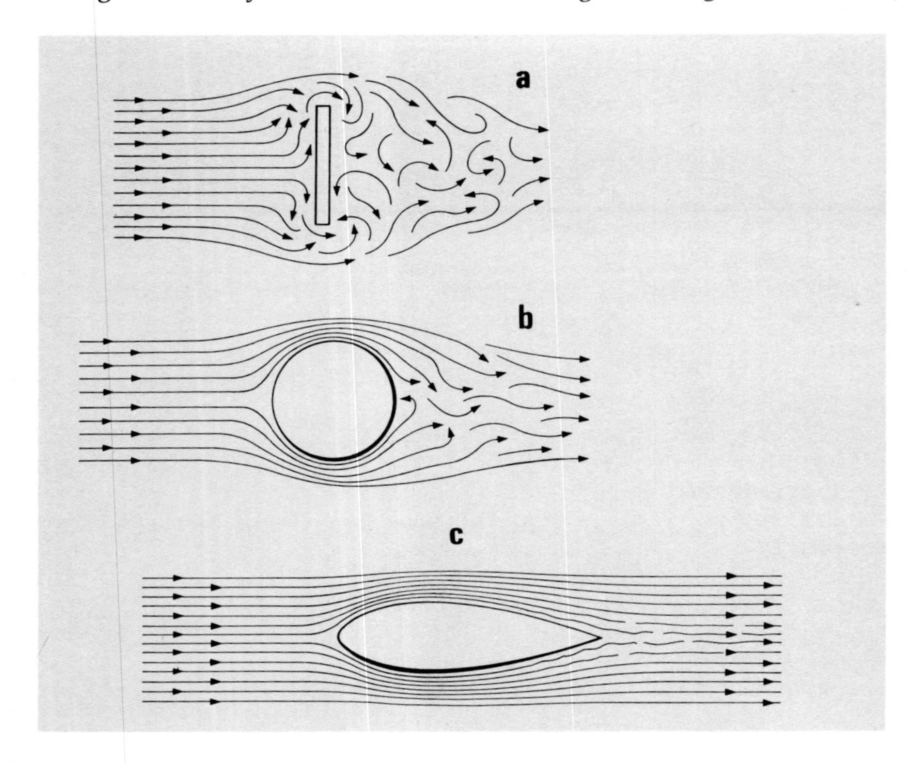


**Below left** The diagrams show the directions of airflow around moving objects. The streamlined shape (c) produces much less flow disturbance, and therefore less drag, than the flat plate (a) or ball (b).

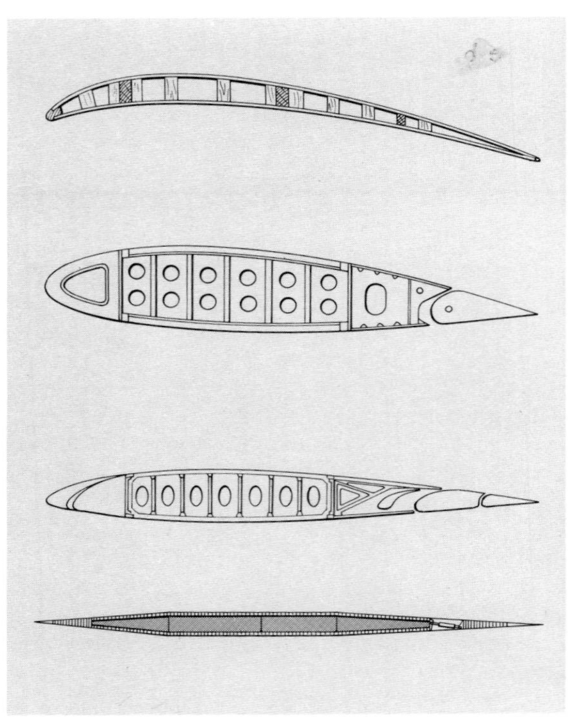

**Below** Progress in wing design. The cross-section profiles show (from top to bottom): Blériot XI (1908), highly cambered, wooden structure; Avro Lancaster (1941), thick-section, light alloy; subsonic Boeing 727 (1962), slimmer-section, light alloy; Mach 3 North American XB-70 (1963), very thin, welded stainless-steel honeycomb structure.

The Boeing 727 has leading-edge slats on the outboard portion of the wing, Krüger flaps on the inboard leading edge, and trailing-edge flaps to boost lift for landing and take-off. Like most jet transports the 727 has engines fitted with thrust reversers to assist the wheel brakes after landing.

a slight increase in pressure. These changes in pressure produce the overall upward force on the wing known as lift. The reduction in pressure on the upper surface makes a bigger contribution to lift than the pressure increase on the lower surface. The changes in pressure are greatest towards the leading edge of the wing; the overall lift force acts at a point called the *aerodynamic centre*, which is about a quarter of the way back from the leading edge.

Five basic factors affect the lift a wing will generate. Two of these, wing-section shape and wing area, are under the control of the designer. Two others, the angle of attack (the angle at which the wing is inclined to the airflow) and the speed with which the air flows over the wing, are controlled by the pilot. The fifth factor, the density of the air, varies with altitude and from place to place (*see* Chapter 6) and is beyond human control.

In general, thicker wing sections give more lift at a given speed, but at high speed they may increase drag to unacceptable levels, so they are used mainly on slower aircraft. Lift is directly proportional to wing area: the bigger the wing, the more lift it will generate. The designer of a plane intended to carry heavy loads has to make a compromise between a small but sophisticated wing with high-lift devices or a large, simple, but heavy wing.

Lift is directly proportional to the square of the airspeed. If the speed is doubled, lift is increased fourfold; if it is trebled, lift increases ninefold. Higher air density brings an increase in lift. Conversely, at low densities pilots have to compensate by flying faster or increasing the angle of attack. The reduction of lift with falling density is one of the reasons why the take-off and climb performance of aircraft is often worse than normal at airports located in the tropics or at high altitudes. In these conditions the problem of lift may be increased because low air density also reduces the power output of most aero-engines.

There is a limit to the amount of additional lift that can be gained simply by increasing the angle of attack. Lift usually increases up to an angle of between 16° and 20°. If the angle is then steadily increased the lift will remain constant for a degree or two and then quite suddenly will fall. This point is called the *stall*. Stalling is caused by a complete breakdown of the airflow over the upper surface of the wing. Instead of remaining smooth and streamlined, and 'attached' to the wing surfaces, the airflow breaks away, forming a region of swirls, eddies, and small vortices. These destroy the favourable pressure distribution and effectively kill most of the lift.

Thin wings with sharp leading edges tend to have more sudden stalling characteristics than well-rounded, thick wings. On some aircraft, for a variety of reasons, one wing tends to stall before the other, and this leads to 'wing drop'.

The stall is usually preceded by the onset of progressively worse buffeting which warns the pilot to decrease the angle of attack. When there is no such warning, sensors may be fitted to detect excessive angles of attack and sound a warning horn in the cockpit. Under normal circumstances, with adequate height, the stall is not in itself dangerous: the loss of lift usually causes the nose to pitch downwards, reducing the angle of attack; after a dive to recover speed, the aeroplane returns to normal flight.

Stalling close to the ground, however, can be fatal. The danger is greatest at high angle of attack after take-off or before landing. The stalling characteristics of certain aircraft, including some with rear-mounted jet engines and high T-tails, are unacceptable because the turbulent air from the stalled wing flows back over the tail and destroys the power of control. Such aircraft may be fitted with a warning device which shakes the control column as the stall is approached and an automatic 'stick pusher' which pushes the control column forward to reduce the angle of attack.

At a given airspeed, the maximum lift a wing can develop is largely determined by the angle of attack at which it stalls. Devices have been developed which either delay the stall, allowing the effective angle to be increased, or boost the lift at a given angle. Stall-delaying devices are usually fitted to the leading edge of the wing and smooth the flow of air over its upper surface. The first of such devices was the Handley-Page *slat*, which moves out above and parallel to the leading edge, like an auxiliary wing. This creates a small slot above the front of the wing through which the air speeds up as it flows over the upper surface. On some thin wings the whole of the leading edge is hinged and droops for take-off and landing to give a better match to the approaching airflow. Most modern jet transports use a combination of slats and leading-edge *flaps*, known as Krüger flaps, for the highest possible lift at high angles of attack.

Devices fitted to the trailing edge can greatly increase lift at a given angle of attack, but they do not delay the stall. Plain trailing-edge flaps, used since the early days of flying, involve the rear portion of the wing moving downward about hinges fitted approximately parallel to the trailing edge. With split flaps only the under surface of the wing is hinged. More complex devices introduced in the 1930s include the Fowler flap, which is rather like a split flap but first moves rearwards along tracks in order to increase the area of the wing. Developments of the Fowler include the double- and triple-slotted flap, which use auxiliary flaps (slats) to speed the airflow over the flap leading edge and prevent break-away.

Power from the engines can also be used to increase wing lift. For example, the slipstream from a propeller increases the speed of the air

**Above** In the early days of aviation, when few aircraft flew at more than 160 km/h (100 mph), designers were interested less in minimising drag than in providing adequate lift and a strong, light structure. The Fokker Dr. I triplane of World War I reflected this attitude. (The replica shown here is powered by a radial engine rather than a rotary, as in the original.)

blowing over the wing – and lift increases as the square of airspeed. Four large propellers blow air over the wing of the de Havilland Canada Dash 7 to improve its short-take-off-and-landing (STOL) performance. The principle is taken much further on the Boeing YC-14, which uses its turbofan engines to blow exhaust over the top surface of the wing, and on the McDonnell Douglas YC-15, whose under-wing engines blow exhaust onto two-section slotted flaps. Other aircraft, such as the Hawker Siddeley Buccaneer, direct high-pressure air from the engine to blow through narrow slits over critical parts of the wing and flap to prevent flow separation and to increase lift.

## Drag

Powered aeroplanes use engines to overcome drag. Drag can be divided into two components: *lift-dependent* drag and *zero-lift* drag. As its name suggests, lift-dependent drag is a function of the lift a wing is generating. It can be minimized by making the wing long and slender and by ensuring that the lift is distributed right across the span in an efficient manner. In practice, long slender wings (or *high-aspect-ratio* wings, as they are called) are excessively heavy for high-speed or highly manoeuvrable aircraft.

Zero-lift drag includes both drag due to the form or shape of the aeroplane and drag due to skin friction. Again, the designer has to compromise between a short bluff shape with low skin-friction drag but high form drag and a beautifully-shaped long design with low form drag but excessive skin-friction drag. Zero-lift drag is proportional to the square of the airspeed, so a doubling of speed brings a fourfold

increase in drag. As the design speeds of aircraft increase, so designers must take ever greater care to ensure that planes have the minimum of drag-creating protuberances.

The study of skin-friction drag has shown that it depends critically on the flow of air within a thin layer – known as the *boundary layer* – close to the surface of the body. If the shape and growth of the layer and the flow of air within it could be controlled, skin-friction drag could be greatly reduced. Boundary-layer control (BLC) using blowing and sucking devices has been tried but, so far, it is limited to research aircraft. (Flap blowing, of course, is a form of BLC; but, as we have seen, it is used only during take-off and at approach and landing, and is designed to increase lift rather than to reduce drag.)

During the early 1940s, when production aircraft began to exceed 645 km/h (400 mph), changes in drag and in airflow were experienced which could not be explained by classical slow-speed aerodynamics. In particular, as the speed of sound (1,224 km/h or 760 mph at sea level) was approached the effects of compressibility and shock waves became important for the first time. This problem could occur even with subsonic aircraft owing to the fact that the lift-generating properties of wings depend on a speeding up of the flow of air over the upper surfaces. It was discovered on some of the quickest piston-engined fighters that air flowing over parts of the upper wing surfaces was approaching supersonic speed. It was clear that for aircraft operating at such speeds designers would have to develop an alternative to the traditional, thick-section wing. The answer was provided by thin-section, swept-back wings, first used by the Germans towards the end of World War II. These reduced the acceleration of the airflow, and thus delayed the onset of shock waves and the problems of compressibility.

By 1955 jet fighters were regularly flying at supersonic speeds and another problem of high-speed flight emerged. Research in the United States showed that the shape of an aircraft's fuselage – or, to put it more exactly, the distribution of area along the aircraft's fore-and-aft axis – greatly influenced performance at supersonic speeds. *Area ruling*, as the find-

**Above** The Messerschmitt Me 108 of the 1930s anticipated many features of modern light aircraft, apart from its tail-wheel undercarriage layout. It also had a family resemblance to the formidable Bf 109 fighter of World War II.

**Below left** The General Dynamics F-16 is designed primarily as a fighter, and its canopy gives the pilot an unrestricted view. One of its most interesting features is the use of electrical signalling, or fly-by-wire (see page 93), rather than mechanical linkages between the pilot's controls and the wing and tail surfaces.

ings of this research are now called, was applied for the first time in 1954 to modify the Convair F-102, a fighter which in its original form had refused to exceed the speed of sound (there really is a 'sound barrier' if aircraft are an unsuitable shape).

Although many military aircraft, especially fighters and reconnaissance planes, are capable of supersonic speeds (and some can even exceed Mach 3), the bulk of everyday aviation is still subsonic. Much research in the 1960s and 1970s has been directed towards methods of delaying the onset of the effects of compressibility, or drag rise, at the speeds at which subsonic planes such as jet passenger transports normally operate. One way to achieve this is by so designing the wing that it generates lift right across the chord (the width between the leading and trailing edges) of the wing instead of concentrating lift, as on conventional wings, at a point immediately behind the leading edges. Such *supercritical* wings, as they are called, have a flattish upper surface and often a slightly down-turned trailing edge. They are suitable for use on some fighters, wide-bodied airliners, and executive jets.

Designers can use the reduced drag from a supercritical wing by installing smaller, more economical engines, or they can specify a thicker, lighter wing with less sweepback and space for more fuel, while still reducing drag significantly.

Whereas swept-back wings bring a reduction in drag at high speeds, they are less efficient than long-span 'straight' wings at low speeds: they generate less lift at take-off and landing, and they are less suitable for cruising well below the speed of sound. Soon after World War II the British inventor Barnes Wallis developed the idea of the swing-wing, or *variable-geometry* layout. This allows the designer to match the degree of sweepback to the speed of the aircraft: 'straight' for take-off and landing; intermediate for normal flight; and fully swept for maximum speed. The American General Dynamics F-111 was the first

The American National Aeronautics and Space Administration (NASA) used this modified Vought F-8 Crusader when flight-testing its supercritical wing. The effect of such wings is to delay the onset of compressibility effects and to lessen drag at supersonic speeds.

production jet aircraft to make use of the idea, and it has been followed by the Grumman F-14, the Panavia Tornado, and several Russian combat aircraft.

## Control Surfaces

Conventional aircraft are designed to be stable so that they return to straight and level flight if they are disturbed by a sudden gust of wind or by atmospheric turbulence. Basic stability is provided by the *fin* and *tailplane*, which are sometimes called the vertical and horizontal *stabilizers*. If its nose is displaced upwards an aeroplane rotates about the centre of gravity and the tailplane moves downwards. The tailplane is then at a higher angle of attack, so that it generates more lift and, in raising the tail (and therefore depressing the nose), brings the aeroplane back to its level-flight position. A similar reaction takes place if the nose is displaced down, producing a negative tail angle and the correct restoring force. The fin provides directional, or weathercock, stability by a similar process: displacement of the nose to right or left produces side forces to bring it back into the wind.

An aircraft with a tailplane and fin located well behind the wing is therefore naturally stable, and this is one of the principal reasons why such a layout has become the generally accepted standard. But other configurations are possible. For instance, one can design an aircraft in which the basic forces of lift, thrust, drag, and weight act in such a way that the horizontal tailplane must be replaced by *canards* (foreplanes), as on the Russian Tupolev Tu-144 (the Russian equivalent of Concorde) and the Saab Viggen (Thunderbolt) fighter. Canards, which look like small wings near the front of the fuselage, have one distinct advantage over tailplanes: they contribute significant lift to the aircraft at both take-off and landing, allowing the use of shorter runways than might otherwise have been possible. Indeed, the Viggen is capable of operating from stretches of highway about 500 m (1,640 ft) in length.

Soon after World War II the British inventor Barnes Wallis (inset) proposed the variable-geometry or swing wing to reconcile the conflicting needs of swept wings for high-speed cruise and 'straight' wings for take-off and landing. Models of his Swallow design, shown here, were successfully flight tested. The need for its wing-mounted engines to swivel to align with the direction of flight posed problems that designers later solved by burying engines in the fuselage.

## Stability v. Controllability

The design of every aircraft, from the Wright *Flyer* to the latest supersonic fighter, has involved a compromise between two desirable but conflicting qualities: stability and controllability. Too much stability gives an aircraft inherent safety and an excellent straight-line performance but makes it unresponsive to the controls and very difficult to manoeuvre. Too much controllability makes an aircraft highly manoeuvrable but exhausting to fly because the pilot must constantly make adjustments to the controls to maintain a given course.

As we have seen, the tailplane and fin confer inherent stability on an aircraft; but they also contribute to its controllability. On some aircraft the entire tailplane and fin may be rotated in order to change their angles of attack. More commonly, however, only the rearmost section of each is movable, being attached to the unmovable front section by a hinge. On the tailplane the horizontal hinged section, which moves up or down, is called the elevator; on the fin the vertical hinged section, which moves from side to side, is called the rudder. The effects of elevator and rudder movements, which provide control in pitch and yaw, are complemented by roll-control surfaces called ailerons, which are usually built into the outboard trailing edges of the wing. Ailerons resemble the conventional flaps that are mounted farther inboard, but unlike flaps they are able to move upwards as well as downwards. Positioning the ailerons near the wing tips provides the greatest possible rolling power for a given control movement. Ailerons, of course, move differentially; as one goes up, the other goes down.

A number of recent supersonic aircraft show interesting variations on conventional tailplane movements and designs. One such variation is the *taileron* for roll control, in which the tailplane on one side of the fuselage rotates to increase the angle of attack while the tailplane on the opposite side rotates to lessen the angle. The lift forces therefore act in opposite directions and cause the aircraft to roll. Some other aircraft are fitted instead with wing spoilers (movable surfaces that reduce the lift on one or other wing), while still others use a combination of spoilers and tailerons. Tailless aircraft, such as the Concorde, have control surfaces on the trailing edge of the wing known as *elevons*, which act as both elevators and ailerons.

In most aircraft the wing tips are higher than the wing roots, and the wings are said to have dihedral; conversely, wings, with tips lower than the roots have anhedral. Both arrangements are related to the need for stability if an aircraft starts to move downwards and sideways – a manoeuvre known as sideslipping. Generally speaking, aircraft with low, 'straight' wings require dihedral, giving additional stability. High, swept-back wings, however, confer a degree of inherent stability that militates against sideslipping, so such wings often have anhedral to reduce stability.

## The Pilot's Controls

A pilot controls an aeroplane's movements with a control column (or joystick), rudder pedals, and a throttle. Moving the control column forward depresses the elevator; this increases the lift on the tailplane and causes the nose to pitch down. Pulling the control column backwards raises the elevator, giving a downward force on the tail to raise the nose. Moving the column to the right (or, on some aircraft, turning a 'yoke' or wheel to the right) depresses the aileron on the left wing tip and raises the aileron on the right wing tip. This increases the lift on the left and reduces the lift on the right, causing the aircraft to roll to the right. The opposite happens when the column is moved to the left. Pushing the left rudder pedal causes the rudder to move to the left. This generates a force to the right on the fin, causing the nose of the aircraft to yaw to the left. Pushing the right rudder pedal produces a yaw to the right.

Within limits, the pilot of a powered aeroplane uses the throttle to control speed when flying level. An increase in thrust, either from a jet or from a propeller, above the setting needed for steady flight will give an excess of thrust over drag, causing the aeroplane to accelerate. Drag depends on the square of the speed, however, so as speed builds up so does drag. After a short while the aeroplane will settle down again at a speed corresponding to the new throttle setting, with thrust equal to drag. If the throttle setting is reduced, drag immediately exceeds thrust and the aircraft will decelerate until it reaches the speed at which the two forces are equal once more. In this sense, an aeroplane is speed stable: a change in throttle setting, whether for a higher or lower airspeed, soon results in an equilibrium being reached between the forces of drag and thrust. If the throttle is cut altogether, the aeroplane will not be able to maintain straight and level flight. Drag will soon slow the plane and the pilot will have to depress the nose to avoid stalling. The result is gliding flight, with continuing loss of altitude.

## Co-ordinating Control Movements

The pilot of a powered aeroplane manoeuvres by making co-ordinated movements of all the controls as well as adjustments to the throttle.

He increases or lessens the inherent lifting ability of the wing by moving various devices on the wing's leading and trailing edges. The lift generated by the wing itself is controlled by using the elevators to alter the angle of attack, or by adjusting the throttle setting to change the airspeed.

An aircraft's response to the various control movements is complex, however, and much of the art of flying lies in correct anticipation of secondary effects. For example, pushing open the throttle results in a build-up of airspeed. But the additional speed produces the secondary effect of greater lift, so that unless the control column is eased forward to reduce the angle of attack the aircraft will begin to climb. On the other hand, certain control movements, if made in isolation, will not have the desired effect. Sideways movement of the rudder, for example, will not make the aircraft turn unless it is also made to bank by using the ailerons.

During many manoeuvres, including a simple turn, the pilot uses the lift force to move his aeroplane in the required direction. To begin a turn, the pilot rolls the aeroplane into the direction of turn so that the lift force, rather than being vertical, is inclined towards the centre of the turn. The lift force now not only has to support the downward-acting force of the aeroplane's weight but must also pull the aeroplane in a circular path. In fact, the lift needs to be increased and the aircraft will begin to lose height unless the pilot uses the elevators to increase the wing's angle of attack and so generate more lift. But the turning problem does not end there. Greater lift and a higher angle of attack increase drag, so the pilot must raise the throttle setting a fraction to maintain the correct speed.

Even when the aircraft is banking correctly, with the necessary increases of lift and power, it will not turn smoothly unless the rudder is

The Dassault Mirage G.8 (1971) was one of a number of military aircraft to demonstrate Wallis's swing-wing principle. Although the French air force has not ordered production models of the G.8, both the Americans and Russians have several swing-wing combat aircraft in service.

applied to prevent it from either sideslipping into the turn under the action of gravity or skidding out of the turn owing to centrifugal force. Although co-ordinated turns become second nature to experienced pilots and can be made almost without reference to the instruments, modern aircraft usually have a turn-and-slip indicator which shows whether turns are being made correctly. In a plane with an open cockpit an expert flyer can judge the accuracy of a turn simply from the feel of the slipstream on his face.

A simple stall causes the nose to drop, with an inevitable loss of height. The aeroplane recovers when speed is increased and the angle of attack is reduced. Unless the pilot takes care to keep the aeroplane straight, however, a stall may develop into a spin. Spins are sometimes entered accidentally, and they were a frequent cause of fatal crashes in the early days of flying. A spin develops if one wing drops during the approach to a stall. The angle of attack of the downgoing wing is increased by its downward motion and it stalls. The opposite wing may remain only partially stalled, so that the plane enters a steep spiral path with both forward and downward speeds quite low. The plane is now said to be in a state of *autorotation*, either with the nose pointing downwards or, in an especially dangerous 'flatspin', in an almost horizontal attitude. With some aircraft designs it may be difficult if not impossible to recover from this state. Other aircraft may recover if the pilot centralizes the controls or applies rudder in the direction opposite to the spin. The design and relative positions of the tailplane and fin are critical to good spinning characteristics. Even inverted spins are possible in some aeroplanes.

## Aerobatics and Gliding

Aerobatic manoeuvres require keen reflexes in the pilot and close co-ordination of the various control movements. The best aerobatic aeroplanes do not need to be particularly fast but they must be structurally sound and they require a large wing area and an engine that is powerful relative to their weight.

The loop is one of the simplest aerobatic manoeuvres, although the pilot of a low-powered machine may first have to enter a shallow dive to build up speed. When sufficient speed has been obtained the nose is raised by pulling back on the control column. As the pilot reaches the top of the loop, in an inverted position, he checks to ensure that the plane is not rolling, and then carefully enters the back (downward phase) of the loop by gradual adjustment of the control column, foot pedals, and throttle setting.

There are basically two types of roll, the slow roll and the barrel roll. In the slow roll the pilot attempts to roll the plane around its longitudinal fore-and-aft axis; in the barrel roll the plane describes a helix – a large, open spiral parallel to the ground. Both manoeuvres require much more skill than the loop. Flick manoeuvres usually require the pilot to enter a deliberate stall by abruptly increasing the angle of attack. In the flick roll, for example, the plane is made to rotate rapidly about its longitudinal axis in a stalled condition.

Although most aerofoil (wing) sections are not symmetrical, since their top surfaces are shaped differently from their lower surfaces, they can produce lift when upside down if set at a greater angle of attack. It is therefore possible for many semi-aerobatic aeroplanes to fly inverted – with lift still equal to weight – provided that gravitational force does not starve the engine of fuel and oil. The designer of a specialized, fully aerobatic plane makes a

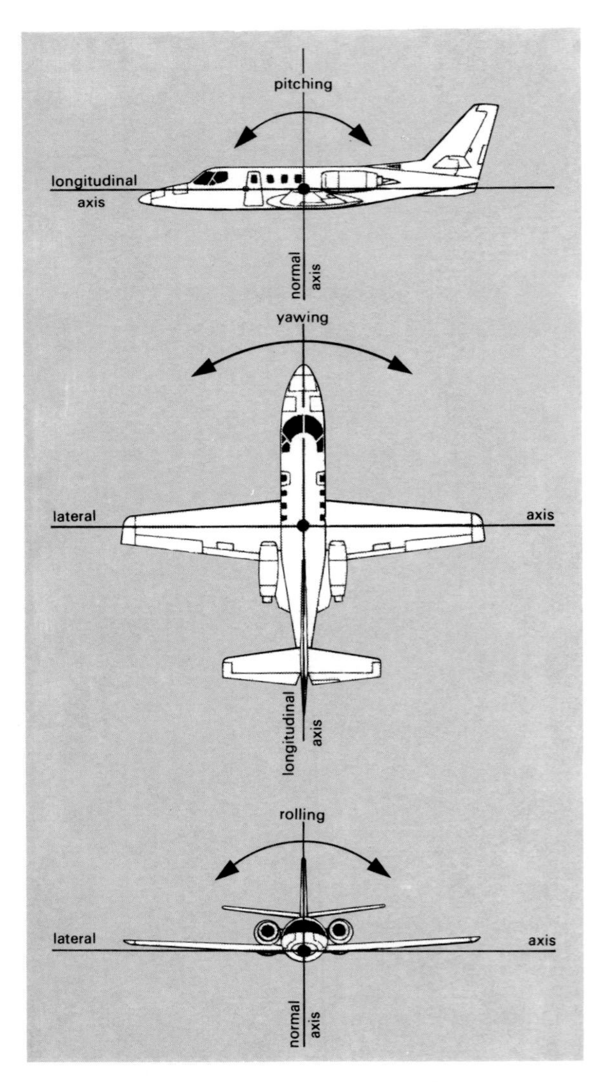


**Left** The three basic movements of an aircraft (from the top): about the lateral axis (pitching); about the vertical axis (yawing); and about the longitudinal axis (rolling).

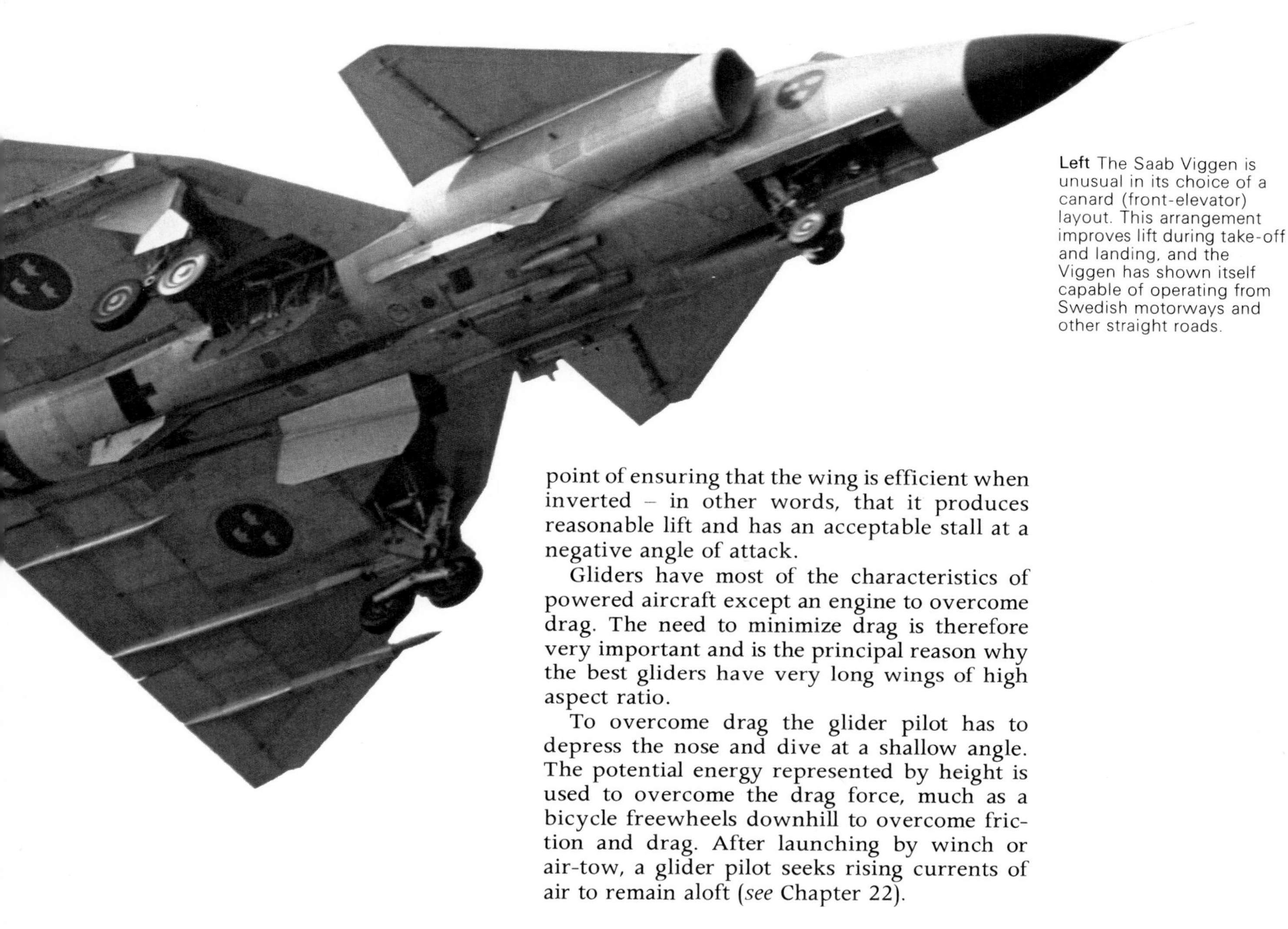

**Left** The Saab Viggen is unusual in its choice of a canard (front-elevator) layout. This arrangement improves lift during take-off and landing, and the Viggen has shown itself capable of operating from Swedish motorways and other straight roads.

point of ensuring that the wing is efficient when inverted – in other words, that it produces reasonable lift and has an acceptable stall at a negative angle of attack.

Gliders have most of the characteristics of powered aircraft except an engine to overcome drag. The need to minimize drag is therefore very important and is the principal reason why the best gliders have very long wings of high aspect ratio.

To overcome drag the glider pilot has to depress the nose and dive at a shallow angle. The potential energy represented by height is used to overcome the drag force, much as a bicycle freewheels downhill to overcome friction and drag. After launching by winch or air-tow, a glider pilot seeks rising currents of air to remain aloft (*see* Chapter 22).

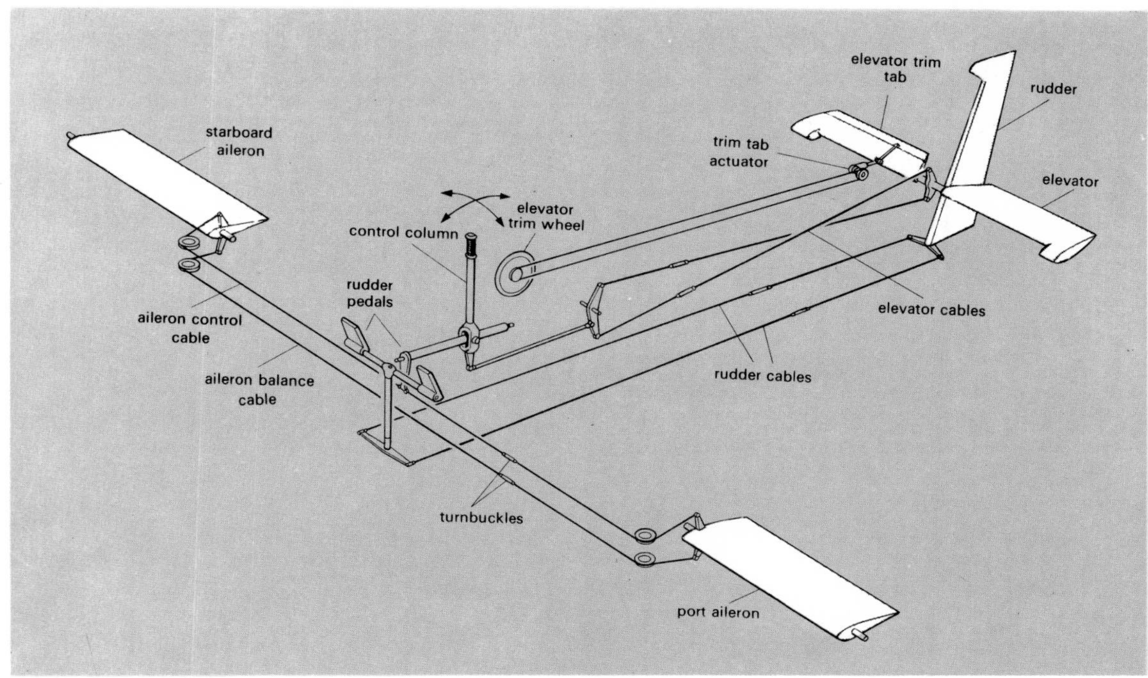


**Right** The cable-operated control system typical of a conventional light aeroplane, showing location of levers, pulleys, and cable turnbuckles.

# 8 Airframes

The history of aircraft can be viewed from many angles. For the aeronautical engineer and designer it is a story of progress in techniques and materials, with advances in one opening the way to developments in the other. Over the years structural materials have ranged from wood, wire, and fabrics to high-strength light alloys, special steels, titanium, fibre-reinforced plastics, and much else; construction methods have developed from externally braced structures to unbraced cantilevers and monocoque (frameless) shells. Yet designs and materials typical of almost every stage in the evolution of aircraft can be found in service today: many successful light planes are still made of wood; tubular space-frame designs of a kind similar in principle to that used in Louis Blériot's pre-1914 monoplanes are still in production; and biplanes based on designs almost 50 years old are still regarded as the best machines for aerobatic and other sports.

The design and manufacture of any structure as complicated as an aircraft involve a series of compromises. For example, the need for great size, which may involve a heavy structure, may conflict with an equally pressing need for lightness in the interests of economical operation; or the need for great range, which will require allocating a lot of space and weight for fuel, may shrink the payload beyond acceptable limits. These problems were as difficult to solve in the early days of aviation as they are today. In addition, however, the pioneers were greatly restricted by the materials at their disposal and by the techniques available for cutting, forming, and joining them. The Wright brothers built their gliders and *Flyers* from scratch, using the facilities of their small bicycle workshop; they even designed and built their own engines with the aid of a mechanic, Charlie Taylor. For their airframes they selected wood as the primary structural material, bracing it with wire and covering it with fabric. Another notable pioneer, Alberto Santos-Dumont, constructed his frames from bamboo.

The Wright *Flyers* were canard biplanes (with front stabilizers instead of tailplanes) powered by two pusher propellors mounted behind a single engine and driven by bicycle chains. Wing warping – that is, pulling on the bracing wires to twist the outer sections of the wings – was used to control rolling. Although remarkably manoeuvrable for their day, these canard biplanes were a design *cul-de-sac*. By the outbreak of World War I the monoplane or biplane with a conventional tailplane and fin, using ailerons for roll control, and with a tractor propeller in front of the engine, was becoming the accepted configuration. There were successful exceptions, of course, but the trend was clear. As speeds increased, streamlining became more important to reduce drag and planemakers began to enclose more of the frame. As manoeuvrability grew in importance, so the structure had to cope with loads not only greater in magnitude but also from more varied directions.

The single-seat Blériot XI monoplane, which made the first crossing of the English Channel by an aeroplane in July 1909, had a thin cambered wing, braced by wires running from the upper and lower surfaces to vertical posts extending above and below the fuselage. Only the centre portion of the fuselage was covered. The tailplane and fin did not carry separate control surfaces but were all-moving. Roll control was provided by wing-warping rather than by ailerons. The design was successful and relatively easy to fly by the standards of the day, and Blériot claimed to have built more than 800 aircraft before 1914.

World War I produced a host of classic designs including the Sopwith Pup and Camel, S.E.5a, DH.2, Bristol F.2B fighter, Fokker Dr.I triplane and D.VII biplane, Nieuport, and Albatros D.Va. Too late to take part in the war, the Vickers Vimy twin-engined bomber flew for the first time in 1918 and in June 1919, piloted by Captain John Alcock and Lieutenant Arthur Whitten-Brown, it became the first aircraft to cross the Atlantic. The 3,043 km (1,890 miles) flight was completed in 16 hr 28 min, and the Vimy weighed 6,035 kg (13,290 lb) compared with the 230 kg (507 lb) of the Blériot XI monoplane which had crossed the Channel 10 years earlier. The Vimy's biplane wings had a fairly robust section, were strutted, and had ailerons on both upper and lower surfaces. The structure was principally wooden but, unlike the Blériot XI, it was completely covered in fabric. Although it had twin fins and a biplane tail, it had rudders and elevators rather than the all-moving surfaces of the earlier French design.

## Junkers and Fokker

Fabric-covered wooden structures braced by wires were not, however, the way ahead. Ideas for a monoplane with a thick, cantilever wing (that is, a wing supported only at its root in the fuselage) were first put forward in Germany by Professor Hugo Junkers (1859–1935) in 1910, and a French aircraft, the Antoinette Military monoplane, was built and flown using this concept in 1911. During World War I Junkers employed wings made of a series of tubes running spanwise, braced internally with metal, and covered with a corrugated aluminium-alloy skin. The skin was not fully stressed as it was on later designs, and it created additional

**Right** The Blériot XI of 1908 had a wire-braced timber-framed fuselage (a) typical of its day. The smaller drawings show (b) a joint between longitudinal and transverse frame members; and (c) a wire-brace fitting on one of the wooden wing spars.

**Above** A Sopwith Schneider float seaplane (1914) under construction at Kingston-upon-Thames. Based on the Sopwith Tabloid, the seaplane won the 1914 Schneider Trophy.

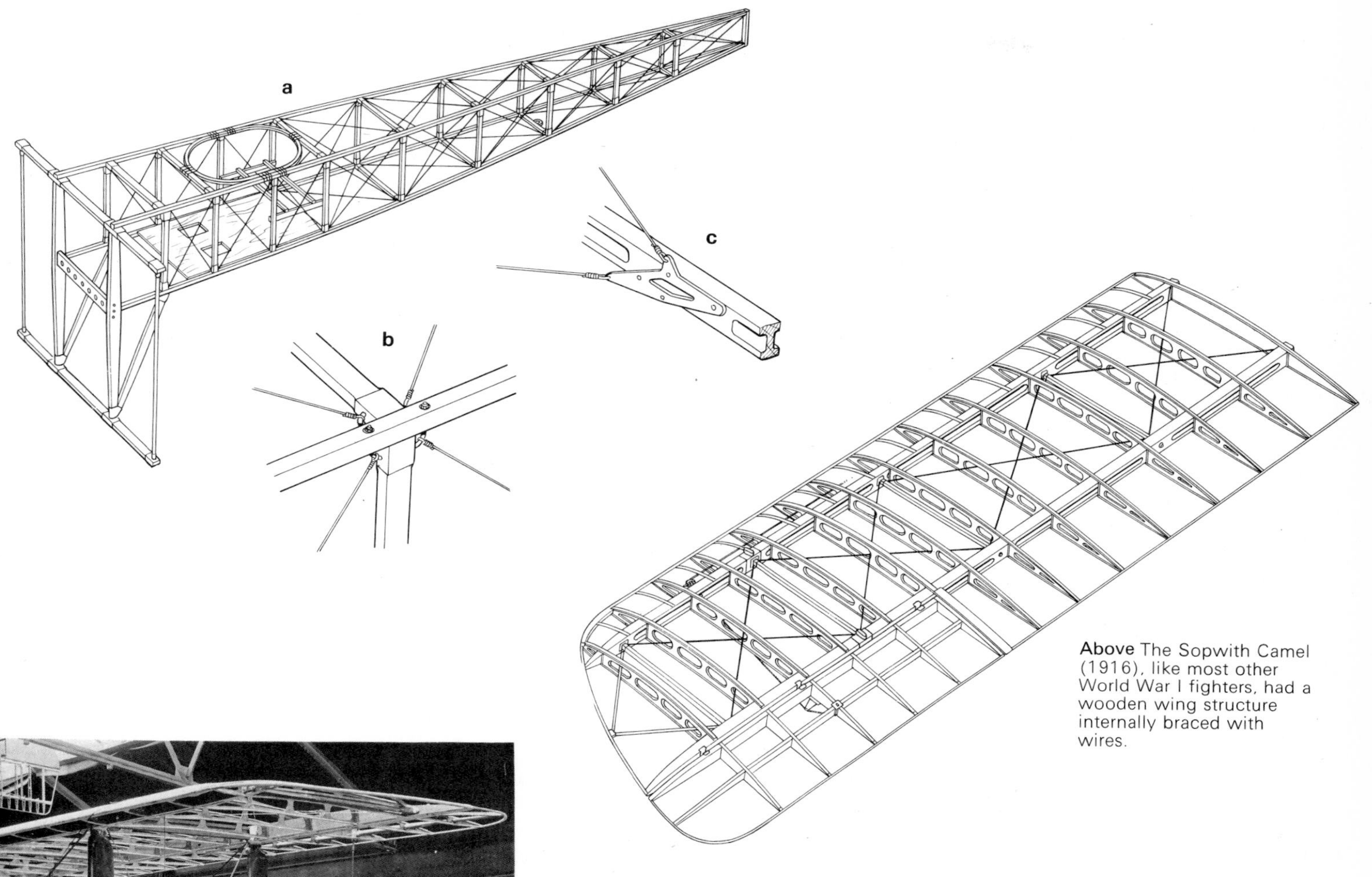


**Above** The Sopwith Camel (1916), like most other World War I fighters, had a wooden wing structure internally braced with wires.

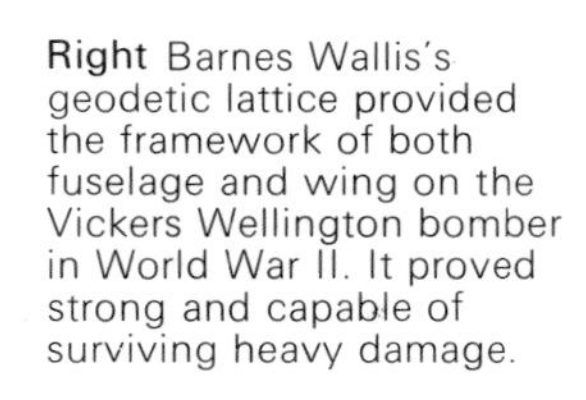

**Right** Barnes Wallis's geodetic lattice provided the framework of both fuselage and wing on the Vickers Wellington bomber in World War II. It proved strong and capable of surviving heavy damage.

drag; but it proved much more durable and practical than its contemporaries. Having applied the principle to a number of small military aircraft during the war, Junkers flew the F13 low-wing civil transport monoplane in June 1919. This single-engined aircraft cruised at 140 km/h (87 mph) and could carry four passengers; its maximum take-off weight was 1,750 kg (3,865 lb). About 350 were built, and in many ways the F13 was the most significant aeroplane of its era.

The Dutch engineer Anthony Fokker (1890–1939) was another leading designer of the period who built military aircraft for the Germans in World War I. Fokker favoured a fuselage made of welded steel tubes covered with fabric, and wings made of wood but with a load-bearing plywood skin. After the war he applied the techniques to a high-wing transport monoplane, the Fokker F.II, which flew for the first time in September 1920. Like the Junkers F13, the F.II was powered by a single engine and could carry four passengers.

Multi-engined aircraft had been built before the 1920s but it was pressure from the airlines which emerged immediately after the war (*see* Chapter 18) which really provided the spur to their development. The airlines wanted aircraft which were safe to fly if one engine failed in flight. Earlier multi-engined types had not been able to maintain height with one engine out of action. The requirement was also reflected in a British Air Ministry specification issued in 1922 for a three-engined type for use in the demanding conditions in Iraq, the Persian Gulf, and on routes to India. Like the airlines, the Ministry recognized that the power plants then available would not give a twin-engined aeroplane the ability to fly on one engine but might enable a three-engined aeroplane to fly on two. The Ministry specification resulted in the Armstrong Whitworth Argosy which flew in July 1926. A less-demanding specification from Imperial Airways led to the de Havilland DH.66 Hercules of the same year. However, it was three-engined adaptations of the basic Fokker and Junkers single-engined monoplanes which really led the way. Fokker's F.VIIA/3m and the Junkers G23 were flown in 1925 and were built in significant numbers. The G23 was first built in the Soviet Union owing to postwar restrictions on German aircraft manufacturers, and it influenced the thinking of Andrei Tupolev, the Russian designer. Tupolev's ANT-9, which first flew in 1929, was a three-engined, high-wing monoplane based on Junker's principles.

The lift generated by the wings of pre-World War I monoplanes was carried to the fuselage by wires, as on the Blériot XI; the wings themselves were too weak to carry the bending load to the roots. The biplane offered a light yet rigid 'cellule' with the two wings, strutted and wire-braced along their span, providing a structure like a girder bridge which could carry bending loads. The monoplanes developed by Fokker and Junkers were a breakthrough. They had unbraced wings which reduced drag and promised big reductions in weight on large aeroplanes. The wing was thick enough to act as a cantilever beam and take the bending due to lift along the span to the fuselage. Although made of wood, the Fokker wing was in some ways the more advanced as it made fuller use of the skin as part of the load-carrying structure; moreover, Junkers' wing generated more drag than he had anticipated.

In spite of Fokker's success with wooden frames and skins, metal was eventually to become the dominating material in the construction of airframes. It is almost impossible now to determine who first used light aluminium alloys for this purpose. In any case, it is clear that quite a number of designers envisioned this use of aluminium alloy from aviation's very infancy. We know, for instance, that as early as 1910 it was used by Louis Breguet in France, while a little later Dr Claude Dornier in Germany developed partly stressed smooth aluminium-alloy skins.

In America Lockheed, Ryan, and Bellanca developed a series of efficient, high-wing braced monoplanes in 1925–30, one of which, the Ryan NYP *Spirit of St Louis*, was used by Charles Lindbergh for his epic flight from New York to Paris in 1927. The fuselages of the Lockheed series were of monocoque (single-shell) stressed-skin construction, at first of wood and later of metal. William B. Stout designed the all-metal Ford Tri-Motor in 1926, using a high-wing layout similar to that of the Fokker series but employing a metal structure with a corrugated skin based on Junkers' philosophy.

## The Birth of the Modern Airliner

During the 1930s the main centre of innovation in the design of transport aircraft passed to the west coast of the United States, and specifically to the two great firms of Boeing and Douglas, whose fortunes were put on a firm basis by the rapid growth of the American domestic airlines. In February 1933 the Boeing 247 made its maiden flight, which was followed five months later by that of the Douglas DC-1. Although neither aircraft was entirely successful, their rapidly introduced successors, respectively the 247D and the DC-2, were bristling with new ideas and can fairly be called the ancestors of the modern airliner.

**Below** The fuselage of the Douglas DC-3, which first flew in 1935, was built up from light alloy frames and stringers, which stabilized the thin light-alloy skin and allowed it to carry stresses. Although fuselages became circular in cross section when pressurization was introduced, the same basic method of construction is still used on modern commercial jets. On the DC-3, wing bending was carried by its three wing spars built up with thin webs and mass booms at top and bottom; torsional (twisting) loads were carried by the wing skin stabilized internally by ribs and stringers.

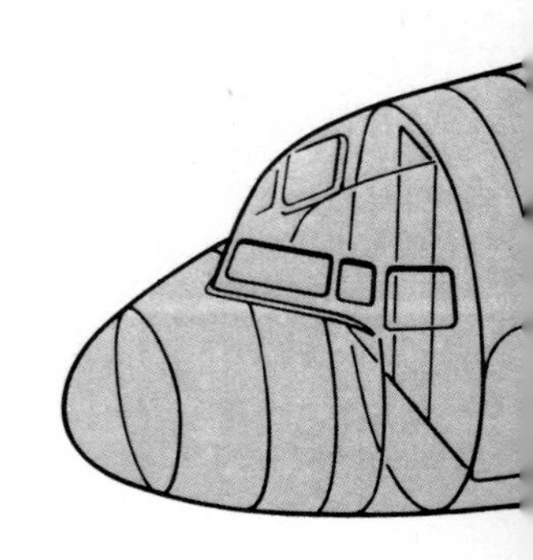

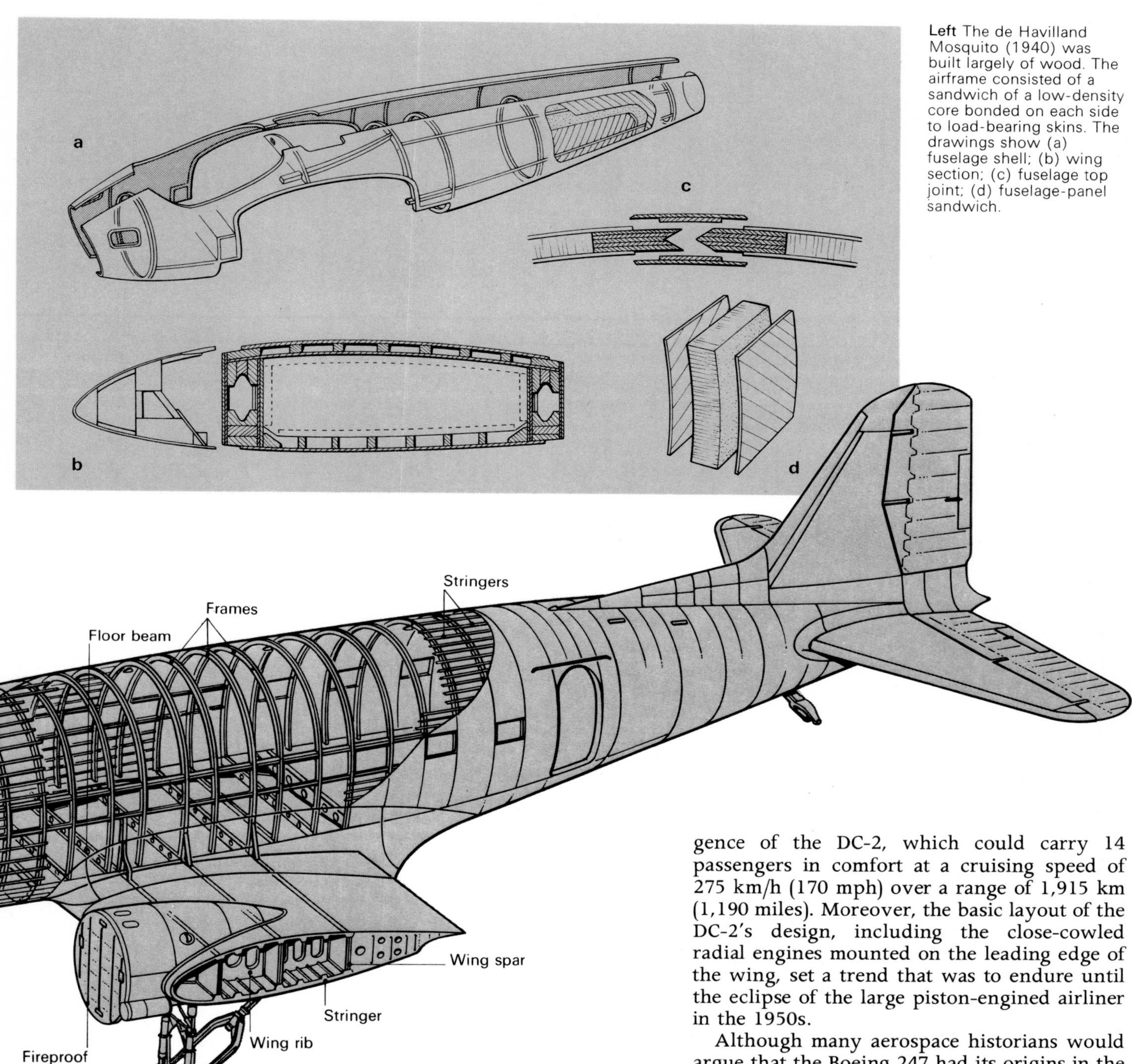


**Left** The de Havilland Mosquito (1940) was built largely of wood. The airframe consisted of a sandwich of a low-density core bonded on each side to load-bearing skins. The drawings show (a) fuselage shell; (b) wing section; (c) fuselage top joint; (d) fuselage-panel sandwich.

Both aircraft were twin-engined, all-metal, low-wing, cantilever monoplanes with stressed-skin structures and retractable undercarriages. Both had *variable-pitch* propellers, allowing the blades to be set to match the speed of flight and giving both aircraft the capacity to fly on one engine. Variable pitch also gave improved take-off and landing performance, which was further enhanced on the DC-2 by the design of its wing flaps.

Just 30 years separated the first powered flight by the Wright brothers and the emergence of the DC-2, which could carry 14 passengers in comfort at a cruising speed of 275 km/h (170 mph) over a range of 1,915 km (1,190 miles). Moreover, the basic layout of the DC-2's design, including the close-cowled radial engines mounted on the leading edge of the wing, set a trend that was to endure until the eclipse of the large piston-engined airliner in the 1950s.

Although many aerospace historians would argue that the Boeing 247 had its origins in the Boeing B-9 bomber, and that therefore military aircraft were the pace-setters in design during the inter-war years, this thesis is debatable. The dominant military types as late as the mid-1930s were fighter biplanes. Biplanes are strong and manoeuvrable but their high drag guaranteed their demise in both civil and military aviation. The last of the British fighter biplanes, the Gloster Gladiator, made its first flight in 1934 – only a year before that of the classic monoplane, the Hawker Hurricane. The Fairey Swordfish biplane first flew in 1934, only a year before the maiden flight of the four-engined monoplane Boeing 299 bomber from which the B-17 Flying Fortress was developed. Clearly the aeronautical world was in transition at that time, but it was civil aircraft which had shown the way. Nevertheless, there were some technical spin-offs from the Schneider Trophy series of racing seaplanes that found expression in the Supermarine Spitfire which flew in 1936, although the latter was quite different in form and structure.

At the beginning of the war most front-line military types, including the Hurricane, Spitfire, and Short Stirling heavy bomber, were not of true stressed-skin construction since all had areas which were fabric covered. The end of the war, however, saw the introduction of the gas-turbine-powered machines and the introduction of new generations of all-metal aircraft.

During the war the Americans had continued with the development and production of piston-engined transports. Over 10,000 twin-engined Douglas DC-3s (Dakotas) were produced, together with 1,200 of the four-engined DC-4s, which first flew in 1942. Lockheed introduced the L.049 (Constellation) in 1943; the DC-6 – a development of the DC-4 – appeared in 1946. The layout of these three last four-engined aircraft was much the same, the main departures from the classic DC-3 formula being the adoption of a nose-wheel undercarriage and pressurized fuselage.

By the end of World War II in 1945 light-alloy stressed-skin construction had been confirmed as the best method of building the basic structure of an aeroplane. There were, of course, exceptions to the general rule, most notably the de Havilland Mosquito and the jet-powered Heinkel He 162 Salamander, both made of wood, and the Vickers Wellington, made in a geodetic lattice of light alloy covered with fabric. The Mosquito and Salamander had stressed skins; wood rather than metal was chosen because light alloy was in short supply. The Mosquito, in particular, with its airframe made from two load-bearing wooden skins bonded, sandwich-like, to a less-dense core, was notable for its low weight and great strength.

In the Cranfield A4 (which was designed purely for aerobatics) and some other modern light aircraft the fuselage frame is built up from welded-steel tube; the external covering of the fuselage provides streamlining but does not carry any loads. Welded-steel spaceframes are relatively easy to make, although they are heavier than a skin, frame, and stringer design.

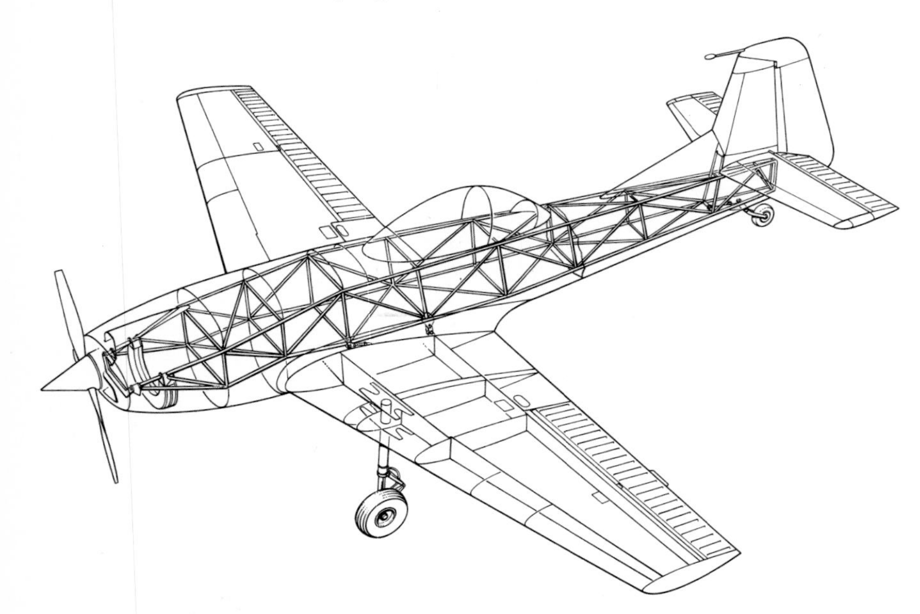

## Airframes in the Turbine Era

The power of jet engines brought a steady increase in aircraft speeds in the years immediately after World War II. As planes flew ever closer to the speed of sound they began to suffer from shock waves and the problem of compressibility (*see* Chapter 7), leading to loss of response in the controls and what appeared to be a dangerous lack of stiffness in some wing sections. The answer was to make wing sections thinner, with swept-back leading edges. Sweep-back, indeed, has the effect of reducing the apparent thickness of a wing. It was soon discovered, however, that thin-section wings have disadvantages of their own: they generate inadequate lift at low speed, and they suffer from lack of stiffness. Low-speed lift was improved by the use of slats and flaps, which have steadily become more complex, and torsional stiffness was increased by thickening the outer skin of the wing.

Many designers concentrated on so-called 'tail-less' jets, which proved to be a mistake. Discounting these, the first post-war swept-wing aircraft were single-engined fighters: the Russian MiG-15, which flew in July 1947, and the North American F-86 Sabre, which followed in October of the same year. Two months later the Americans unveiled the extremely advanced six-engined Boeing B-47. The location of the B-47's engines, suspended from pods beneath the wings, was (like the swept wing) a German idea, and it has since been adopted on almost all the larger jet transport aircraft.

The de Havilland Comet, which first flew in July 1949, was significant as the world's first jet-powered transport; but in terms of airframe it marked little if any advance over the conventional layout of piston-engined aircraft, with thick, only slightly swept wing, unswept tailplane, and engines buried in the wing roots. The first British aircraft with swept wings was the Vickers Valiant bomber of 1951, but this, too, had engines embedded in the wings. In contrast, the Americans continued to develop the thin, swept-wing and podded-engine concept pioneered by the B-47, exploiting it again on the enormous eight-engined Boeing B-52 bomber of 1952. Two years later Boeing flew the prototype of their Model 367–80, with a similar wing and engine layout. Originally earmarked as a military transport, the 'Dash Eighty' was later to become known world-wide as the Boeing 707. Although this aircraft came five years after the Comet – and made its first commercial transatlantic flight only in 1958 – it was the first civil jet-powered liner to bring together all contemporary aviation technology in a practical form, and it enjoyed a phenomenal success for 20 years.

The early career of the Comet, which might have given Britain a decisive lead in jet-powered transport, was blighted by two mysterious disasters in 1953 and 1954 in which the planes seemed to have blown up in mid-air. At least one of these disasters, it was later discovered, was due to explosive disintegration of the fuselage. Like other modern airliners, the Comet had a pressurized cabin and it emerged that, after a period of continuous service, the repeated application of pressurization loads had caused *metal fatigue* in the corner of a small but poorly designed aperture at one point in the structure. From this point failure spread quickly throughout the fuselage structure.

Most metals, especially certain alloys, suffer fatigue (that is, progressive weakening) if they are subjected to repeated loads. Although such loads may be well below those which a structure is designed to withstand, repeated application of the loads over a period of time will cause an accumulation of damage leading to cracking and then to fracture.

The Comet disasters led to some concentrated research into metal fatigue. The findings taught aircraft designers some important lessons. Certain alloys, for instance, are now no longer used in load-bearing structures, the permitted overall stress levels have been reduced, and much greater attention is now given to avoiding design details that could lead to local concentrations of stress. Fatigue cannot, of course, be entirely eliminated, so all aircraft structures are now designed to one of two fatigue categories: *fail safe* and *safe life*. A fail-safe structure is one in which the strength is so distributed that the structure will not fail even if a crack develops in one part of it. In a safe-life structure the designer specifies a fixed life for each component, after which it must be replaced even though it may show little or no evidence of fatigue.

Although the Russians introduced the twin-engined, swept-wing Tupolev Tu-104 jet liner in 1955, the first major innovation in jet-powered transport after the Boeing 707 came from France in 1959, when the Sud-Ouest (now Aérospatiale) Caravelle entered service on short- and medium-range routes. The Caravelle was unique at the time in locating its two engines in pods near the rear of the fuselage, with the tailplane mounted behind and above them on the fin. This layout gave a cleaner, more efficient wing, a better take-off performance, and a much quieter cabin. The Caravelle's layout was adopted on the four-engined Vickers VC10 in 1961, the BAC 1-11 and the Ilyushin Il-62 in 1963, and the DC-9 in 1965. Since then there has been a variety of permutations on engine layouts. In the early 1960s came the de Havilland (now Hawker Siddeley) 121 Trident and the Boeing 727, both with two rear-fuselage engines and a third engine mounted at the base of the fin. This layout also introduced the T-tail assembly, with the tailplane and elevators mounted at the top of the fin. By the end of the decade under-wing engine pods had come back into favour, finding most notable expression in the massive Boeing 747 and its derivatives, which ushered in the era of wide-bodied transports. The big three-engined McDonnell Douglas DC-10 and Lockheed Tri-Star, both of which first flew in 1970, represent a third configuration, having two wing-mounted engines and the third engine at the base of the fin. The Boeing 737 and the very advanced Airbus A300B, both twin-engined, have one pod under each wing.

Turboprop-powered aircraft, with jet engines driving conventional propellers, have tended to follow the airframe and engine layout of the later piston-engined airliners. So far only the Russians have built a swept-wing turboprop, the Tupolev Tu-95 and its civil deriva-

Boeing 727s under construction in the company's enormous plant at Seattle, Washington. More than 1,500 examples of this aircraft have been sold, and production well into the 1980s is assured. Long production runs enable constructors to spread overheads and bring unit costs down.

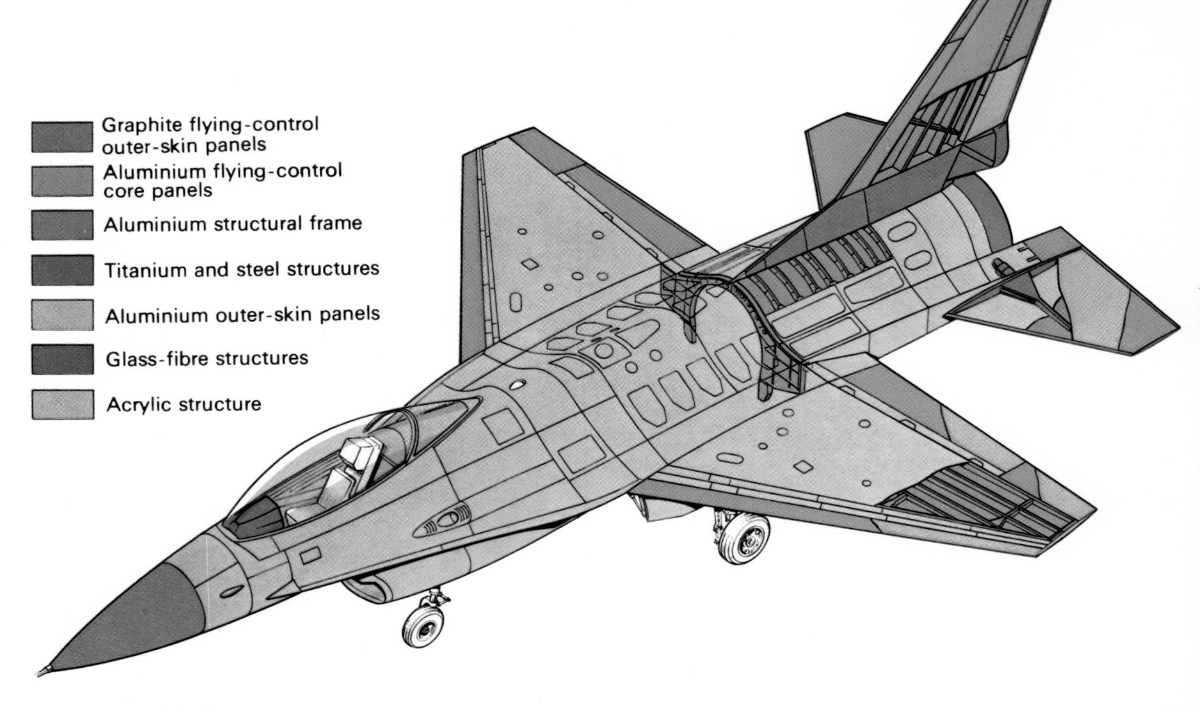


Some of the principal materials used in construction of the General Dynamics F-16 fighter. Designers of modern aircraft, especially military types, are turning increasingly to non-metals, such as plastics and fibre-reinforced materials, for use in airframe structures.

tive, the Tu-114. It is quite possible, in our energy-hungry world, that there will be a revival of interest in turboprop engines owing to their relatively low fuel consumption.

The swept-wing configuration necessary for flight approaching and exceeding the speed of sound has led to two main developments: the variable-geometry wing (*see* Chapter 7) and the delta (triangular) wing; the latter form involves dispensing with the tailplane. Instead of being based on two main spars, as on straight or slightly swept wings, delta wings involve multi-spar construction in which wing and fuselage become fully integrated. In the Concorde and its Russian rival, the Tu-144, for instance, main wing spars and fuselage frames are built as one piece.

Supersonic flight poses the problem of kinetic heating due to friction (*see* Chapter 6). One of the reasons why Mach 2 was chosen as the cruise speed for Concorde and the Tu-144 was that at higher speeds the level of kinetic heating would have ruled out the use of structures made primarily of light alloy – and this would have added greatly to the cost of the aircraft. The Mach 3 supersonic transport (SST) planned by the Americans, but abandoned in 1971, would have used titanium as the primary structural material and stainless steel in many critical areas. Temperatures due to kinetic heating may vary considerably from one part of the airframe to another. On Concorde, for instance, temperatures at cruise range from about 160°C (320°F) at the nose to just over 110°C (230°F) at the tail, and the designers of the aircraft had to make provision to allow for the safe expansion and contraction of the structural components at different rates and by different amounts.

## Engineering Techniques and Materials

The Douglas DC-2 and Boeing 247D pioneered the construction of airframes from light-alloy sheet and strip that could be bent, pressed, rolled, beaten, or machined into shape. In the 1930s the alloy sheets were usually joined by riveting. By the end of World War II spot-riveting (welding) by electrical means had been perfected, as had methods of bonding together different types of metal with adhesives. Perhaps even more important, it had become possible to make large, fault-free castings from light alloys. Magnesium, in spite of the fact that it corrodes and is a potential fire hazard, was used in a number of large airframe parts owing to its low weight. Forging, which involves hammering metals into shape, and extrusion, in which hot metal is forced through a shaped die, were also widely used because of their capacity to increase the local or directional strength of certain metals.

The coming of supersonic flight and the need for thinner wing sections and stiffer skins led to the development of new metal-working techniques. One of the most important of these is skin milling, which was well established by the mid-1950s. It enables a wing skin to be machined from a solid block of metal instead of having to be fabricated from separate sheets and strips. 'Hogging', as cutting from the solid is called, extends the fatigue life of metals. Another significant development, the increasing use of titanium alloys and special steels, was enforced by the need to stress certain structural components to levels beyond the capacity of the lighter alloys that had been used hitherto. The extreme hardness of some of the newer

**Right, above** A Hawker Siddeley Hawk trainer undergoes vibration and flight-load tests at the British Aerospace factory at Kingston-upon-Thames. This kind of test rig offers a much more thorough examination of an aircraft's strength and endurance than component testing or the most rigorous test-flights can provide. It has enabled the company to claim a longer structural life for the Hawk than for its competitors.

**Right, below** A fully automated skin-milling machine in operation. Although the capital costs of such machines are very high, airframe components produced by this method have weight, fatigue-life, and other advantages over the conventional method of building up structural panels piece by piece.

The Inflatobird, which has an inflatable wing, is one of a series of experimental airframe designs flown by the Goodyear company. With its pylon-mounted engine above the wing, single main wheel, and wing-tip outriggers, the Inflatobird resembles a powered glider.

metals used in airframes has occasioned the development of special techniques for detailed cutting and shaping, while new methods of welding have had to be devised for bonding radically different types of material.

Although light alloys are still found in many different types of aircraft, designers are making ever-increasing use of new plastics and fibre-reinforced composite materials in airframes. Composites based on carbon or graphite fibre have nearly two-thirds the tensile strength of steel at about one-sixth the weight, and three times the strength of aluminium alloy at half the weight. Their properties are, however, highly directional and their strength in compression and at right angles to the orientation of the fibres is low. This means they are very suitable for structures subjected to tension in only one direction, such as rotors and spars, but that they have to be built up from laminations rather like plywood if loads are likely to vary in magnitude and direction.

Designers of gliders and helicopter rotors make considerable use of glass and carbon fibres, and glass-reinforced plastics are specified on many subsonic transports in non-critical areas. Honeycomb-structured materials made of light alloy, stainless steel, and even paper are an ideal filler for lightweight 'sandwiches': the honeycomb stabilizes the thin sheets of metal, allowing them to be highly stressed without bending or buckling. Some aircraft, particularly light planes and gliders, now use rigid plastic foam to stabilize structures against buckling.

While British designers have tended to favour carbon-reinforced plastics for major structures, the Americans prefer composites based on fibres of boron or graphite; all are light, strong, and stiff, but carbon fibres are cheaper. The Americans have long used composites in vital parts of the structures of fighters. Uncertainties about their long-term properties and about the legal consequences of any accident that could be blamed on failure of the new materials have so far deterred designers from using them for the structures of any civil transport. Concorde has brake discs made from special carbon fibre, and they result in a weight saving of more than 910 kg (2,000 lb) per aircraft compared with the steel equivalent. McDonnell Douglas expects to specify composites for the multi-spar-wing torsion box of the AV-8B Advanced Harrier to be built in co-operation with the Hawker Siddeley division of British Aerospace.

Most designers would agree that composites promise a big reduction in the weight of aerospace structures. This saving can be used directly to carry more payload, to reduce fuel consumption, or to reduce the size of aircraft needed to carry a given payload. Composites clearly have a future in airframe design.

# 9 Engines

There have been several distinct avenues of aero-engine development, but all have their origins in engineering ideas which pre-date successful manned flight. The piston engine reigned supreme until the 1940s, when gas turbines began to take over as the accepted powerplants for large-capacity or high-speed aircraft. The quest for still higher speeds and altitudes brought an interest in the ramjet and rocket but, in the event, their application was limited to missiles and, in the case of the rocket, to vehicles designed to fly outside Earth's atmosphere. Considerations of cost have so far prevented the gas turbine taking over as the accepted source of power for the lightest and cheapest aircraft, while considerations of fuel economy and low-speed performance have kept the ramjet away from conventional aircraft entirely. Nevertheless, the process of engine development is continuous, and smaller turbines are becoming acceptable, just as higher speeds and altitudes may make ramjets more attractive. No one knows with certainty how long aviation will be able to rely on relatively cheap oil-based fuels, and the next few years could see the start of a serious search for an alternative. Piston engines, turbines, and ramjets could all run on an alternative fuel, so powerplants may change in detail rather than in principle.

The Wright brothers developed their own powerplant because contemporary motorcar engines were too heavy and motorcycle engines were not powerful enough. They adopted water cooling and a crude form of fuel injection for their *in-line* four-cylinder, four-stroke petrol engine, which weighed about 82 kg (180 lb) and produced about 12 hp. Even though steam engines had been used to power models and electric engines were a theoretical possibility, the Wrights correctly dismissed them both as being too heavy and impractical. However, the Wright engine was arguably not the most efficient petrol engine of its day, even though it powered the world's first man-carrying aeroplane.

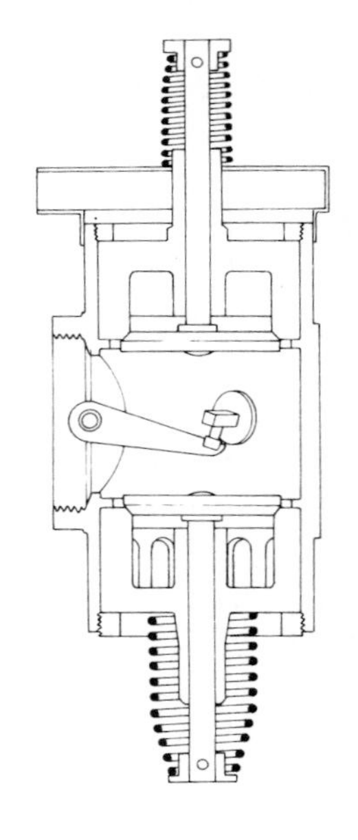

## Rotary and Radial Engines

In 1902 the American Charles M. Manley was commissioned by Professor Samuel Pierpont

**Below and right** The Wright *Flyer* engine, developing about 12 hp, weighed 91 kg (200 lb). Both the engine and its two pushers propellers were designed and built by the Wright brothers.

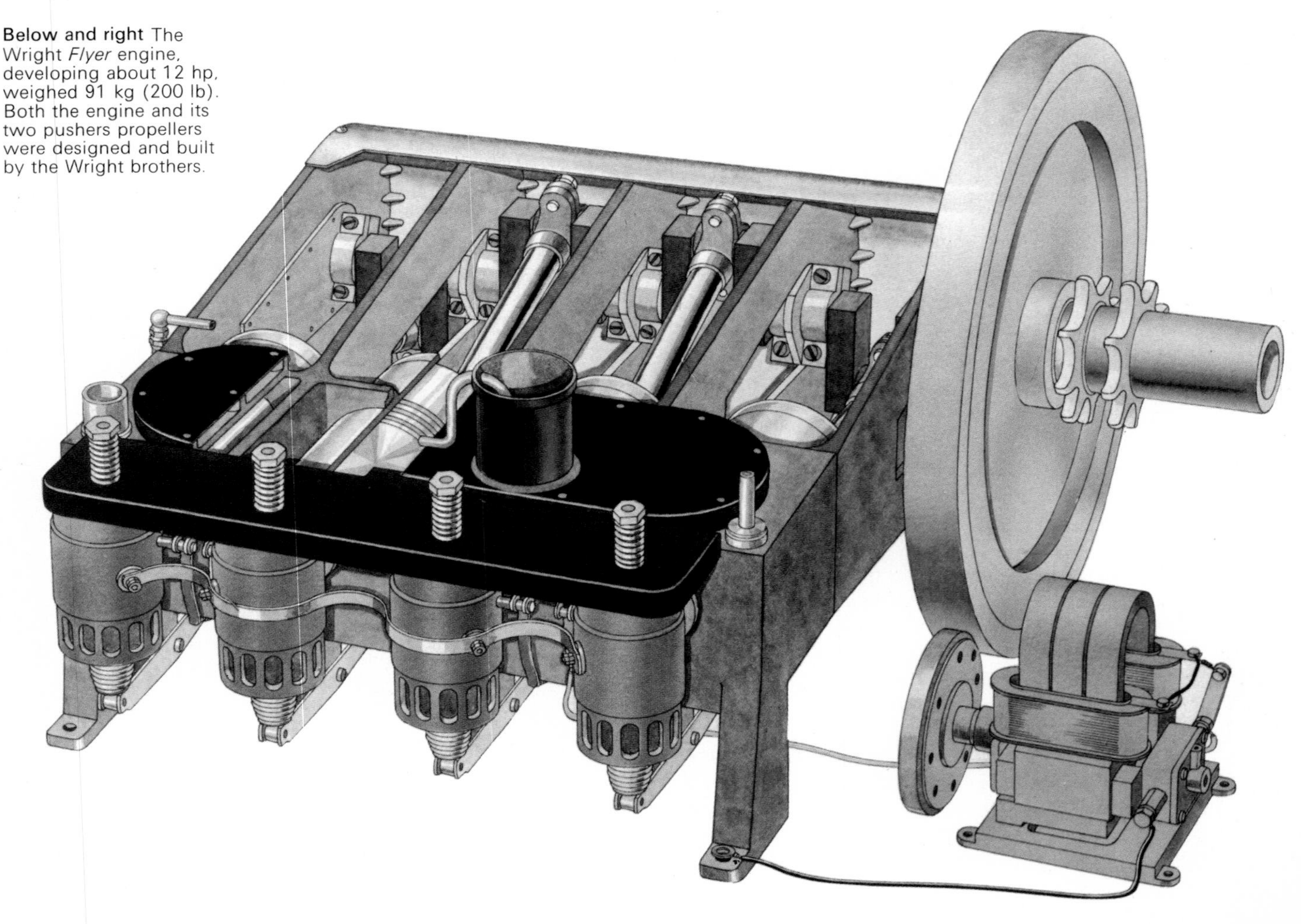

Langley to build an engine for his single-seat aeroplane called the *Aerodrome*. Manley chose a radial layout with five cylinders spaced at regular intervals, like the spokes of a wheel, around a central crankcase. The engine was water-cooled and produced 52 hp for a dry weight of 57 kg (125 lb). Whereas the Wright engine began to run roughly after just a few minutes, Manley's was able to achieve 10 hours of non-stop operation without any loss of power. Many of the features of the Manley engine became standard for radial engines during the next 50 years of aviation. The *Aerodrome* failed to fly, however, and Manley's excellent engine was not adopted for any other aeroplane.

By 1910 just about every layout had been tried for petrol-fuelled aero-engines, and there were about 70 manufacturers of engines in Europe alone. Some designers favoured air cooling because it made engines lighter, simpler, and more reliable; others chose water-cooled arrangements in an attempt to reduce drag. In 1912 the Kaiser offered a prize for the best German aero-engine to be built to a specification which included a power output of between 50 and 115 hp and a weight of not more than 6 kg/hp (13.2 lb/hp) including an allowance for fuel for seven hours' running. The competition, one of the first to recognize the importance of fuel consumption, attracted 26 companies and 44 engines, including three two-strokes. The fact that the winner was a four-cylinder, in-line, water-cooled, 100 hp Benz, which achieved little success after the competition, was not really important. What the *Kaiserpreis* showed was the wide variety of designs available and the growing importance of a layout known as the rotary. The rotary, of which 14 examples were entered, looked superficially like a radial, but its crankshaft was bolted to the airframe while the crankcase and cylinders were bolted to the propeller and rotated as a unit. The rotary had numerous disadvantages, but its rotation kept the cylinders cool and its inertia gave smooth running and damped out vibration. Overheating and vibration were the two principal drawbacks of all early aero-engines. Soon after war broke out in 1914 the rotary established itself as the premier fighter engine.

Perhaps the best known rotary was the French Gnome. It was well-made, reasonably reliable by contemporary standards, and served as a model for all later designs. On the original two-valve, seven-cylinder models the fuel-air mixture was drawn through the rear of the fixed crankshaft into the crankcase and passed to the combustion chamber through a valve in the crown of the piston. Lubrication was pro-

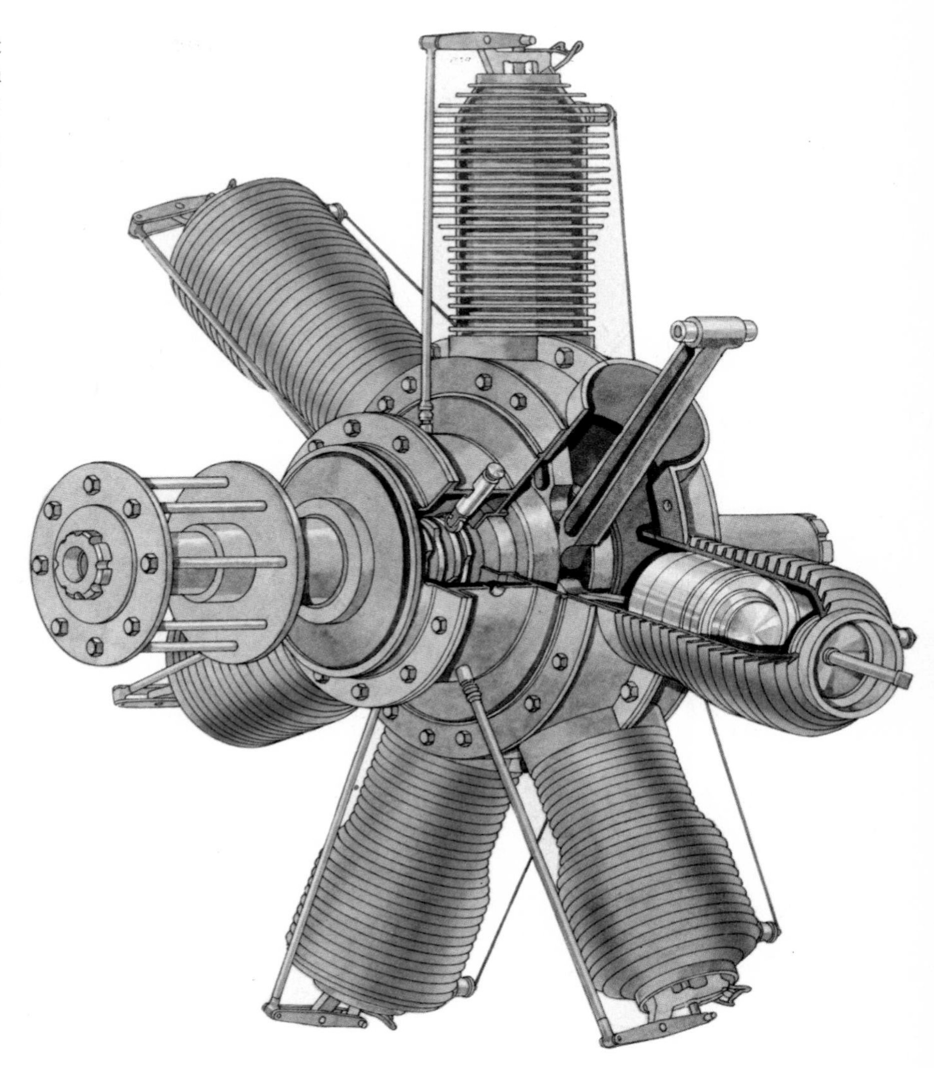

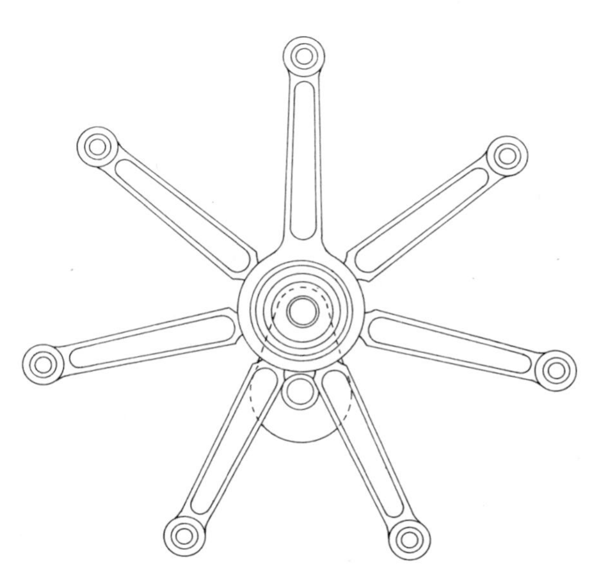

**Above and left** The Gnome rotary engine of 1908, designed by Laurent and Gustav Séguin. The engine's seven spoke-like cylinders turned with the propeller, so cooling themselves even when the aircraft was at rest. Remarkably smooth-running, this first Gnome developed about 75 hp and was the most widely used engine of its day.

vided by castor-based oil in the crankcase, which mixed with the fuel and burned. The engine exhausted through a single valve on the top of the cylinder head – and tended to fling partially burnt castor oil in all directions. Rotary engines were usually cowled (except for a small segment to let in cooling air) in order to reduce aerodynamic drag and cut down friction due to windage. A cowling cut down the amount of castor oil sprayed around, but did not reduce its characteristic aroma. Because there was no satisfactory way of controlling the mixture on the two-valve Gnome, the pilot did not have a throttle; the only power control was provided by an ignition cut-out, 'blipped' in and out for landing.

Just before the war, the *monosoupape* (single-valve) Gnome was introduced. This used ports in the cylinder liner rather than a valve in the piston to allow the mixture to pass into the combustion chamber. It boosted power and, to some extent, improved reliability. In addition, it allowed the pilot some control over power output. During the war the British car designer W. O. Bentley was asked by the Admiralty to help develop a series of aero-engines. The result was the Bentley family of rotaries. Power outputs ranged up to 230 hp, the highest reached by any production rotary, and in this form the B.R.2 powered one version of the Sopwith Camel. The Bentley-powered Camel proved both the strengths and weaknesses of the large rotary. In skilled hands its manoeuvrability using the torque reaction of the engine was legendary; in inexperienced hands it could be lethal to its own pilot.

It became increasingly clear that there was a limit to the size to which rotaries could be developed, and by the end of the war this line of advance was at an end. The air-cooled static radial and the in-line V engine were, however, beginning to prove their worth, and over the next three decades they were to be constant competitors. While one provided simplicity and low weight, the other promised low frontal area and at first lower drag. In the United Kingdom Rolls-Royce began to develop a series of V engines based on the Hawk, Eagle, and Falcon of World War I. In the 1920s Bristol surpassed these engines with the Jupiter, the first of a succession of radials which was to continue to the end of the piston era. During World War I the Americans showed a preference for in-line engines, ordering no fewer than 22,500 units of the V-12 Liberty engine from six automobile factories. Production, however, barely got under way before the end of the war owing to numerous changes to the design. Later, the success of the Bristol Jupiter encouraged the Americans to take a more serious interest in the air-cooled radial.

The Sopwith Camel, which first flew in 1916, was powered by Clerget, Le Rhône, Gnome, and Bentley rotary engines developing between 100 and 150 hp. In dogfights, Camel pilots exploited the rotary's torque reaction to make sudden and very tight right-hand turns. (The aircraft shown here is a modern replica powered by a 165 hp Warner Super Scarab radial engine.)

Like Bristol, the Wright company came into the radial business via a take-over. In 1923 Wright bought the highly-successful Lawrence company and put the latter's new 200 hp radial into production as the Whirlwind. Wright did not, however, show great interest in higher-powered engines. One of the company's leaders, Frederick B. Rentschler, left to form Pratt & Whitney Aircraft. The P & W Wasp, a modern radial like the Jupiter, ran in December 1926. Wright responded to the challenge, but were beaten by Pratt & Whitney for an important US Navy Order. Wright soon launched a new 500 hp-plus engine called the Cyclone, and a commercial rivalry was established which carried the US engine industry through the next three decades. Ford and Fokker Trimotors and the Douglas series of airliners, starting with the DC-2, relied on engines from both Wright and Pratt & Whitney.

## In-line Engines

While the radial engine was developing rapidly, the other mainstream of development, the liquid-cooled V, was also making swift progress. It was the challenge of pure competition, however, rather than commercial reward or military strength that encouraged this development. In 1919–31 the Schneider Trophy races proved that, in the existing state of the art, the low frontal area of a liquid-cooled engine brought big advantages when speed was important. It also proved that fuels, combustion efficiency, and supercharging were as important to high power output as was the development of mechanical strength. Britain won the trophy in 1927 with a Napier Lion-powered Supermarine S.5 seaplane at a speed of 453.3 km/h (281.6 mph), in 1929 with a Rolls-Royce R-powered Supermarine S.6 at 528.9 km/h (328.6 mph), and in 1931 with a Rolls-Royce R-powered Supermarine S.6B at 547.3 km/h (340.1 mph). During the nine months before the 1931 race the output of the R engine had been increased by 450 hp, to give an output of 2,350 hp, and taught Rolls-Royce a great deal, not least about supercharging.

Superchargers, which on aero-engines date back to 1910, are basically a device for feeding the engine air at higher than normal pressure. The use of a supercharger offsets the decrease in the density of atmospheric air with altitude (*see* Chapter 6). It therefore increases the altitude at which an aeroplane can fly, and it also boosts performances for take-off, allowing greater payloads to be carried. Higher cruising altitudes bring better fuel economy and greater range. The supercharger technology Rolls-Royce developed in the 1930s was to aid Britain in World War II.

By the late 1930s the Americans had perfected a series of air-cooled radials producing outputs of between 1,500 and 2,000 hp, but these were principally transport-aircraft engines designed for steady cruising power rather than for combat at full-throttle. The US armed services began to recognize the need for liquid-cooled combat engines; but after extensive research it emerged that their best fighter engine was a conventional air-cooled radial, the Pratt & Whitney R-2800 Double Wasp developing 2,000–2,800 hp.

The Germans produced both air-cooled radials and liquid-cooled Vs, but they tended to concentrate on the latter. Initially development had been restricted by the Versailles treaty of 1919; but in 1934 Daimler-Benz launched its DB600 series of inverted V-12 engines, which were used in the Messerschmitt Bf 109. Later this was developed via the DB601 into the 1,900 hp DB603, while Junkers provided competition with the Jumo 211 and 213. BMW rivalled both with its excellent 801 radial. In all these engines the fuel was injected directly into the combustion chambers, a system later adopted by other companies. Another feature of German aero-engine production was the development of common-standard

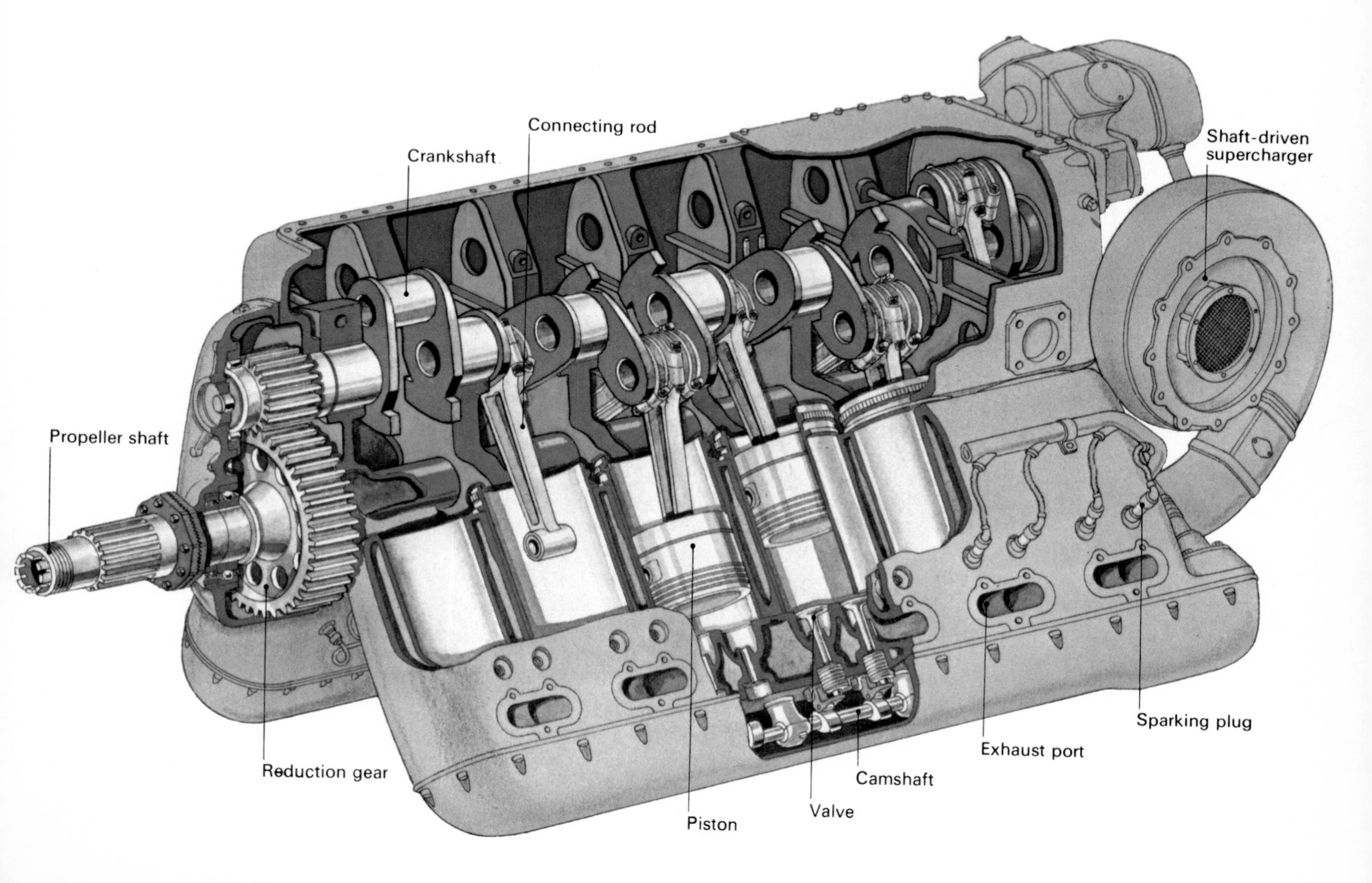


The Daimler-Benz DB 601A 12-cylinder, inverted-V, liquid-cooled engine, developing about 1,100 hp, powered the Messerschmitt Bf 109E, one of the classic fighters of World War II. It had overhead camshafts and, like all piston aero-engines, two sparking plugs for each cylinder.

The close-cowled nacelles of its four 525 hp Gipsy King engines contributed to streamlining on the elegant de Havilland D.H.91 Albatross. The airliner first flew in 1937; its career was blighted by chronic weaknesses in its wooden structure.

V or radial installations, allowing either type of engine to be used on aircraft such as the Junkers Ju 88 and the Focke-Wulf Fw 190.

The British engine produced in the largest quantity was the Rolls-Royce Merlin, a water-cooled, in-line V-12. This started life in 1934 as a prototype giving barely 625 hp at sea level: by the end of World War II bench runs of 2,640 hp had been demonstrated on special fuels. This remorseless improvement of the Merlin typified the aggressive attitude of Rolls-Royce.

After the war the Americans continued to develop big radials for civil use and carried off most of the airline market. The only significant competition came from Bristol, whose Hercules and Centaurus sleeve-valve radials were adopted for several new transports. But the end of the large piston engine was in sight. Wright attempted to hold back the tide for a short period by applying the compound principle of extracting energy from the exhaust with three blow-down turbines geared to the crankshaft. The Turbo-Compound added an extra 550–700 hp to the output of the R-3350 Cyclone and cut fuel consumption at a given rating. The Douglas DC-7 and Lockheed Super Constellation achieved transatlantic range with this engine, but they were the last of the large piston-engined airliners.

Diesels had been flown intermittently since 1918, and Junkers made thousands of opposed-piston two-strokes before ceasing production in 1942. After 1945 Napier developed a two-stroke diesel linked with an axial-flow turbine – the ultimate compound engine – but abandoned it when neither civil nor military authorities showed sustained interest. The future clearly belonged to the turbines.

Lower down the capacity scale, in-line air-cooled engines had established themselves as reliable and simple powerplants for light aircraft. De Havilland had produced its famous Gipsy series of upright and inverted engines between the wars, but the post-war standard became the horizontally opposed flat-four and flat-six developed by Continental and Lycoming in the United States. Although this class of engine has been refined for five decades, it is possible that the threat from small turbines will spur a further stage of development. An outstanding line of advance will be the coupling of piston engines to a shrouded propeller or ducted fan. This arrangement, sometimes called the *propulsor*, reduces the power losses associated with an open propeller and also considerably lessens noise.

## Fuel Technology

During World War I aero-engines ran on low-grade petrol. During the following 20 years, as performance became more important and racing became a way of developing airframes and engines, more and more research was devoted to fuels. The engine companies found that the use of exotic mixtures allowed greater power to be squeezed from a given engine; and fuel technology was recognized to be as important as mechanical refinement. One way to higher power output was to increase the compression of the fuel in the combustion chamber, but this tended to cause premature detonation, known as 'knocking'. The answer was fuels with greater detonation resistance, or higher *octane rating*. The rating of aviation fuels increased from 50 in 1914–18 to 73 in 1929, 87 in 1935, and 100 in 1939. In Britain 'four-star' petrol, used nowadays by most motorists, has an octane rating of 97–8.

By the end of 1939 the RAF had standardized on 100 octane, to increase the power of, first, the Bristol radials and then of all other combat engines. Work on fuels by the British established that the anti-knock rating was not the same when the engine was running on a lean

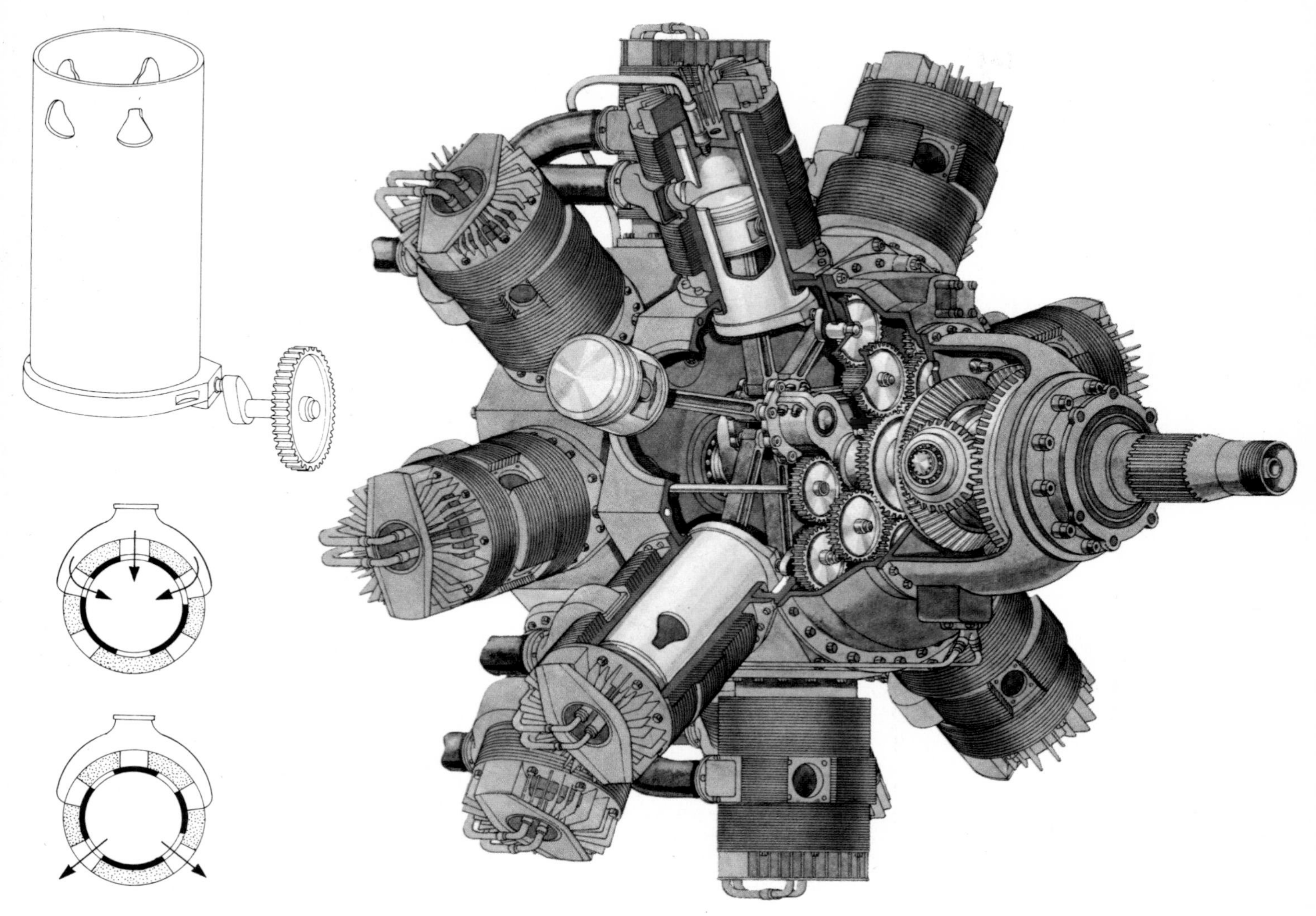

The Bristol Hercules radial engine, first produced in the late 1930s, was widely used in British aircraft in World War II and in its later versions developed more than 2,000 hp. The 14-cylinder, double-row Hercules was one of the first engines to use sleeve valves (upper left drawing) – thin-walled, oscillating sleeves, inserted between the cylinders and pistons. Movement of each valve was timed so that ports in the sleeve coincided at the appropriate moments with inlet and exhaust ports in the cylinder wall (lower left drawings).

mixture (with a high proportion of air) as it was when running on a rich mixture. This led to dual-rating fuels, such as 100/130, in which the first number represented the lean rating and the second the rich; other fuels used in the war included 115/145 and 100/150.

## Generating Thrust

During the summer of 1944 the RAF urgently needed more powerful engines to enable its fighters to catch the Fi 103 flying bomb (known as the V-1, or 'doodlebug'). The Fi 103 was rail-launched (although air launching was employed after the Allies had over-run the launching sites) and was powered by a single pulsejet. This engine was basically a long tube open at both ends. Air entered at the forward end and passed through a grid of non-return butterfly valves and fuel-injection nozzles. The mixture of diesel oil and air was ignited by a spark, whereupon the non-return valves closed automatically owing to the increased pressure, and the hot gas was forced to flow along the pipe and out of the rear end. The valves were then free to open and the process was repeated. The ignition frequency was around 3,000 cycles per minute, giving a deafening motorcycle noise which at a distance was a throbbing rumble. The Fi 103 needed a catapult launcher because its pulsejet produced little thrust below 320 km/h (200 mph).

Ramjets also make use of a simple open-ended duct. They do not have the non-return butterfly valves in the intake but rely on a rise in pressure due to ram effects to choke the intake. Unfortunately they produce no thrust at low forward speeds, and have to be travelling close to the speed of sound to be efficient. So far their application has been limited to missiles, helicopter tip-jets, and experimental aircraft; but in the future they may be combined with jets or rockets to give a hybrid powerplant for supersonic aircraft. A rocket shrouded by a ramjet is known as a ram-rocket.

All aero-engines generate thrust by changing the velocity of their working fluid. Some use a propeller to accelerate atmospheric air, others (such as the pulsejet, ramjet, and turbojet) pass air through the engine, heat it, increase its speed, and exhaust it through a nozzle. Rockets are unique in that they are not air-breathing and carry their working fluid with them. They use a chemical reaction, often involving burning, to generate large volumes of gas, which passes out through a nozzle at high speed. Because the working fluid is given a momentum in one direction, the rocket and the vehicle to which it is attached experience a thrust in the opposite direction.

Many different methods of producing thrust were proposed and flown in model form in the century before the Wright brothers' first powered flight in 1903. Some used steam engines to drive propellers or rotors, others used compressed air or electricity, and at least one used blank cartridges to flap the wings of an ornithopter. But although some of these models were successful, inasmuch as they flew at all, only the petrol engine promised the power-to-weight ratio necessary for a man-carrying, heavier-than-air machine.

## Propellers

A propeller works rather like a small wing (*see* Chapter 7), with its blades inclined to the airflow at a low angle of attack. The pressure generated by the blades rotating in a circle accelerates the air rearwards and results in a forward force on the blades. More sophisticated *variable-pitch* propellers change their angle of attack relative to the airflow to generate the maximum amount of thrust for a given power, the angle being fine (small) at full power at low speeds (as at take-off) and coarse at high speed.

The amount of power a propeller can usefully absorb and convert into thrust depends on the number and shape of its blades, its diameter, and the speed at which it rotates. As engines increased in power during World War II it became increasingly difficult to absorb the power. Propellers which were too large, or rotated too fast and had high tip speeds, ran into problems of compressibility as the tips approached the speed of sound. A few engines were coupled to *contra-rotating* propellers which were mounted one behind the other and geared to turn in opposite directions. This arrangement had two advantages: it tended to keep the slipstream straight, rather than flowing in a spiral; and it eliminated propeller torque (a tendency of the plane to roll in a direction opposite to that of the propeller's rotation), which made some small but powerful single-engined fighters tricky to handle.

## Jet Propulsion

It was obvious that propellers would not be suitable for very-high-speed aircraft, and it

The RFB/Grumman American Fanliner uses a Wankel rotating combustion (RC), or rotary, engine to drive a shrouded pusher propeller immediately aft of the wing. Like the turbine, the RC engine has yet to prove a threat to the small air-cooled piston engine on light aircraft.

also became obvious that piston engines were getting more and more complicated. The answer was to look for a basically different type of propulsion system that was lighter and could run on less-exotic fuels. Although ideas for jet engines had been mooted in the 18th century, practical proposals began to be taken seriously only in the 1930s. Much of the practical inspiration for jet-powered aircraft came from two Britons, Dr A. A. Griffith of the Royal Aircraft Establishment at Farnborough, and Frank Whittle, a young RAF officer. Griffith carried out both theoretical and practical work beginning in 1926. This led to the running of an axial compressor at Farnborough in 1936 and to the involvement of the engineering company Metropolitan-Vickers in 1937. The first Metrovick axial turbojet ran in 1940. Frank (later Sir Frank) Whittle first suggested the use of jet propulsion for aircraft in 1929, and the first engine designed by him was tested in 1937. Rolls-Royce took up the Whittle design, which used a double-sided centrifugal compressor, and ran their first engine in 1942.

The Germans also took an interest in jet propulsion, and it was to be a Heinkel design, the He 178, which made the first flight of a true jet-propelled aeroplane in August 1939. Heinkel had helped young Pabst von Ohain to run his first centrifugal jet in 1937. At the same time Junkers had begun work on an axial design. The Messerschmitt Me 262, powered by two Junkers axial engines, was to be made in far greater quantity than any other jet before the Korean War of 1950–3. It flew in 1942, the same year as the American Bell P-59 Airacomet and one year ahead of the Gloster Meteor. The first British jet was the Gloster E.28/39, flown on a Whittle engine in May 1941.

The principle of the jet engine is simple. Air is passed through a compressor and its pressure raised. Fuel is then added and burnt in a combustion chamber, producing a high-velocity hot gas which flows out through a turbine. The turbine extracts just enough energy to power the compressor, to which it is attached by a shaft running through the centre of the engine. The rest of the energy in the gas is used to provide thrust by expanding at high velocity through an open jet nozzle.

From the early days there have been two types of jet engine. The first, favoured by Whittle, uses a centrifugal compressor which takes the air from the engine intake and compresses it by spinning it outwards across both faces of a rapidly-rotating disc finned on both sides. The second type compresses the air as it

**Above** The horizontally opposed air-cooled engine has become the accepted powerplant for most light aircraft. This French Piel CP60 of the mid-1960s has a Continental engine, one of the dominant makes in this field.

**Below** Propellers have developed from the simple wooden two-bladers used on World War I types such as the Sopwith Camel (left) to the four-bladed glass-fibre designs on the de Havilland Canada Dash 7 (right). The variable pitch propellers of the Dash 7's PT6A-45 turboprops are driven through a gearbox to keep the blade speed down, so reducing noise.

Frank Whittle's original jet engine (seen here in its much-rebuilt final form in 1944) had ten separate combustion chambers. Although Whittle was the first to propose a practical jet engine, his basic configuration using a centrifugal compressor has been superseded by the axial compressor type for most large gas-turbine aero-engines.

moves along the length of an axial compressor looking like a series of fans mounted one behind the other. The centrifugal-compressor type was the better understood (it had been used in superchargers for piston engines) and it is robust and simple; but the axial compressor is slimmer and can work at much higher pressure-ratios, aiding fuel economy. Gradually the use of the axial-flow arrangement has become universal except in the smallest jet engines used in general aviation.

Initially it was very difficult to make the gas turbine efficient enough to run and produce worthwhile power or thrust. It also proved difficult to develop materials capable of standing up to the high stresses at high temperatures (almost white heat) in the combustion chamber and in the turbine. Nevertheless, the basic concept was simple and promised to do away with all the gears, cams, valves, and superchargers associated with the piston engine; it also ran on low-grade fuel similar to kerosene (paraffin). High-temperature alloys and advanced cooling techniques are the key to gas-turbine performance, and they continue to be a major area of research.

At the end of World War II development in Germany ceased and the United Kingdom was the undisputed leader in turbine-engine technology. Work began in the United States and Soviet Union based on data and batches of engines supplied by the British, or captured from the Germans. While Britain had been preoccupied with combat aircraft, the Americans had maintained their lead in transport-aircraft design based on the air-cooled radial. Britain attempted to make use of its gas-turbine technology to break the American dominance and began work on a family of turbine-powered aircraft. It is one of the politico-industrial ironies of the period that British manufacturers were obliged, on government instruction, to pass on much of their hard-learned gas-turbine technology to their rivals.

The world's first jet-powered airliner was the de Havilland Comet, powered by four centrifugal-compressor de Havilland Ghost engines, which made its maiden flight in July 1949. But the British realized that the turbojet would not be suitable for all types of transport.

Whereas the propeller works by giving a large mass of air a small increase in velocity, the turbojet works by accelerating a small mass of air to a high velocity. It is a basic law of propulsion that the former method is more efficient than the latter, so one of the first postwar developments was the turboprop. This engine extracts much more energy from the jet and uses it to drive a propeller. The first such engine, the Rolls-Royce Trent, was flown on a converted Meteor in September 1945. Rolls-Royce followed with the turboprop Dart to power the Vickers Viscount on its maiden flight in 1948; amazingly, the Dart is still in production.

Turboprop engines helped to overcome the high fuel consumption of the early jets, but they did little to relieve the inherent problems of the propeller. Whittle had taken out a number of patents covering adaptations of the pure-jet principle, and his company (Power Jets) and Metropolitan-Vickers ran two engines driving fans at the end of the war. However, it was not until the Rolls-Royce Conway engine was developed in the late 1950s that the *bypass* principle was applied to a production engine. In this system, part of the air passing through the compressor is ducted around the outside of the 'core' engine and either expelled through separate nozzles or re-introduced into the exhaust duct. Designers found the arrangement works best if the compressor and turbine are each divided and connected to two different concentric shafts. With this *two-spool* arrangement, as it is called, the first few stages of the compressor, the low-pressure compressor, are driven by the last stages of the turbine. The final part of the compressor (the high-pressure spool) is powered by the initial, high-pressure stages of the turbine.

A typical split, or bypass ratio, between the air flowing through the bypass duct and through the core of the engine is one to one. The Conway had a timid ratio of only 0.3:1. It spurred Pratt & Whitney and General Electric to modify their existing turbojets into turbofans of a 1:1 ratio.

By the early 1960s designers were ready to take bypass technology a stage further and proposed powerplants with bypass ratios of about 5:1. With these engines, the air not passing through the core is accelerated by a single-stage (or $1\frac{1}{4}$- or $1\frac{1}{2}$-stage) fan and no attempt is made to mix the two flows. The first high-bypass-ratio turbofan was the General Electric TF39 of 18.6 metric tons (41,000 lb) thrust used to power the Lockheed C-5A Galaxy military transport which flew in 1963. The TF39 was followed by the Pratt & Whitney JT9D for the Boeing 747, the General Electric CF6 for the McDonnell Douglas DC-10, and the

**Above** The Fokker F.27 Friendship is powered by two 2,050 hp Dart engines. The high-wing layout of the F.27 provides good ground clearance for the propeller blades but calls for a long-legged undercarriage.

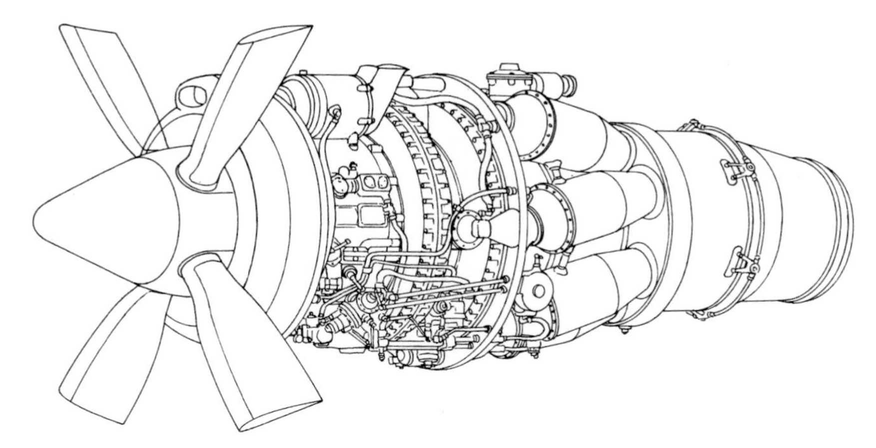

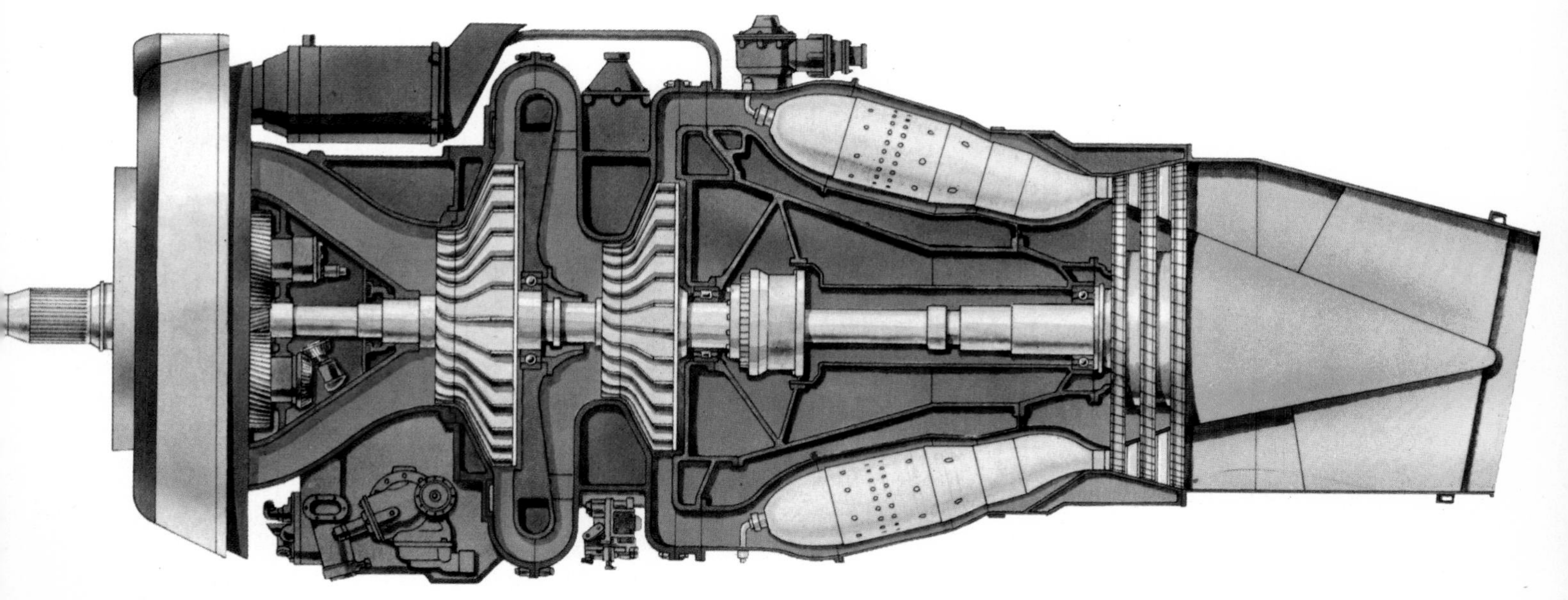

**Right and below** The Rolls-Royce Dart, which made its first bench run in 1946, was the earliest turboprop to go into service. The Dart remains in production after more than 30 years, and over 7,000 have been delivered.

The McDonnell Douglas DC-10-30 trijet has two under-wing engines in pods and a third in a unique straight-through mounting at the base of the fin. The Americans pioneered the use of underwing-podded engines in the Boeing B-47 bomber of 1947, although the idea had been developed by the Germans toward the end of World War II.

Rolls-Royce RB.211 for the Lockheed TriStar. Of these engines, the RB.211, which has a three-shaft or three-spool layout, is possibly the most advanced, although it has been out-sold by both its US rivals. All three manufacturers are now offering versions with over 23.0 metric tons (50,000 lb) thrust.

Turbojet engines were used first in fighters, and designers were soon called on to provide additional thrust for dash or acceleration. It was realized that thrust could be boosted by burning extra fuel in the jetpipe. This technique, known as *reheat*, or afterburning, is not efficient in terms of fuel consumption but it gives a big increase in thrust for a small increase in engine weight. The need to match the shape of the jetpipe exit to the flow requirements of the exhaust gases brought the complication of the variable nozzle. On both civil and military aircraft devices have been fitted to the jetpipe to divert the direction of flow and provide reverse thrust for braking.

Higher speeds also brought complications to the intake, which must ensure that air reaching the compressor is travelling at an acceptable speed (usually less than half the speed of sound) and is free from turbulence. For aircraft able to operate over a wide range of speed and altitude the intake system may need to be very complex. The intake is vital to propulsive efficiency because it may, at supersonic speeds, provide a greater pressure rise than the compressor itself. In supersonic aircraft, therefore, the engine becomes a minor item between large and complex variable inlets and nozzles.

Over the last decade aircraft noise has become a major social issue. It was more by accident than by intention that the high-bypass-ratio turbofans – the JT9D, CF6 and RB.211 – greatly reduced noise; and research is

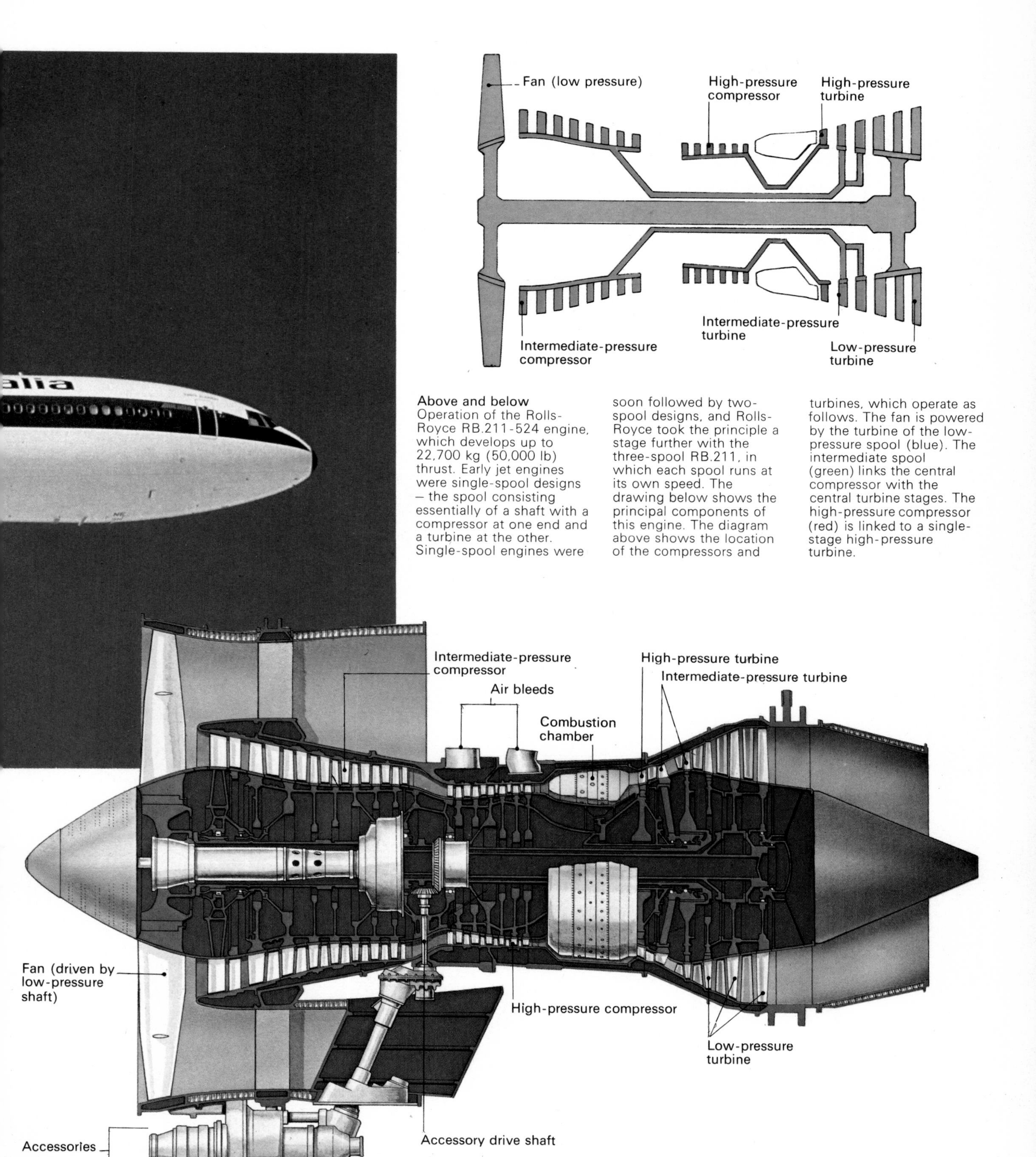


**Above and below**
Operation of the Rolls-Royce RB.211-524 engine, which develops up to 22,700 kg (50,000 lb) thrust. Early jet engines were single-spool designs – the spool consisting essentially of a shaft with a compressor at one end and a turbine at the other. Single-spool engines were soon followed by two-spool designs, and Rolls-Royce took the principle a stage further with the three-spool RB.211, in which each spool runs at its own speed. The drawing below shows the principal components of this engine. The diagram above shows the location of the compressors and turbines, which operate as follows. The fan is powered by the turbine of the low-pressure spool (blue). The intermediate spool (green) links the central compressor with the central turbine stages. The high-pressure compressor (red) is linked to a single-stage high-pressure turbine.

now in hand to ensure that the next generation are even quieter. It has proved particularly difficult to silence older jets – including the engines of the supersonic Concorde and Tu-144 – because much of the noise is generated by the high-velocity exhaust. Reducing exhaust velocity on these designs is impossible without sacrificing cruise thrust. In any case, the Concorde and Tu-144 rely on engines with low frontal area and high exhaust velocity to give Mach 2 performance. Some noise reduction has been achieved on subsonic aircraft by introducing sound-absorbent material in the intake and nozzle, and often by fitting a multi-lobe exhaust nozzle to promote mixing. Supersonic aircraft will probably remain impossible to silence, unless they can be re-engined with a so-called *variable-cycle* installation, which acts like a turbofan for take-off and landing and like a turbojet or low-bypass-ratio engine for supersonic cruise.

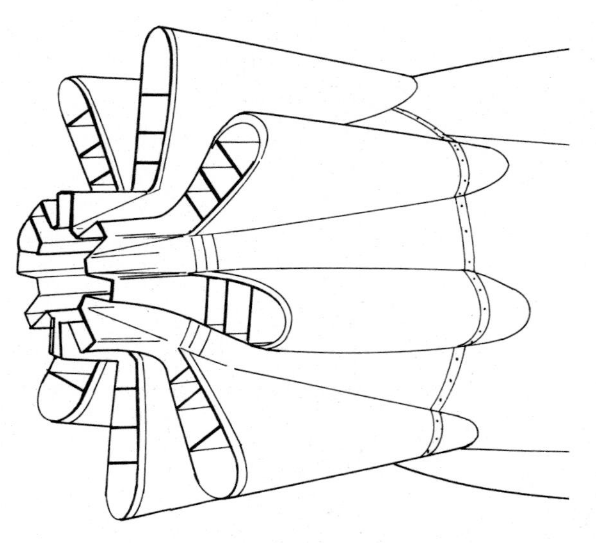

**Left** A multi-lobe nozzle on a jet engine. The long, wavy outline of the nozzle helps to lessen engine noise by speeding the mixing of the jet efflux and the outside air.

**Below** Reheat (or afterburning) boosted the thrust of each of the two Rolls-Royce Avons in the English Electric Lightning from 5,108 kg (11,250 lb) to 6,524 kg (14,370 lb). Reheat greatly increases fuel consumption and so shortens the range of an aircraft, and is therefore used only for a few seconds at a time.

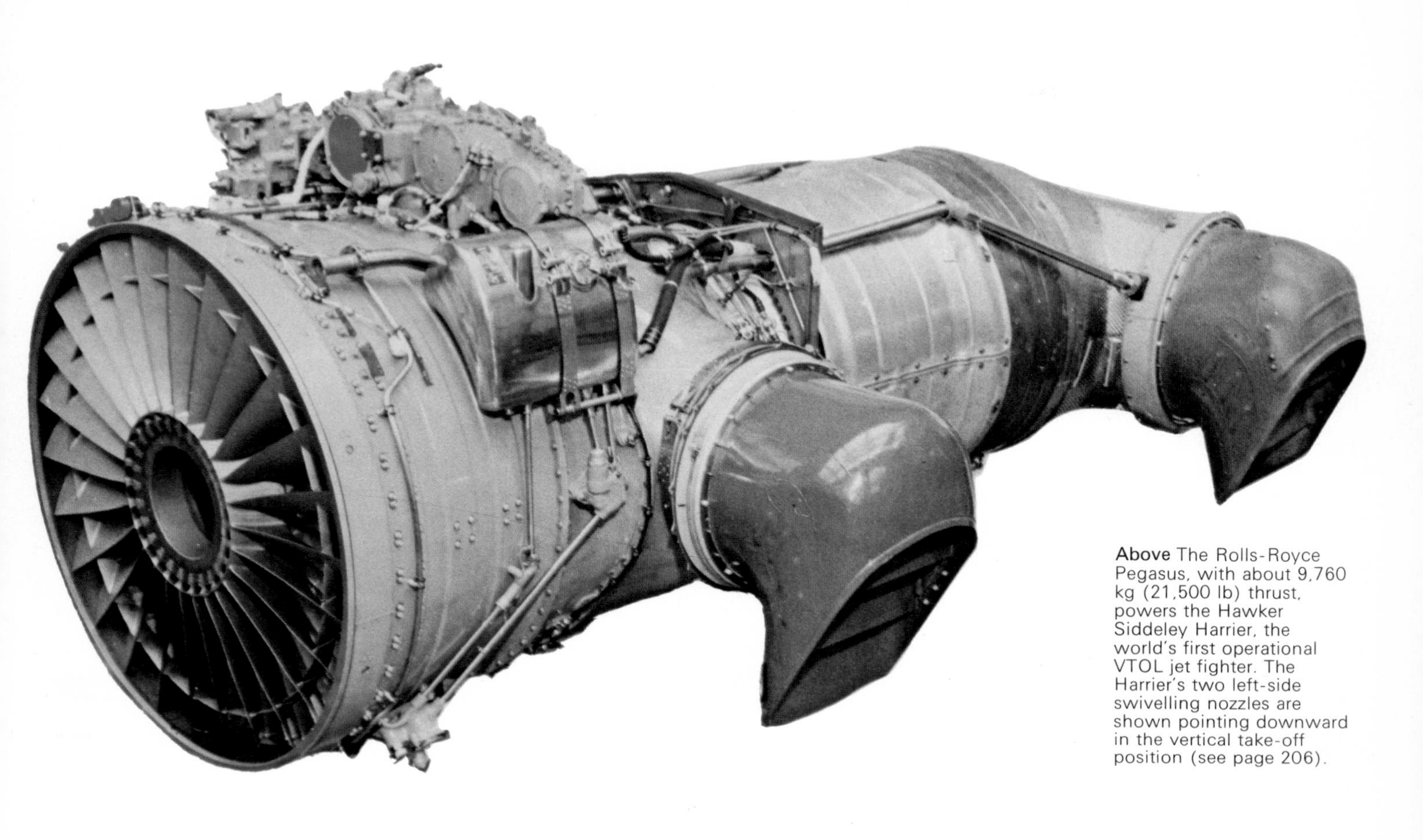

**Above** The Rolls-Royce Pegasus, with about 9,760 kg (21,500 lb) thrust, powers the Hawker Siddeley Harrier, the world's first operational VTOL jet fighter. The Harrier's two left-side swivelling nozzles are shown pointing downward in the vertical take-off position (see page 206).

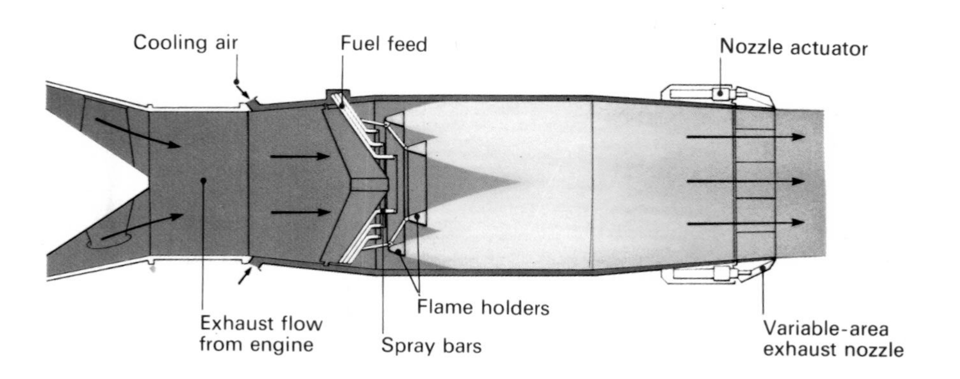


**Left** Reheat, giving a temporary boost to thrust, is achieved by burning extra fuel in the jet pipe. It is used on many military aircraft and on Concorde and the Tupolev Tu-144.

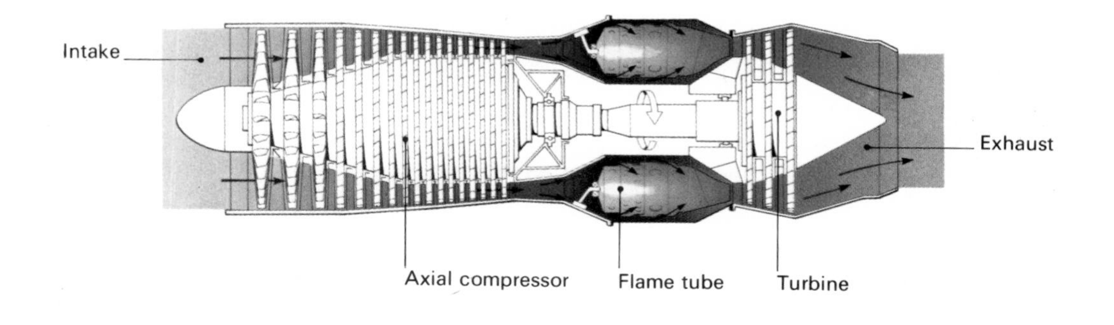


**Right** The passage of working fluid through a typical gas-turbine engine. Air entering the intake (at left) is compressed in an axial compressor, then mixed with fuel and ignited. Combustion increases the energy of the air before it flows through the turbine blades. The turbine extracts just enough energy to drive the compressor before the exhaust passes at speed through the jet pipe.

# 10 Systems

Early aeroplanes had few systems; they were basically an airframe plus an engine. Most had neither instruments nor engine controls. One of the greatest achievements of the Wright brothers was to develop a workable flight-control system and live to tell the tale. Today's advanced aeroplanes are made up of a myriad systems and sub-systems, many of them interdependent and most operating without the direct control of the pilot.

Aircraft themselves may be part of a larger system. A fighter, for instance, is only an element in an air-defence system which includes ground-based radars, computers, control centres, missile bases, and airborne radar carried by large early-warning aircraft. The airliner is only part of a transport system which extends from the ticket counter to baggage collection at the destination airport. A helicopter may be part of a weapon system bringing together ground-based personnel, the helicopter, and a missile designed to knock out tanks.

On all but the smallest modern aeroplanes there are probably systems distributing electric, hydraulic, and possibly pneumatic power. The power system is usually driven from the engine through a direct mechanical off-take. When the engine is not running, power has to be supplied by a trolley-mounted ground power unit (GPU) or from a built-in auxiliary power unit (APU). Flight-rated APUs are now fitted to most airliners and many military types. They add weight and complexity, but bring independence from ground supplies during check-out and starting.

## Engines

The engine relies on systems of its own, including a starter, a fuel system, and a control system. The starter may use electrical power provided by a ground trolley, an APU, or a built-in battery. Some military engines use a cartridge starter which generates sufficient gas to drive a turbine-driven starter-motor. Most modern airliners use air-turbine starters fed by a hose from a GPU or APU; when one engine is running it feeds air to start the others.

The engine control system may automatically compensate for changes in pressure, temperature, and Mach number, and may include an auto-throttle to help the pilot flying in difficult conditions. The need to control closely the intake and nozzle conditions of all supersonic aircraft has tended to increase the importance of engine-related systems. The fully variable Concorde and Tu-144 intake systems are vital to the operation of these aircraft, while the hinged inlets on the McDonnell Douglas F-15 smooth the airflow at high angles of attack. At the other end of the scale, the design of an intake and cooling system for an air-cooled light-aircraft engine is often a matter of trial and error.

Simple aircraft may have a fuel system using gravity feed much like a car, while others require a fully-pumped arrangement. Concorde has a particularly complicated system and uses the fuel not only as a heat sink, pumping it round to cool various equipment and airframe parts, but also to provide *trimming*. At supersonic speeds the centre of the lift force moves, and fuel must be pumped between nose and tail to re-balance the trim. Fuel management is complicated because fuels vary in density. Fuel is usually loaded according to weight, but fuel-flow meters and contents gauges measure in volumetric units. The potential heat energy of a fuel is more closely related to its weight than to its volume.

In-flight refuelling capability is now essential for military types. Most aircraft use a probe thrust into a flexible hose trailing from the tanker, but the USAF Strategic Air Command uses a telescopic boom guided by a tanker crew member into a receptacle on a passively waiting receiver aircraft.

## Flight Control

The flight-control system translates commands of the pilot into a movement of the control surfaces, causing the desired response from the aeroplane. A simple system may involve nothing more complicated than wires, pulleys, and levers to link the control column and rudder pedals to the ailerons, elevators, and rudder. Larger aircraft and higher speeds often result in forces at the control column which are beyond the convenient strength of an average pilot. Under these circumstances the control surfaces need to be moved by some form of aerodynamic or powered assistance, or by fully powered controls. Most propeller aircraft have manual controls; so do many executive jets, and even a small number of larger jets. Often the rudder, tailplane, or ailerons are manual and the rest powered. As the power system may fail, most aeroplanes with powered controls also have manual reversion or some form of powered back-up. A typical scheme is to use three completely separate hydraulic systems to drive each powered surface, so that two can always over-ride a system that fails.

Changes in speed bring changes in control forces and in the response of the aircraft, and the control system makes these changes acceptable to the pilot. In addition, with a fully powered system there may be no actual 'feel' at the control column at all. The control system may have to introduce artificial 'feel', taking account of aircraft speed and atmospheric

density, to prevent the pilot over-controlling and over-stressing the airframe.

Most aircraft use a mechanical linkage to transmit commands from the cockpit to a point close to the control surface. At this point a hydraulic, electric, or possibly pneumatic jack or actuator may take over. Some modern systems, such as on the Panavia Tornado and General Dynamics F-16, use electric signalling known as fly-by-wire (FBW), or signalling using fibre optics – carrying pulses of light – called fly-by-light. Both weigh little and are easily routed around a complex aircraft. Fibre optics are less likely to interfere with, or be affected by, other electrical equipment.

FBW brings two big advances. First, mechanical linkages can be omitted altogether, and the F-16 pilot has a small side-stick rather than a long, central control column. The pilot's seat can thus be reclined to give more comfort and more resistance to black-out under high-'g' conditions. Second, the electrical signals can be

**Above** The cockpit of the three-engined Junkers G.31 airliner of the 1920s. Flying instruments such as compass, turn-and-slip indicator, and altimeter were in front of the captain on the left; engine instruments were in front of the right-hand seat. Cables to control the ailerons can be seen running down the control column; throttles for the three engines are between the seats, as on a typical modern airliner.

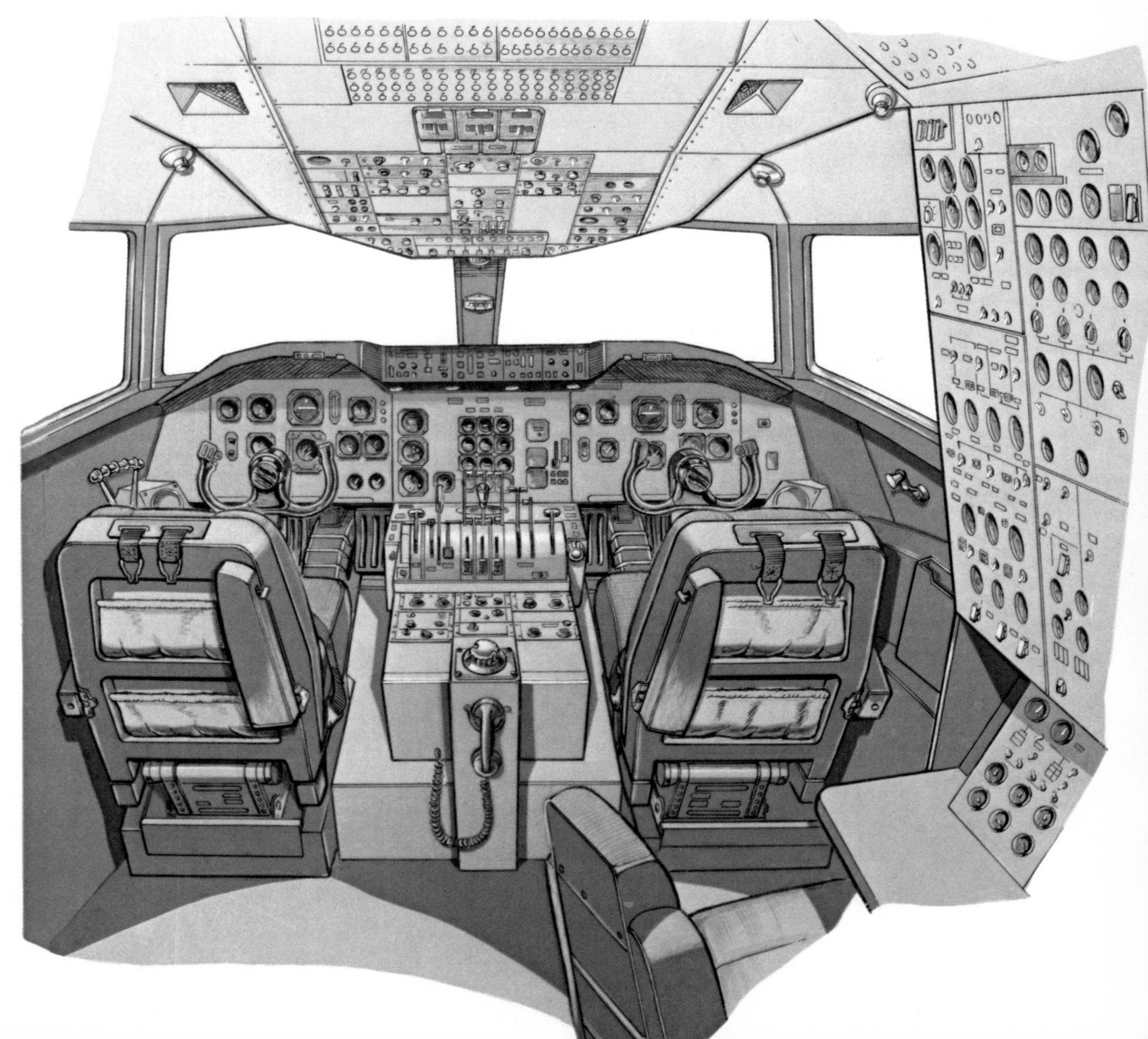

**Right** The cockpit of the McDonnell Douglas DC-10, like other modern airliners, has the captain on the left and the first officer (co-pilot) on the right; the engineer sits behind and to the right. Primary flight-control instruments are placed in front of each pilot, with engine instruments located between. Weather-radar scopes are on the extreme left and right of the panel. Controls for the flight systems (which automatically control speed, altitude, and so on) are on the glare shield, while controls for other avionics (including communications) are behind the throttles. Lighting and cabin-system controls are in the roof panel. The engineer's station has more detailed instrumentation to monitor the fuel, engine, hydraulic, electrical, and a variety of other systems.

On large modern aircraft the forces required to move the control surfaces are far beyond the muscle power of the pilot. The Boeing 747 relies on hydraulic power to actuate the flying controls. It uses no fewer than four independent hydraulic systems – shown here in red, yellow, blue, and green – each powered by a different engine. The primary flight controls for pitch, roll, and yaw are actuated by all four systems; each of the other flight controls is powered by at least two of the four systems, so that if one engine fails or one system breaks down, the others continue to provide adequate control power.

computer-processed to give the pilot the impression that the aircraft is more stable than it really is. Low natural stability brings good manoeuvrability.

On all but the simplest light aircraft flaps are usually moved by powered screw-jacks. Spoilers and lift dumpers can also be considered as part of the flight-control system.

By 1940 simple autopilots had been developed to control altitude, speed, and direction, so that the pilot could fly 'hands off'. The latest autopilots can be programmed to fly set manoeuvres, to fly within more detailed constraints including Mach number, or they can be coupled to follow commands that reflect data derived from ground-based aids.

The first retractable landing gears had to be wound up by hand; now they are hydraulic, electric, or pneumatic. A very few civil aircraft, such as the Britten-Norman Trislander and Boeing 727, are offered with optional rockets which can be fired if a main engine fails on take-off. The Hawker Siddeley Trident 3B is unusual in having a turbojet take-off booster installed at the base of the fin. Arrester hooks, long used on naval aircraft, are now fitted to land-based combat aircraft as an emergency system in the event of brake failure. Landing parachutes are also common on military machines, and various kinds of emergency barriers and overshoot beds are commonly fitted to military and also to some civil runways.

## Pressurization

The efficiency with which a pilot can work falls off at altitudes above about 3,000 m (10,000 ft), and pressurized oxygen is needed above 5,500 m (18,000 ft) (*see* Chapter 6). It also becomes uncomfortably cold at high altitude. Face masks fed by an oxygen system and warm clothing are only a partial solution; the better answer is to provide pressurization and air conditioning. Since the early 1940s most large airliners and warplanes have been pressurized. Pressurization is now offered even on some general-aviation aircraft. Compressed air is often supplied directly from a bleed, in the case of a turbine-engined aircraft, or from a compressor in the case of a piston engine. Electronic and other systems generate heat and need to be cooled.

Low outside-air temperatures bring problems of icing. A reliable detector is essential, to control hot air ducted to critical areas, electric

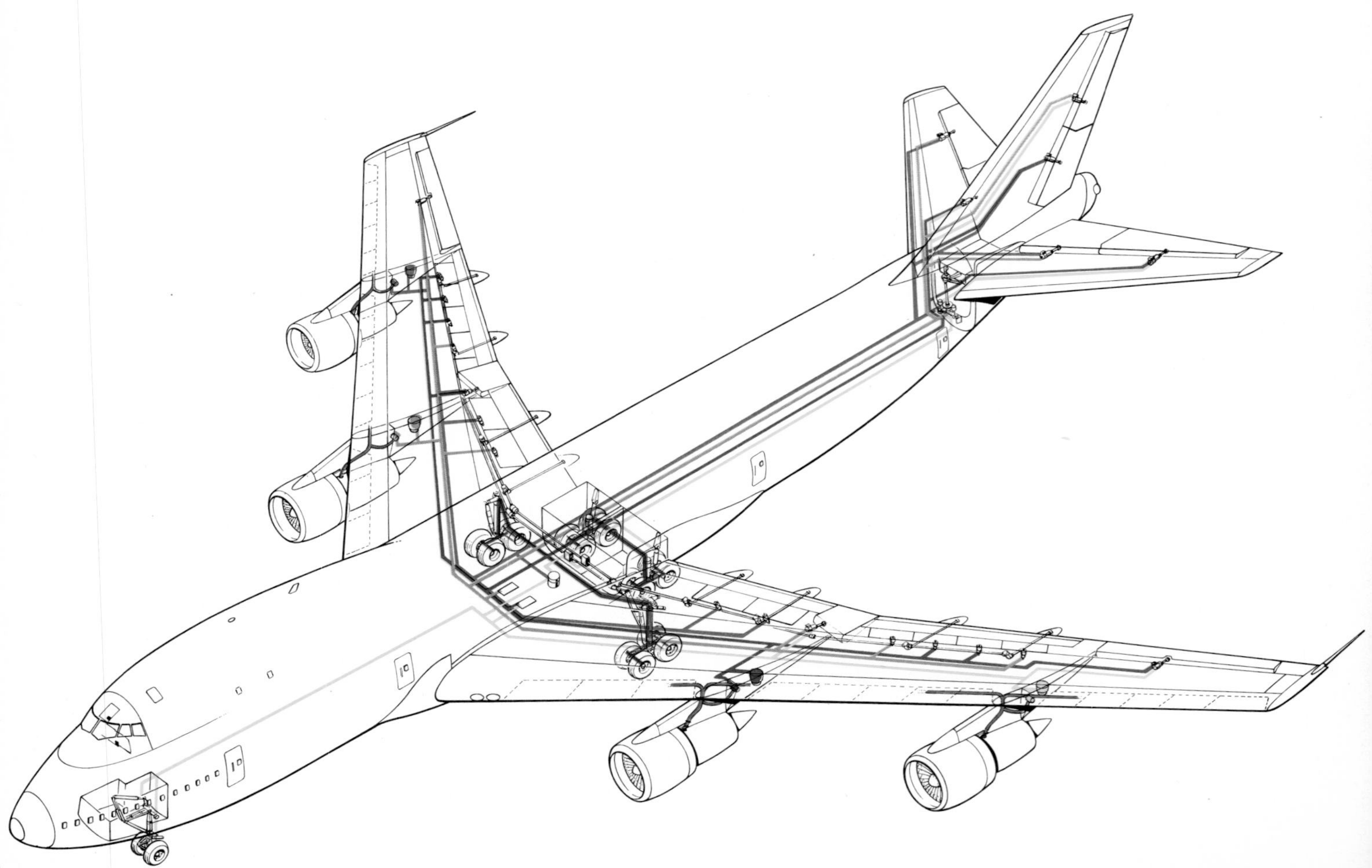

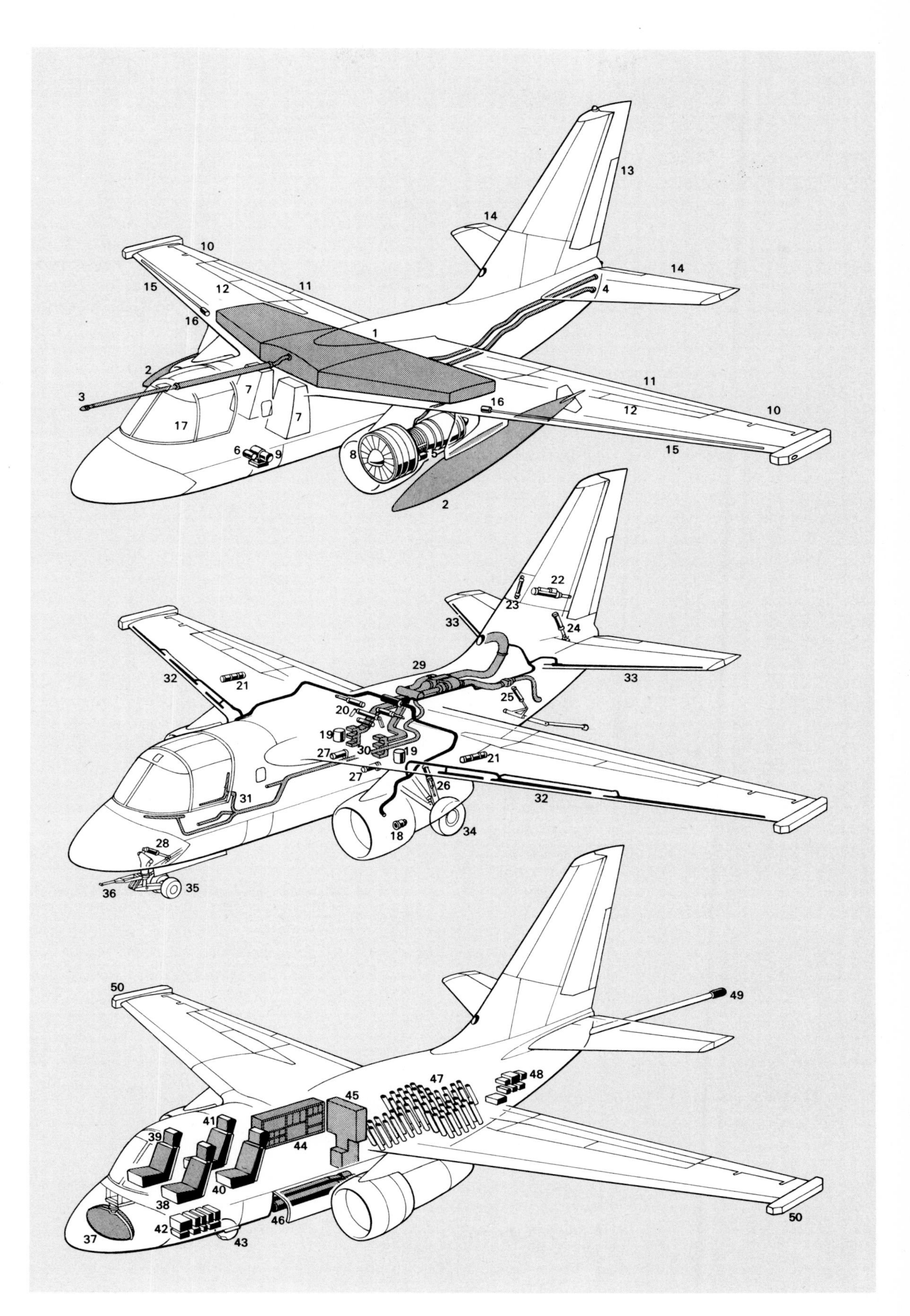

Systems on the Lockheed S-3A Viking, a long-range carrier-based submarine destroyer.

**Top** Fuel system, electrical system, and basic flight controls: 1 Wing-integral fuel tanks, 2 Auxiliary fuel tanks, 3 Inflight refuelling probe (retractable), 4 Fuel dump and vent pipes, 5 Engine-driven electrical generator, 6 Emergency generator, 7 Electrical load-distribution panels, 8 Engine (General Electric T-34), 9 Auxiliary power plant, 10 Ailerons, 11 Slotted flaps, 12 Spoilers, 13 Rudder, 14 All-moving tailplane with flaps, 15 Drooping leading edges, 16 Leading-edge electrical actuators, 17 Electrically heated windscreen.

**Centre** Hydraulic system, pneumatic system, and landing gear: 18 Engine-driven hydraulic pump, 19 Hydraulic accumulators, 20 Aileron flap and spoiler servo controls, 21 Wing-fold rotary actuators, 22 Rudder actuator, 23 Tail-fold jack, 24 Tailplane actuator, 25 Arrestor-hook jack, 26 Main-landing-gear jack, 27 Weapons-bay door actuator, 28 Nose-wheel jack, 29 Environ-mental control system, 30 Avionics cooling ducts, 31 Cabin air-conditioning ducts, 32 Wing leading-edge de-icing ducts, 33 Tailplane de-icing ducts, 34 Rearward-retracting main landing wheel, 35 Rearward-retracting nosewheel, 36 Catapult strop attachment.

**Bottom** Operational personnel and equipment: 37 Search radar, 38 Pilot, 39 Co-pilot, 40 Tactical-control officer, 41 Sensor-control officer, 42 Forward avionics bay (port and starboard), 43 Forward-looking infra-red scanner (FLIR), 44 Mission-control avionics (port and starboard), 45 Univac-control computer, 46 Weapons bay (two torpedoes each side), 47 Sonobuoy launch tubes, 48 Aft avionics bay (port and starboard), 49 Magnetic-anomaly detector (MAD) boom, 50 Wing-tip sensors.

heating, de-icing fluid pumped through a porous strip, and pneumatic 'boots' which are inflated and deflated to keep breaking up any ice that forms.

## Instrument Panel and Displays

The pilot's instrument panel presents data from a variety of sensors giving an immediate read-out of all important flight conditions. Back-up data on the rate of climb or descent, rate of turn, and altitude relative to the horizon are also shown. On powered aircraft engine speed is prominently displayed; other indicators show fuel state, engine temperatures and pressures, and condition of the electric, hydraulic, pneumatic, and other systems.

Many advanced aircraft carry radar. On combat aircraft radar is part of a system for finding targets and aiming weapons. Many use a head-up display (HUD) which presents data to the pilot on a transparent plate mounted behind the forward windscreen. It is focussed at infinity, so that symbols are clear to the pilot when he looks through the windscreen. The HUD saves the pilot having to look down at the instrument panel. The pilot selects data for navigation or attack, such as airspeed, altitude, closing speed to target, distance to go to target, and distance to go to weapon release.

On an advanced aircraft the navigation and attack sub-systems will take data from a ranging device, based on a laser or on a forward-looking radar or infra-red (FLIR), to calculate distance to the target, and will combine this with the known speed and direction to calculate the weapon-release point. The pilot will be helped to fly towards the target and told exactly when to release the weapon by information on the HUD. The HUD does not rely on visual sighting, and the pilot can make an accurate attack without even seeing the target. On a fighter the HUD may show the trajectory that shells from the cannon would follow if the firing button were pressed. The computer makes allowance for the velocity of the shells, the effect of gravity and drag, and the speed and direction of the aircraft. When the theoretical trajectory crosses the target aircraft, the pilot is told to fire.

## Electronic Counter-measures

The introduction of radar before World War II tipped the balance of aerial warfare in favour of the defender. However, it was soon discovered that attacking aircraft could confuse a radar system by releasing clouds of chopped metal foil, which acted as billions of reflective targets. The foil, known by the code names *chaff* and *window*, remained airborne for long periods because of its light weight. Chaff was the first stage in what was to become the fast-moving science of electronic counter-measures (ECM). No combat aircraft is now safe without an ECM system.

ECM can be divided into two groups: passive, such as chaff, and active, such as jamming devices. A passive device may analyse the signals from an enemy radar. This then alerts the pilot and activates the jammer. Jamming has to be carried out with care, because the enemy may use the emission from the jammer as a beacon for homing. The most active form of ECM could be considered to be missiles which home on enemy radars.

Chaff has now become more sophisticated and is supplied as a mixture of foil of different lengths – or actually chopped to length in the air – chosen to reflect the frequencies used by the enemy. Radars have also become more sophisticated and are able to change their frequency many times a second – a capability known as frequency agility – to avoid active jammers. This is called ECCM (electronic counter-counter-measures).

A major peacetime activity of the armed service is electronic intelligence (Elint), the gathering of information about the early warning systems, ECM, and ECCM of potential enemies. The frequent probing of Britain's defences by the Tupolev Tu-95 Bear is a reminder of how seriously the Soviet Union views this task. In war, particular aircraft are given the task of providing ECM cover either with built-in equipment or with external pods.

Not all counter-measures are electronic. Many small anti-aircraft missiles are designed to home on the infra-red (IR) heat of an engine's exhaust. A simple form of infra-red counter-measure (IRCM) is for an aircraft to release flares to provide decoy targets.

While counter-measures help to confuse the enemy, a sub-system known as 'identifunction friend or foe' (IFF) provides an insurance against being shot down by one's friends. Voice transmissions or visual recognition are too slow and unreliable. Therefore, before a missile is fired at an unidentified aircraft an automatic transmitter asks the aircraft for what amounts to a password. If the correct code is received, the firing sequence is inhibited.

## Safety

In almost all aircraft, safety and emergency systems come into action as soon as anything goes wrong. They range from simple fire extinguishers, in the engine nacelles and bomb bays, to the ground-proximity warning system (GPWS) which alerts the pilot if the aeroplane is closing with the ground too rapidly. Some emergency systems assist crew-escape. Most

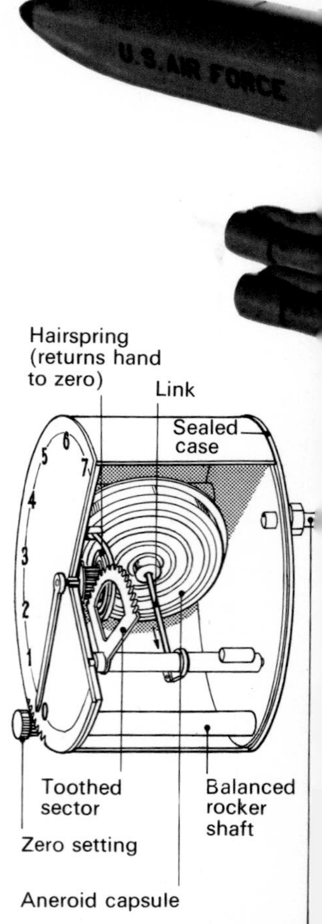

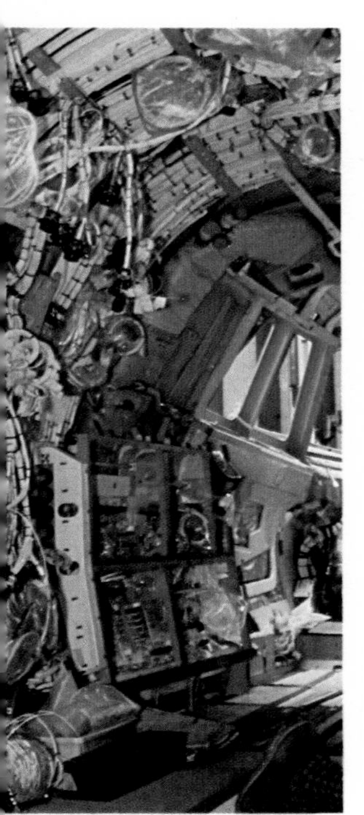

military aircraft have ejection seats to fire the crew clear of the aircraft. The General Dynamics F-111 is equipped with an emergency escape module which includes the complete cockpit. This module can be fired from the aircraft at zero altitude, and even under water, and serves as a survival shelter. Many airliners are provided with an inflatable escape chute for use after an emergency landing. The normal doors are also supplemented by emergency exits according to rules laid down by the safety authorities. The aircraft manufacturer has to demonstrate that a full load of typical passengers can get out from only half the exits in less than 90 seconds. Passengers are provided with

**Left** Concorde during wiring-up. Electrical systems on large modern aircraft are so complex that routine checks are possible only with the help of automatic testing equipment.

**Below** The Boeing E-3A AWACS (Airborne Warning and Control System), based on a 707-320C transport. Its rotating-radome radar has a range of more than 400 km (250 miles).

life-jackets and an emergency oxygen supply, fed through face masks dropped from the hat-racks or seat backs, if pressurization fails.

A cockpit voice recorder (CVR) and a crash-proof flight-data recorder (FDR) are mandatory on airliners and common on combat aircraft. If something goes badly wrong and the aircraft crashes, they provide vital evidence for accident investigators. The crash recorder is also played back on a regular basis to check piloting standards and to aid maintenance.

On transport aircraft all vital systems are duplicated or 'paralleled'. For example, if there should be hydraulic failure, there is at least one alternative hydraulic supply or a mechanical back-up. If one instrument fails there is another like it on the co-pilot's panel, or the data can be gathered from information displayed elsewhere.

Modern aircraft are more than just pretty shapes. They are only as good as their systems, and capability is no longer measured simply in terms of performance. For example, when the Trident first flew in January 1962 it was not merely another jetliner. It was the first airliner in the world to have blind-landing capability. Three independent hydraulic systems work in parallel to operate all flight controls, landing gear, nosewheel steering, brakes, and windscreen wipers; and the electrical system also has three separate channels. Tridents were cleared to land under conditions of 30 m (100 ft) vertical visibility and 400 m (1,300 ft) horizontal visibility in February 1969, and under conditions of 3.7 m (12 ft) vertical and 270 m (880 ft) horizontal in May 1972.

**Left** The simple altimeter is sensitive to atmospheric pressure, which decreases with altitude. Outside air is fed into the sealed case by a tube. Any change in pressure causes the aneroid capsule to expand or contract, and this movement is communicated, via a sensitive mechanism, to the pointer on the dial. The simple altimeter shows height above a pre-set zero setting (sea-level, for example). Modern radio altimeters, by contrast, measure the distance between an aircraft and the ground below – which may be a valley at sea-level or a mountain top.

## Computers

Systems technology is the vanguard of aerospace advance, and airborne computers are influencing aircraft design. Rather than each system having its own calculator and logic device, it is now easier to centralize all the computing functions with a master computer, in turn, co-ordinating all the systems. Even small general-aviation aircraft may use a central computer in the not too distant future.

A centralized computer could provide the crew with several large television-like screens on which they could call up selected data. Important information could be displayed all the time, but secondary data would be displayed only when needed or when one of the safety/emergency systems triggered a warning. Navigation data could be provided in a pictorial form, showing the destination airport, the desired track, and the actual track. The pilot would no longer be faced with a bewildering array of instruments, many of them provided only for monitoring. The new display would improve reliability and reduce cost and weight.

A second function of the computer would be to provide artificial stability. Advanced military and civil aircraft already have a degree of stability augmentation. A more positive intervention by the computer, particularly when coupled with FBW signalling, would allow conventional stability requirements to be greatly relaxed; the tail, for instance, could be smaller, saving weight and drag.

The Rockwell B-1 bomber (cancelled in 1977) was to have used so-called active controls to smooth the ride through turbulent air. Sensors in the fuselage would have warned of gusts (sharp up- and down-currents in the atmosphere) and the control surfaces would have reacted to counteract their effect on the aircraft. If this idea was taken to its logical conclusion, the structures of aircraft would not need to be so strong, saving both weight and cost.

# War in the Air

# 11 Reconnaissance

The history of military aviation began on a familiar theme, which was repeated off and on for more than a century. Enthusiasts and visionaries kept talking and writing about the way flying machines might revolutionize warfare, while the people in authority – the generals and government officials – poured scorn on the whole idea. Admittedly, some of the enthusiasts let their imaginations run away with them, and many early writers frightened their readers with 'flying ships . . . manned by a race of savages from the North' and with predictions about how 'our ship, descending out of the air to the sails of sea ships . . . may over-set them, kill their men . . . and this they may do not only to ships but to great buildings, castles, cities . . .'. The possibilities seemed limitless.

Most of these scenarios were pure fiction when they were first published, and impossible of attainment. But one role, reconnaissance, did appear to have some practical possibility. Finding out what the enemy was doing had always been an important part of warfare, and it must have been thousands of years since scouts first discovered that you can see farther by climbing a tree or a hill. The proposition that the law was universal, and that a man with a spyglass might do even better if he could fly, was published in 1784 (possibly much earlier). The upshot was that the French regular army in the wars that followed the 1793 Revolution contained the world's first 'air force'. Called Les Aérostiers (Aeronauts), they comprised a cavalry force with the addition of horse transport carrying large balloons and hydrogen-gas plant. At the Battle of Fleurus on 26 June 1794 a Captain Coutelle spent the day reporting every move by the enemy forces while he hung suspended under a tethered Aérostier balloon. He is credited with playing a vital part in enabling General Jourdan to defeat the formidable Allied forces. Later he rose above the battle of Ourthe, near Liège, and helped the French to defeat the Austrians.

Subsequently balloons played a part in many campaigns and battles, including the American Civil War (1861–5), the Franco-Prussian War and Siege of Paris (1870–1), the Boer War at the turn of the present century and, above all, World War I. But the balloon had drawbacks as a reconnaissance platform. It was large and clumsy, and needed even bulkier equipment to generate or store its gas. (Curiously, the much simpler hot-air balloon was not used in warfare.) It either had to be tethered to the ground, so restricting its range and height, or it had to be left to the mercy of the winds, which might blow it into the hands of the enemy or away from the battle entirely. And the development of rifled guns meant that a balloon near enough for its observer to see much with binoculars was also near enough to be shot down.

Potentially the aeroplane was much more useful, but the early flying machines were so flimsy and so poor in performance that this potential was difficult to discern. Until World War I was in progress most senior officers believed that the flying machine had no role in war. This was partly because their disciplined minds found it hard to evaluate anything so completely new, but the opinion also stemmed from the fact that they were aware only of the feeble early aeroplanes which often could not stagger into the air at all. Those that did fly could not carry much and were extremely unreliable. So there was some basis for the scorn of the armies and navies for what was widely held to be a means of locomotion unfit for a gentleman.

A prophetic water-colour, *Military Aviators on the Eastern Front*, published in a French magazine in August 1910. The airmen have a camera, note-pad, and two-way system of speaking tubes. The aircraft is based on a contemporary Voisin or Farman pusher.

**Previous two pages**
Tactical air power has had a profound influence on land warfare. Here ground-attack Hawker Typhoons destroy German armour during the breakout from the Falaise Gap, Normandy, in August 1944.

Thus, during the first decade of the 20th century, the only aviation enthusiasts who succeeded in interesting the military were the Wright brothers, who sold aircraft to the US Army and helped train the first military heavier-than-air pilots. But by 1910 the proven capability of the aeroplane, in crossing the Channel and the Alps, caused some countries to think again. In the autumn of 1911 Italy went to war with Turkey over the disputed possession of Tripolitania and Cyrenaica in North Africa, and sent several aircraft to support the Italian army. On 22 October 1911 a Captain Piazza took off in a Blériot to reconnoitre Turkish positions at Al Aziziyah (near Tripoli), and after an hour returned from the first-ever military aeroplane mission. He had found he had his hands full making notes and drawing sketches in addition to flying the Blériot, and later – after the world's first bombing raid (*see* Chapter 12) – he took up a camera and photographed enemy encampments in the pioneer photo-reconnaissance (PR) sortie. Even this was a full-time task because the camera was almost as big as a modern television set.

## World War I

By the summer of 1912 the Italians had done much to make air warfare a reality, with photo reconnaissance (including ciné film taken from an airship) and aerial bombing. But the rest of the world took little notice. On 13 April 1912 Britain formed the Royal Flying Corps without any clear idea of what it was to do. The main type of aircraft, the B.E.2, was specially designed to be highly stable in flight, so that it could virtually fly itself while both the pilot and observer scanned the ground, made notes, and took photographs. The infant RFC had a lot of fun in its two years of peaceful flying before the outbreak of World War I, but in spite of professional skill and enthusiasm it went to war on 4 August 1914 in an almost unbelievably primitive state. It knew virtually nothing about aerial fighting (as will be described later), and not much about reconnaissance. In the first weeks of the war the pilots flying reconnaissance missions usually got lost and had to come down and ask where they were; at least twice pilots found they had landed in territory overrun by the enemy. And sheer lack of experience led their observers to make all kinds of incorrect reports about what they imagined they had seen. 'Columns of enemy troops' were later found to have been civilians, or even patches of tar or shadows on a road.

Gradually the RFC and other young air forces learned what to look for: new footpaths, moving shadows, concentrations of road or rail trucks, fresh digging, the glint of metal, unexplained smoke, and rifle fire directed against the aircraft. Cameras developed swiftly, and soon a range of specially designed aerial cameras were in use, fixed to point vertically downwards, or at a known angle to one side, and with the film plates changed automatically between one exposure and the next. From mid-1915 all troops in the front line were continually supplied with detailed mosaics built up from hundreds of overlapping photographs, and later with detailed maps, showing the enemy defences ahead in the most complete definition. Reconnaissance airmen also learned a quite different technique: 'spotting' for the artillery. Having with their photographs shown the artillery where to aim, the spotter aircrew could watch where the shells were bursting and tell the gunners how to correct their aim. Eventually this was done by imagining a large clock-face centred on the target. By Aldis signal lamp, or by radio, or even by dropped written messages, the spotter crew could tell the gunners '150 yards, seven o'clock' (or whatever the correction might be) to get an almost immediate bull's eye. Inevitably, spotters became prime targets for enemy fighters.

**Below left** One of the first purpose-designed reconnaissance aircraft was the German two-seat Aviatik B.I. of 1914, with water cooling radiators on the sides of the fuselage. Many Aviatik B.IIs, with slightly longer range, were built in Germany and Austria from 1915 onward.

**Below** A floral greeting for the pilot of a Rumpler-built Taube in 1914. (He wears the ribbon of the Iron Cross, unusual so early in the war and obviously gained in a previous exploit.) A Taube reconnaissance plane brought warning of a Russian advance at Tannenberg in 1914, and another dropped small bombs on Paris; but by 1915 these flimsy 1910-designed monoplanes had become obsolete.

## Progress in Photography

Between the wars solid progress was made with cameras and film. Ordinary film was sensitive to light at the blue end of the visible spectrum, so that red, orange, or yellow tended not to show up at all (for this reason, RAF roundels in old photographs often appear to have no red centre). Panchromatic film was invented to record all colours, and special filters were added to cameras to give clearer results, mainly by filtering out blueish haze. Infra-red film was developed that could record objects clearly over great distances, although with unnatural black, grey, and whitish tonal values. Cameras grew ever larger, to give better and sharper pictures from increasing altitudes. The first were operated by hand – no easy job with frozen fingers inside bulky gloves – but later cameras were power-driven by windmills and then by electric motors. Multi-lens cameras were produced which could take a central picture surrounded by five, six, or seven others which, when adjusted in a special optical device, gave complete all-round coverage.

Triple-camera installations took one vertical and two oblique pictures, so that as the aircraft flew on a straight course it produced overlapping photographs of a strip of land up to 10 miles wide. Although a few specialized reconnaissance aircraft were built between the wars, most cameras were fitted into regular bombers and army co-operation aircraft (*see* Chapter 15). Clandestine (spying) photo-reconnaissance was done by many specially fitted aircraft.

Before and during World War II a remarkable Australian, Sidney Cotton, working in Britain, developed single-handed great advances in technique, and masterminded the creation of Photographic Reconnaissance Unit (PRU) as part of RAF Coastal Command. Gradually PRU received aircraft different from those previously used. They were special versions of the fastest combat types, with exceptional speed, altitude, and range. The most important were the Supermarine Spitfire PR.XI and the de Havilland Mosquito PR.IX, XVI, and 34, all unarmed but almost impossible to intercept. They flew lone missions all over Europe, bringing back pin-sharp pictures of enemy activity or the results of Allied air attacks (which in the early days of World War II were usually ineffective). Occasionally missions had to be flown not at extreme altitude but at tree-top height. For example, German radars were first spotted and photographed by a Spitfire flying at full throttle 'right down on the deck'. For such a mission low-angle, oblique cameras were needed, with the shutter, film, and aircraft speed precisely matched to give clear, undistorted pictures.

## Specialized Aircraft

The vital role of reconnaissance was increasingly reflected in special versions of aircraft or even special types of aircraft equipped with nothing but cameras to accomplish their mission. Some were extreme-altitude machines (which in turn led to a spate of extreme-altitude interceptor fighters), while others relied upon speed. A few were merely converted bombers, neither specially high-flying nor fast but carrying many cameras over great distances; several were fitted with a photo-processing laboratory. The chief Allied strategic-reconnaissance conversions of bombers were based on the Boeing B-17 and Consolidated B-24, with range adequate for many missions in the western Pacific. These were far out-performed in 1944 by the F-13A version of the Boeing B-29 Superfortress. At about the same time the Luftwaffe introduced the first jet-powered reconnaissance aircraft, the Arado Ar 234B. It was fast enough to fly with impunity over Britain, northern Italy, and the western battle-front (where intelligence coverage for Germany had previously been a blank: the Luftwaffe did not bring back pictures of the D-day invasion build-up, or the preparations for the invasion itself from a week before 6 June 1944, because no reconnaissance aircraft had penetrated the British fighter screen). But the Ar 234, like most early jets, had a short range.

Jet propulsion was a 'natural' for reconnaissance aircraft. Some of those built after 1945 were among the biggest and heaviest military aircraft ever used, owing to their need to carry a great load of fuel for long ranges. Biggest of all was the Convair RB-36, with 10 engines and a crew of 22 to fly the monster and operate its 14 cameras. Although it was much faster, the Boeing RB-47 had to be refuelled for inter-continental missions, and this was true of almost all the new crop of camera platforms except the Boeing B-52 Stratofortress and the Soviet Tu-95 Bear, which are still in use after 25 years. The B-52 was originally designed so that its bomb bay could accommodate a reconnaissance capsule containing cameras, electronics, various systems, and a crew of two extra men, but this was eventually abandoned.

## Electronic Reconnaissance

Reconnaissance is undertaken to find out about the enemy, and by 1942 it had rather suddenly become of great importance to find out facts of an electronic nature. Where were the hostile radars? On what frequency did they operate? Did enemy night fighters carry radar? If so, of what type, and how did they communicate with the ground controllers? The RAF formed

1474 (Wireless Investigation) Flight to find answers. Its aircraft were the first in the world to fly the so-called *ferret* mission whose purpose is electronic reconnaissance. This often calls for sustained courage of the highest order. For example, on 3 December 1942 a Wellington ferret 'trailed its coat' over the bases of Luftwaffe night fighters to try to discover details of a device called Emil Emil which the German machines used to help them intercept RAF bombers. Alone in the dark, the crew expected at any moment to be blasted by cannon shells – and they were. Eventually the brave crew ditched in the Channel, all of them wounded, after having sent back all the required details of what turned out to be Lichtenstein FuG 202 fighter radar.

In the 20 years following World War II electronic ferreting was of enormous importance. The scale of operations can be guessed from the number of aircraft, mainly of the US Air Force and Navy, that were shot down. Some were across alien frontiers, while others were reported to have been over international waters or in other harmless places, but in this kind of mission one cannot argue too forcefully over legalities. Aircraft types included the Consolidated P4Y Privateer, Boeing ERB-47, Douglas EB-66, and the Lockheed U-2. This last, a specially designed ultra-high-altitude aircraft, made flights over Communist territory for four years until, on 1 May 1960, a U-2 piloted by Gary Powers was brought down by a Russian surface-to-air missile (SAM) battery.

Today reconnaissance missions are flown by an unprecedented variety of aircraft. For strategic missions at the highest possible altitude and speed nothing can equal the US Air Force Lockheed SR-71A Blackbird, which at present holds the world absolute speed and sustained-altitude records at 3,524 km/h (2,185 mph) and 25,929 m (85,069 ft). One of these remarkable blue-black monsters flew from New York to London in 1 hour 55 minutes in 1974. Running the SR-71A rather close is the Soviet Union's MiG-25 Foxbat reconnaissance version, though this has a much shorter range.

Tactical reconnaissance is flown either by tactical attack or fighter aircraft carrying multi-sensor pods on external weapon pylons, or by special versions with the equipment housed internally; the latter cannot fly combat

De Havilland's final version of the reconnaissance Mosquito was the PR.34, which served mainly after World War II. Development of the high-altitude Rolls-Royce Merlin engine with paddle-blade propellers resulted in an aircraft able to take off at over 11,350 kg (25,000 lb) weight with 5,760 l (1,267 gal) of fuel (part of it in a bulged belly compartment), and fly more than 4,830 km (3,000 miles) at about 515 km/h (320 mph); its top speed was 685 km/h (425 mph).

**Left** The Mitsubishi Ki-46-11 was one of the most successful Japanese reconnaissance aircraft of World War II. Entering service in 1941, it had a top speed of 604 km/h (375 mph), a service ceiling of about 10,800 m (35,500 ft), and a range of 2,400 km (1,490 miles).

**Left** The Arado Ar 234, which entered service with the Luftwaffe in September 1944, was the world's first jet-powered reconnaissance bomber. Very fast for its day, the reconnaissance version had a maximum speed of 742 km/h (460 mph) and a range of 1,630 km (1,013 miles), and was powered by two Junkers Jumo turbojets each developing 900 kg (1,980 lb) thrust.

missions. The sensors (search equipment) may include forward-looking, vertical, and oblique cameras (with various kinds of film sensitive to different electromagnetic wavelengths to defeat attempts at camouflage); infra-red (IR) linescan, resembling a camera but taking pictures showing temperature variation (useful for seeing which factory chimneys are warm, vehicles whose engines are running, and places where vehicles were parked a few seconds previously); and side-looking airborne radar (SLAR), which can give an amazingly fine picture of a different kind over great distances. Lasers can also provide extremely accurate reconnaissance information in the form of pictures or measurements.

## RPVs and Satellites

In the past 20 years an increasing proportion of reconnaissance flights have been made by pilotless vehicles. The smallest and simplest are drones light enough to be carried by a man, which can fly over a battlefield on a pre-programmed course and, after recovery, be readied for a second mission in a short time. Larger remotely piloted vehicles (RPVs) can fly for 10 to 20 hours at great heights, carrying a full range of sensors; they are much smaller and cheaper than manned aircraft and do not, of course, put pilots' lives at risk. A special class of unmanned vehicles is the ultra-quiet platforms that can study battle areas without being detected. Another class is tethered platforms, or free-flying remotely piloted helicopters (RPH), which can operate from a small truck and promise to have civil as well as military applications.

Such are the developments in cameras and other sensors that since 1960 an increasing number of reconnaissance satellites have kept surveillance on Earth activities from space. All of them are in orbit, constantly following a track which takes them over all the parts of the land or oceans they wish to see, unlike comsats (communication satellites) and many other types which hover over a single chosen spot. By far the most numerous are the Soviet Cosmos series, of which a high proportion of well over 950 so far launched are used for military purposes. Most of these weigh about 5,900 kg (13,000 lb) and carry many sensors; some are manoeuvrable. Largest of the much fewer US Air Force reconnaissance satellites are the Lockheed Big Bird series, weighing about 11,350 kg (25,000 lb) and about 15 m (50 ft) long. These pass over every point on Earth twice each 24 hours and have exceptional capabilities. Their use has immeasurably improved the quality and flow of data, which no longer has to be ejected and recovered on the earth in capsules but can be sent in a stream of digital data by a television link.

**Below** In 1940–60 the USAAF and USAF operated several long-range reconnaissance aircraft derived from bombers. One of the best was the Boeing RB-47E, an early example of which is seen here on test. Its nose lacked the bomber version's radar/autopilot bombing system, and it had a remotely controlled, airconditioned camera bay instead of a bomb bay. Six General Electric turbojets gave it a top speed of 980 km/h (606 mph), a ceiling of 11,600m (38,000 ft), and a range of about 6,450 km (4,000 miles).

**Left** Undoubtedly the most mysterious aircraft of recent times, the Lockheed U-2, which went into operational service in 1956, was designed to fly clandestine spy missions over unfriendly territory at very high altitude – about 25,900 m (85,000 ft) in most versions. Especially rare is this much-enlarged version, the U-2R, in which the length has been increased from 15 m (49 ft) to 19 m (63 ft), giving it a greater payload than earlier types.

**Centre** The Russian Tupolev Tu-95 'Bear' (here escorted by a McDonnell Douglas Phantom) is a reconnaissance version of a former bomber that first flew in 1954. Its four Kuznetsov turboprop engines give it a maximum speed of 870 km/h (540 mph) and a range of about 12,550 km (7,800 miles). Another version of the Tu-95 is an AWACS with a saucer-shaped radome.

**Right** A Teledyne Ryan Compass Cope 'R' remotely piloted vehicle (RPV) undergoing a test flight in 1974. One of two versions produced for the USAF, this aircraft later set an endurance record, flying for more than 24 hours at altitudes of over 16,770 m (55,000 ft).

# 12 Bombers

In 1910 men of the Signal Corps, US Army, dropped a home-made bomb, with satisfying results, from a Wright *Flyer* beside an air meeting in San Francisco. Although this was the first time a bomb had been dropped from an aeroplane, the first bombing mission proper was, like the first reconnaissance mission, flown by the Italians against the Turks in 1911. It was on 1 November that Giulio Gavotti took off in his Rumpler-built Taube (Dove) of the Squadriglia di Tripoli and let go four small bombs against Turkish positions at Ain Zara and Tagiura. The bombs caused great alarm, and brought an angry protest that the Italians had committed 'an outrage against a hospital'.

Unlike other countries, Italy just went ahead and used aeroplanes in war. In the following year, the Italians used quite large numbers of aircraft for reconnaissance, bombing, and troop harassment in the Balkan war. Many of the pilots fighting the Turks were not in fact Italians but Russians, Bulgars, Hungarians, and others, including some Britons. Dozens of primitive aircraft were used, and the armament was contrived on the spot. Bombs were made by various methods, such as adding fins to artillery shells and fitting fuses to cans of explosive; there were also purpose-made bombs delivered from an Italian factory. Some were dropped by hand over the side of the cockpit, and others were hung on the outside and released by withdrawing a pin or pulling a string; one great Russian aviator tied string loops to his bombs, hung them on his feet, and released them with a few sharp kicks!

## Airships

This was clearly about as primitive as air attack could be, although any kind of bombing from the air had a large effect, even on disciplined troops. But the only aircraft capable of carrying really effective bomb loads before World War I were airships. The first to be used in war were the Italian P.2 and P.3, of Forlanini design, which first bombed Turkish forces on 10 March 1912. Germany's Zeppelin and Schütte-Lanz ships were much larger, and the Imperial German Army and Navy both explored military airship operations in 1911–4. By 1914 the Army had begun to drop bombs, but thought of the airship chiefly in terms of its role over land battlefields. Peter Strasser of the Navy set his sights higher and regarded the airship as potentially capable of long-range strategic operations. Both services had bad luck and fatal crashes, but they persevered against terrible

The first great four-engined bombers in history were those of the IM (*Ilya Mourometz*) series built by young Igor Sikorsky at the Russo-Baltic Wagon Works in 1914–7. They came in various versions, and equipped the EVK (the Squadron of Flying Ships), one of whose bombers is seen here.

difficulties, without benefit of a textbook of air warfare to guide them.

The Army was first off the mark in World War I, dropping 250 kg (550 lb) of bombs on the Belgian city of Liège only two days after the start of the war, on 6 August 1914. The airship was hit by the world's first anti-aircraft fire and crashed on landing. Over the next few weeks several other Army ships were lost, one of them being destroyed by fire on 8 October 1914 in an audacious bombing attack by the British Royal Naval Air Service (RNAS). But this was not the first aeroplane bombing attack of the war.

## World War I: the Early Raids

At the start of World War I there was no established air-defence system anywhere. Despite this, it was no mean exploit when on 30 August 1914 a German, Lt Hiddessen, flew his frail little Taube monoplane all the way to central Paris, where he dropped a small load of grenades and leaflets. The Taube was a 1910 design, stable and pleasant to fly but hardly up to playing an effective role except as a reconnaissance platform. There was only one purpose-designed bomber among the western Allies and another in the East. The western one was the French Voisin pusher, another aircraft with a long heritage and (with 70 hp) even less powerful than most Taubes; but the Voisin, built of steel, was extremely strong and capable of considerable development. As well as a crew of two, in a pusher nacelle, the Voisin accommodated about 60 kg (132 lb) of bombs, a load that increased in later and more powerful versions. Gradually the French Aviation Militaire built up the world's first really numerous and formidable bombing force with Voisins and other similar aircraft. The basic concept of the pusher propeller was by now becoming obsolescent, but it was so universally used in France that the general rule among French troops was to fire on all tractor machines; and the pusher Voisin bombers continued in production to the end of the war.

The Allied bombers on the Eastern Front were far bigger and technically ahead of their time. The young Russian designer Igor Sikorsky (1889–1972), assisted by G. I. Lavrov, built his first four-engined biplane in the winter of 1912–3 and it flew in May 1913. In 1914 he flew an even bigger machine, the *Ilya Mourometz*, named after a Russian legendary hero. Powered by four 100 hp engines, this was by far the most capable aeroplane then flying. The Tzar gave the order to build a 'squadron of flying ships', and by 1917 nearly 80 of these great machines were flying. They differed in detail, and many had two pairs of different engines, one pair in the inner positions and the other outboard. Most carried more than 680 kg (1,500 lb) of bombs, and they were defended by up to seven machine guns. Until the Revolution in October 1917 only one *Ilya Mourometz* was lost in action during some 400 bombing raids, and this machine was shot down only after it had itself destroyed three of the fighters attacking it.

Compared with these stately Russian aircraft, British attempts to produce bombers were at first feeble. The RNAS attack on Zeppelin sheds on 8 October 1914 was made by two of the smallest and lightest machines ever to belong to any air force. They were Sopwith Tabloid single-seaters, and although unable to carry much they were outstandingly fast and nimble. For their raid on the German Army airship bases at Düsseldorf and Cologne each Tabloid carried four 9 kg (20 lb) bombs but had no defensive armament. Just over a month later, on 21 November 1914, three slightly larger RNAS aircraft – the first of the famous Avro 504 type to be delivered to the Admiralty – dropped eleven 9 kg bombs on the Zeppelin factory and gasworks at Friedrichshafen. From its base at Belfort, the mission was a major feat involving a penetration of almost 160 km (100 miles) into Germany. Two of the small bombs wrought

Some of the earliest strategic bombing raids in history were made in World War I by the Italians against Austria-Hungary, over rugged mountainous routes that taxed men and machines severely. This Caproni was one of the final, most-produced Ca 5 family, slow but adequate for its purpose. The mid-upper gun position, a kind of metal basket just ahead of the pusher-bladed middle engine, is unoccupied in this picture.

havoc: one badly damaged airship L.7, nearing completion in its shed, and the other destroyed the gasworks, which blew up in a gigantic explosion.

## World War I: the Big Bombers

Nobody had any right to expect such results from such puny bombs. Britain's Royal Ordnance Factories began to make a much bigger bomb weighing 51 kg (112 lb), and in December 1914 the Admiralty issued a challenging specification for a large aircraft able to carry six bombs of this size. The pioneer constructor Frederick Handley Page showed a suggested design to the Director of the Admiralty Air Department, Commander Murray Sueter, who studied it and said he wanted something even bigger. In a classic command he ordered the designer to build 'a bloody paralyser of an aeroplane'. This Handley Page did. The O/100 flew on 18 December 1915, and 46 were supplied to the RNAS, most of them powered by two 250 hp Rolls-Royce engines. It carried not six but 16 of the 51 kg bombs, or eight of 114 kg (250 lb), and a crew of four rather than the stipulated two, the extra men manning up to five machine guns. From this formidable machine was developed the O/400, with engines of 360 hp, of which some 550 were built (plus a further 107 in the USA). The Handley Pages won such renown in hundreds of daring missions that the name 'Handley Page' even got into an edition of the *Oxford English Dictionary,* which defined it as 'a type of large aeroplane'.

The O/400 saw action not only with the RNAS but also with the Royal Flying Corps (RFC), which – to the surprise of many – found the giant Handley Pages could bomb more accurately by night and suffer fewer casualties than the faster, smaller RFC bombers did by day. In April 1918 the RNAS and RFC were amalgamated into the Royal Air Force (RAF), and an Independent Force was formed for the express purpose of mounting a strategic bombing offensive against the heart of Germany and even Berlin. To this end an even bigger bomber had been ordered in late July 1917. This aircraft, the V/1500, was really something. Powered by four of the most powerful engines available – various engines from 375 to 500 hp – arranged in push/pull tandem pairs, it could carry 4,545 litres (1,000 gal) of petrol and up to 30 bombs of 114 kg (250 lb) or one monster weighing 1,500 kg (3,300 lb). But the war ended before any of the long missions planned for the V/1500 could be made.

Italy and Germany both built powerful fleets of bombers in World War I. The Caproni company, started by a wealthy count in 1908, was the first in the world to produce purpose-designed bombers – a consequence of the Italian Army's pioneering use of aerial bombing at a time when other nations ignored the idea. The first Caproni bombers were not very good, because they had three Gnome rotary engines one behind the other in the central nacelle, the rearmost one driving a pusher propeller and the others driving left and right tractor screws. Later versions had three separated engines, and carried bomb loads up to 1,450 kg (3,200 lb). Most had several machine gunners, often with paired or even triple Revelli guns, one gunner having to stand in a metal cage directly above the centre engine a few inches in front of the pusher propeller. With an air temperature below freezing and the inhospitable Alps below, these missions against the Austro-Hungarians were no picnic.

Germany's chief strategic bombers in World War I were the Gothas and the Zeppelin (Staaken) Giants, although Friedrichshafen and AEG models were also used in numbers. Most had pairs of engines of about 260 hp, and on missions against London carried about half the maximum of 12 bombs of 50 kg (110 lb) each. But the Staaken Giants were another matter. Although not the largest German machines of their day, they were impressive enough, with four to six engines and bomb loads of 2,000 kg (4,400 lb). They fairly bristled with Parabellum machine guns and/or 20 mm Becker cannon. They were ahead of their time in their instruments, systems, and general capability. The crew sat in enclosed cabins, had oxygen and radio, and could navigate by advanced radio direction-finding methods also used by the Navy Zeppelins. Two even had additional engines in the fuselage to drive superchargers supplying the main propulsion engines, and thus could maintain higher performance at greater heights. The Luftwaffe was to return to this scheme in World War II.

Yet despite these impressive heavy bombers, most bombs in World War I were dropped by thousands of modest two-seaters over the battlefronts. Gradually bomb racks ceased to be 'Heath Robinson' lash-ups and were made to proper design drawings that were standardized throughout air forces. Bomb sights had derived from a pioneer geometric device invented in 1910 by Riley Scott, a former US Army officer, which contained all the parts needed for fairly accurate aiming. The rest was a matter of refinement, to allow for crosswinds causing drift or other disturbing factors. Sights were used either by special bomb aimers, who often lay prone in the bomber's nose, or by crew members seated in a cockpit looking diagonally down through a glass panel or a plain hole in the floor. Although crude compared with later

**Right** In the early 1930s the Martin Bomber made a sensation by combining all the latest advances – cantilever monoplane shape, stressed-skin construction, retractable landing gear, internal bomb stowage, gun turret, and fully cowled engines with variable-pitch propellers. This US Army Air Corps B-10 version, capable of 340 km/h (211 mph) was faster than most contemporary fighters.

sights, and consisting of little more than a series of graduated scales on hinged mountings, sometimes with an optical telescopic sight as well, these simple devices were good enough for the immediate purpose.

## Between the Wars

One of the most famous single attacks in the history of bombers took place as a peacetime exercise on 21 July 1921 off the Capes of Virginia, United States. The target was one of the largest and latest German battleships, the *Ostfriesland*, which had surrendered almost three years earlier. Widely proclaimed as unsinkable, it became the focal point of a bitter row between the admirals and the infant US Army Air Service. (Precisely the same political battle was going on in Britain, made sharper by the fact that the RAF was a separate service entirely.) The American sailors were determined to promote the idea that no bomber could sink a battleship; the Navy's former Secretary said he would 'stand bareheaded on the bridge' of any battleship under bombing and expect to remain safe. Finally, after trying to stop such a test, the Navy limited the flamboyant Brigadier General 'Billy' Mitchell's air crews to bombs of 270 kg (600 lb) or less. But Mitchell went out with eight Martin MB-2 bombers (which somewhat resembled the Handley Page O/400), each carrying two bombs of 454 kg (1,000 lb). After seven bombs had been dropped the giant battleship rolled over and sank. Later Mitchell was court-martialled because of the way he uncompromisingly campaigned on behalf of air power.

Gradually the rest of the world began to comprehend what Mitchell had meant. One country that cottoned on quickly was the Soviet Union, where a design collective drawn from the Central Aero and Hydrodynamic Institute, and led by Andrei N. Tupolev, built two bomber families that were outstanding in design and accomplishment. The first, flown in 1925, was the ANT-4, called TB-1 (heavy bomber type 1) by the Soviet air force. Powered by two 680 hp M-17 engines, it was an all-metal cantilever monoplane following the philosophy pioneered by the German Junkers company (*see* Chapter 8). In 1929 a TB-1 flew eastwards around the world, making the sectors between Khabarovsk and Seattle as a twin-float seaplane similar to the TB-1P torpedo-bomber seaplane version. In 1931 a TB-1 was successfully tested carrying two I-4 fighters, one on each wing, as an experiment to see if a bomber could carry defensive fighters with it.

Tupolev's next heavy bomber was the most formidable in the world until the late 1930s. First flown on 22 December 1930, the TB-3 was a logical development of the TB-1 to a size needing four engines. The first production version usually had 730 hp M-17F engines, a giant wing of about 40 m (130 ft) span, in-flight access through the wing to the engines, and tandem pairs of main wheels. The bomb load, initially 1,000 kg (2,200 lb), was later increased to 5,800 kg (12,800 lb), the greatest carried by any mass-production bomber of the inter-war period. About 800 of these fine aircraft were built in several versions, and among their accomplishments were the first landings at the North Pole, the first mass parachute descents, the first airlift of a tank, and the first (and only) simultaneous release from a bomber of no less than five fighter aircraft.

In the United States developments in the late 1920s included naval dive bombers and stressed-skin construction. The dive bomber aimed the whole aircraft at the target, diving on it at an angle of from 75° to 90° and releasing the bomb(s) when the pilot judged they would plummet down to hit the target. It demanded a robust structure, but for service aboard aircraft carriers airframes had to be very strong anyway. The US Navy's dive bombers were among the most lethal anti-ship weapons in the world during 1928–50. From the mid-1930s the reborn

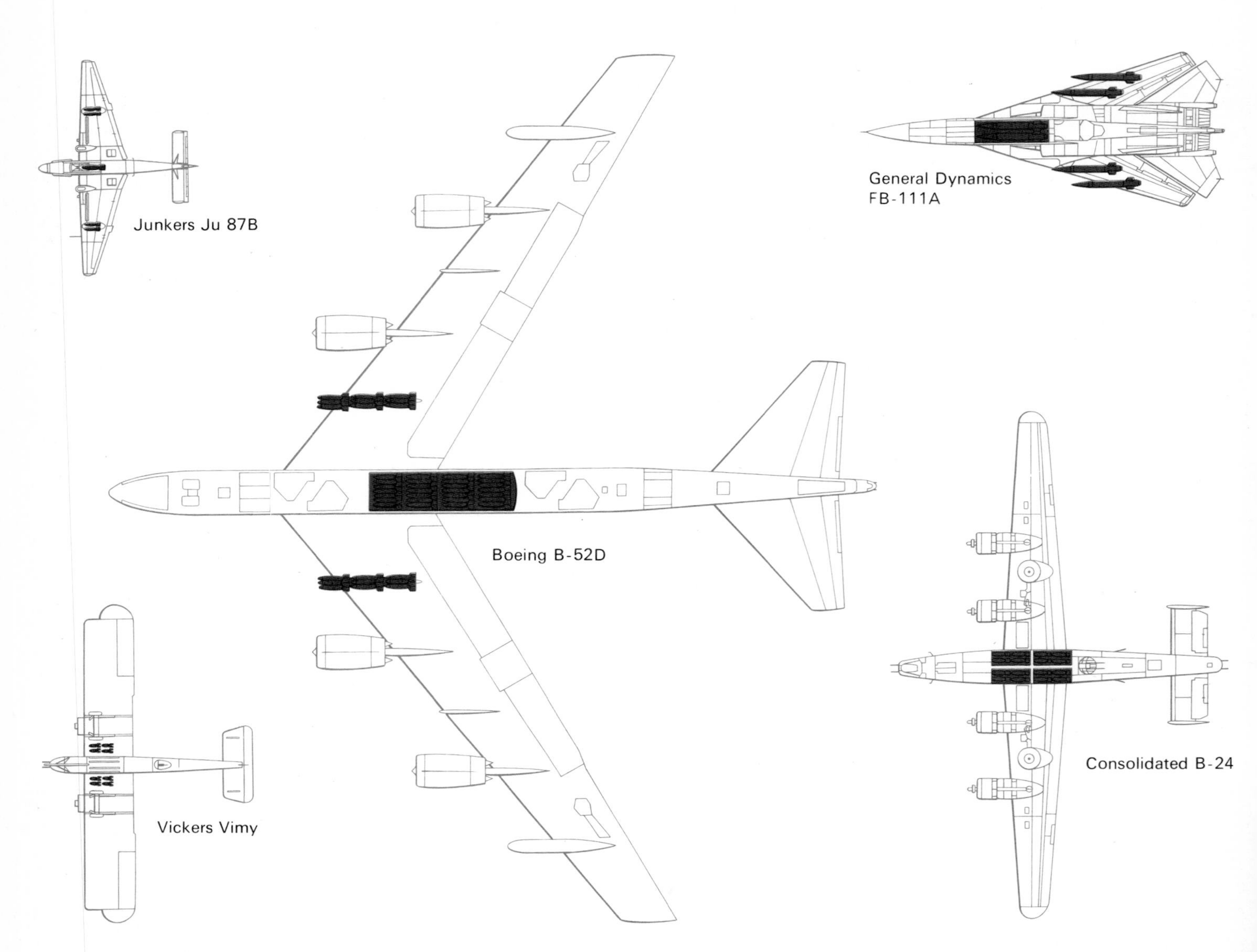


Bomber offensive loads: Vickers Vimy (1918) with four 51 kg (112 lb) bombs (on short-range missions Vimy could carry 22 such bombs); Junkers Ju 87B (1938) with one 250 kg (551 lb) and four lighter bombs; Consolidated B-24 Liberator (1941) with 3,630 kg (8,000 lb) of bombs carried internally or two 1,815 kg (4,000 lb) bombs slung on inner-wing racks; Boeing B52D, modified for conventional warfare (1969), with 31,780 kg (70,000 lb) of bombs, usually of 340 kg (750 lb) each; General Dynamics FB-111A (1969) with maximum of six SRAM missiles.

Luftwaffe relied on the Junkers Ju 87 Stuka dive bomber to clear a path for the German armies before and during World War II, and it made the concept of the *Blitzkrieg* (lightning war) possible. Although not an advanced design, the Ju 87 was shatteringly effective in the absence of strong fighter opposition, and from 1937 (in Spain) until 1942 it did more than any other weapon to smash ground opposition to the Nazi armies and to sink warships and merchantmen in European waters. But in the Battle of Britain it suffered heavily, and thereafter came to be used in various close-support roles (*see* Chapter 15).

## World War II

In 1935 the Luftwaffe had embarked on a programme involving big four-engined bombers for strategic operations, but these were later cancelled in favour of smaller twin-engined machines for large-scale use in support of land campaigns. All were powered by engines of 1,100 to 1,300 hp and initially were defended by three hand-aimed machine guns in the nose, dorsal, and ventral positions. The largest was the Heinkel He 111, with a vast wing able to lift heavy bomb or missile loads; but by 1941 they were becoming outclassed in performance. The Dornier Do 17Z and Junkers Ju 88 were faster, and had a crew of four grouped close together in the nose. Experience in the Battle of Britain rammed home the lesson that they needed heavier defensive armament, but no really good scheme emerged, and the Germans failed to develop a first-rate bomber. The Ju 88, although an outstanding and versatile machine, was in no sense a strategic bomber of the type built in vast numbers by the Allies.

First of the Allied strategic bombers to appear was the four-engined Boeing B-17 Fortress, built in 1935 to a requirement by the US Army Air Corps (as the Air Service had become in 1926) for a new bomber able to hit an enemy invasion fleet. In 1935 an invasion fleet was the only target that seemed plausible to the Americans, who were then geographically too remote to be bombed by any likely enemy. Gradually this remarkable machine progressed to the very forefront, and it was the chief US bomber in Europe in World War II. By 1945 no fewer than 12,731 had been built, in four main versions all powered by 1,200 hp Wright R-1820 Cyclone engines with turbosuperchargers to increase performance at altitude.

The American philosophy was boldly to fly in a massed formation across enemy territory by day, if possible with a fighter escort, and to bomb with great precision using the complex, costly, but extremely accurate Norden bomb-sight. Joined by the Consolidated B-24 Liberator – a later and more advanced aircraft built in numbers exceeding that of any other bomber (19,203 including spares) – the B-17 had a purpose in Europe unlike that of other bombers. Instead of avoiding combat it sought it, and the massed fire of its large-calibre 12.7 mm (0.5 in) machine guns proved lethal to hundreds of Luftwaffe fighters. The objective of a mission was often as much to bring the Luftwaffe to battle as to hit the target, and despite severe casualties – especially when operating without fighter escort – the USAAF (as the Army Air Corps had become in 1941) succeeded completely in its ultimate objectives.

In contrast RAF Bomber Command adopted a policy, forced upon it by lack of an alternative, of general area-bombing of cities by night. Although from 1935 British bombers had been defended by gun turrets, which became better and harder-hitting, they were quite incapable of surviving in daylight against strong fighter defences. At first they could not find their targets, but gradually the science of electronics was harnessed by British engineers to help the bombers navigate, find their targets, confuse hostile radars, and, by 1942, put down special

**Above** The Avro Lancaster, the greatest night bomber of World War II, had a normal bomb load of 6,350 kg (14,000 lb).

**Below** A Junkers Ju 87B Stuka dive bombing over Poland in 1939. The white strips on the wings are dive brakes.

target markers – brilliant flares whose colour codes were changed with each attack – to provide a clear aiming point. The technique of the Pathfinder Force was perfected, the PFF crews having special equipment and experience and leading the main force unerringly to the target and showing them where to bomb.

Having switched to monoplanes shortly before the war, the RAF soon concentrated on four-engined heavy bombers of great carrying power, with bomb loads of around 6,350 kg (14,000 lb). By 1945 some special Avro Lancasters were carrying single monster bombs of 10,000 kg (22,000 lb) which penetrated deep into the ground and shook their targets down like an earthquake. Other 'Lancs' had demolished German dams using a special form of depth charge encased in a spinning cylinder which skipped across the lake and over defensive torpedo nets until it slowly sank against the wall of the dam. Other bombers carried various forms of mines, anti-submarine depth charges, and other stores, while the bomber version of the de Havilland Mosquito went into action in 1942 to demonstrate the feasibility of a bomber fast enough to need no defensive armament.

By 1944 the USAAF was using the Boeing B-29 Superfortress, a bomber so greatly in advance of its predecessors as to render them obsolete. It had a wholly new order of weight, power, complexity, and capability. Its four 2,200 hp Wright R-3350 engines, each with two turbochargers, could take it up to well over 9,150 m (30,000 ft), despite a wing loading of up to 396 kg/m² (81 lb/sq ft), more than double that of other bombers such as the B-17. At this height, at which the B-29 could cruise at over 485 km/h (300 mph), its crew of 10 or 11 could still breathe comfortably in pressurized cabins. Its devastating armament of 12.7 mm (0.5 in) guns and cannon was located in five power-driven turrets, of which four were remotely aimed by gunners looking through sighting stations. In August 1945, by which time as many as 800 B-29s at a time were laying waste to Japan in area-bombing on the RAF pattern, two specially equipped aircraft dropped two atomic bombs, bringing the war to an end.

Originally intended as a fighter, the General Dynamics F-111 (which first flew in 1964) became an outstanding tactical bomber and later, in this FB-111A version, a strategic bomber. Flying here with four 2,273 l (500 gal) tanks on its four wing pylons, it carries two SRAM missiles internally (and four more can replace the tanks). Most of a mission is flown with the wings in the position shown, but for a supersonic dash to the missile-release point they are folded back to a 72.5° sweep.

## Jet Propulsion and Missiles

The technology of the B-29 was the foundation for many subsequent bombers, particularly those of the Soviet Union where the B-29 had been illegally built as the Tu-4. After 1945 virtually all new-design bombers were jets, and the first to go into use – apart from the rather hastily contrived German Me 262A-2a and Arado Ar 234B – was the USAF B-45 Tornado, with conventional piston-type airframe and four turbojets. The English Electric Canberra, delivered from 1951, was again conventional and modest, but these qualities made it a highly useful tactical aircraft which is still being rebuilt and refurbished for customers all over the world. Boeing's B-47, first flown in 1947, was a totally different machine from anything seen before, with swept wings and tail, six podded engines, and a gross weight of 101,700 kg (100 tons). It was one of the first bombers to be refuelled in the air as a routine procedure, using the flying-boom method. It was also typical of the new breed of jet bombers in that it had a crew of only three, the defensive armament being remotely controlled in the tail and many functions being taken over by automated electronic equipment. One of the last survivors of the old school was the Convair B-36, biggest bomber ever in service, with six 3,500 hp pusher engines later supplemented by four podded jets. It had a crew typically of 15, and could stay up for days at a time; but it never went to war.

The most important strategic bomber of the post-war era has been the Boeing B-52 Stratofortress, first flown in 1952. Generally resembling an enlarged B-47, it began life with eight engines in four double pods, giving a total thrust of 31,960 kg (70,400 lb), and was finally developed with much more efficient turbofans with an aggregate thrust of 65,375 kg (144,000 lb). As built the B-52 carried about 12,250 kg (27,000 lb) of bombs, but it soon appeared with long-range missiles on under-wing pylons. The Luftwaffe had pioneered air-to-ground missiles in World War II to hit difficult or heavily defended targets with unerring precision. The missiles carried by the B-52, and by the somewhat smaller British bombers, the Avro Vulcan and Handley Page Victor, were much longer-ranged. The B-52's Hound Dog missiles could fly up to 1,125 km (700 miles), allowing the bomber to stay far from enemy defences. Inside its bomb bay the B-52 could carry a small

**Right** The working life of the Boeing B-52 bomber may be extended by equipping it with ALCM (Air-launched Cruise Missiles), seen here being test-launched from an NB-52G. There is no assurance, however, that these subsonic missiles could penetrate modern defences.

**Below** Although cancelled in 1977, the Rockwell B-1 illustrates better than any other the way a modern bomber defends itself with electronics rather than with guns or missiles. The drawing shows the main elements of the B-1's RFS/ECM (radio-frequency surveillance, electronic counter-measures) system. The system detects hostile radio signals and automatically protects the aircraft against them, making it difficult to locate and hit with missiles. Key to letters: A (right-hand bay) main RF sources and transmitters, B (right side), numbers indicate RF bands of the main transmitter aerials, C (in main-wheel well) additional ECM equipment, D (tail) more transmitter aerials with RF band numbers, E (left side) and F (left-central bay) main computers.

pilotless aircraft, the ADM-12 Quail, which could be released as hostile defences were approached to behave like a B-52 and lure away enemy radars, fighters, or missiles. Various defensive schemes were tried, including fighters launched from the bomber, and later pilotless interceptors were carried to shoot down the enemy fighters. The 2,250 km/h (1,400 mph: Mach 2) Convair B-58 Hustler was intended to carry the Fairchild Blue Goose pilotless interceptor, despite the high speed of the bomber itself. Later the B-52 was to carry the SCAD (Subsonic Cruise Armed Decoy) to confuse or contend with defences.

By the 1960s nearly all bombing was being done by tactical attack aircraft (*see* Chapter 15). There were still large bombers in some countries, but these were being used chiefly as missile launch platforms and for the vital role of electronic reconnaissance. Until 1977 the Rockwell B-1 was developed for the USAF to give credibility to the manned strategic deterrent, carrying the SRAM (Short-Range Attack Missile) of 160 km (100 miles) range, also carried by the B-52 and by the much smaller General Dynamics FB-111A, a slightly longer-ranged version of a tactical bomber. Today the B-1 has been cancelled, and the future of the manned bomber is likely to lie in its role as a mobile launch base for the ALCM (Air-Launched Cruise Missile). Even small cruise missiles, no bigger than traditional naval torpedoes, can carry thermonuclear warheads farther than 1,600 km (1,000 miles).

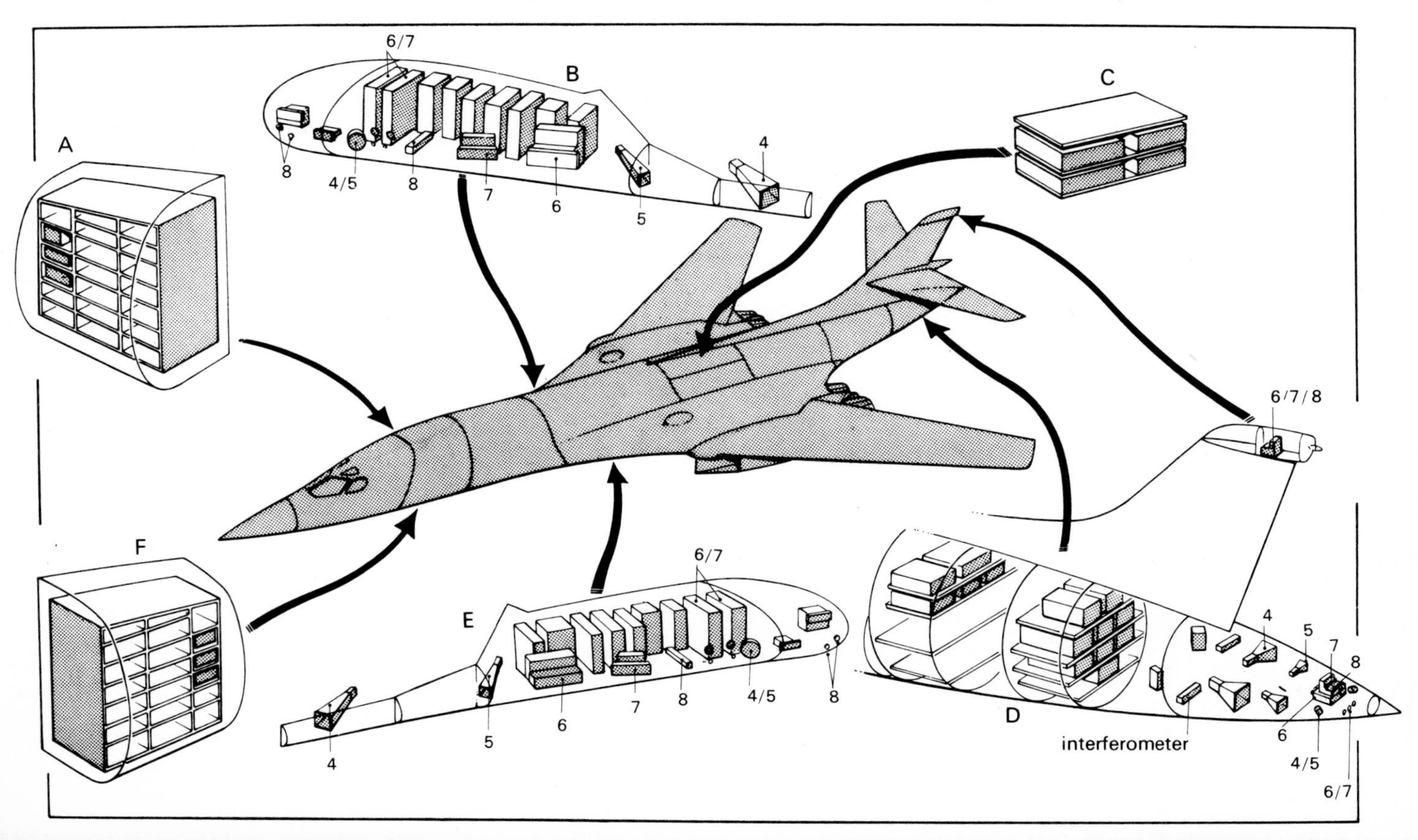

# 13 Fighters

Writers of fiction in the 19th century imagined several kinds of aerial warfare, including balloons or flying machines that could make war on each other. By 1910 several flying machines, in the United States and France, had carried men with rifles. By 1912 three kinds of machine gun (counting the Hotchkiss and Benet-Mercier as two types, though they were basically similar) and a shell-firing automatic cannon had been successfully fired from aircraft. Early experiments with such armaments were bedevilled by problems. On the US Army Wright *Flyers*, for instance, the pilot and gunner sat side-by-side, and as the Benet-Mercier had a rigid clip of ammunition sticking out on one side and a rigid ejector chute on the other, the gun inevitably got in the way of the pilot's control column.

One of the pastimes of the infant Royal Flying Corps in 1912 was to take pot shots at kites. This, too, posed problems. The standard SMLE rifle was too long to carry easily in a cramped cockpit, and as its bullets did no good to structural parts of the aircraft or to people on the ground in the line of fire, such games were frowned upon. But little effort was made to develop aircraft guns, so that when World War I began in August 1914 the RFC had no aeroplane that could be called a fighter and little comprehension of air combat. Yet a few visionary designers were beginning to put their ideas into practice. At least eight schemes for aeroplane armament had been published, and several gun-carrying machines had actually flown. The Vickers EFB (Experimental Fighting Biplane) series had been conceived in 1912 and exhibited the following year with a Maxim machine gun in the nose of its pusher nacelle. The RFC had fitted the new Lewis gun in the front cockpit of a Maurice Farman 'Shorthorn', and the British F.E.2a of August 1913, designed by de Havilland at Farnborough, was planned from the start as a true fighter – but its production in adequate quantities was delayed until 1916, two years too late.

This delay was to cost the lives of hundreds of pilots. The B.E.2 biplanes, which became by far the most numerous type in the RFC, were modern tractor machines, but they were hopeless in any kind of combat. The observer sat in the front cockpit, between the wings, with hardly any arc of fire and surrounded by wires and struts. Some observers carried hand-loaded rifles and 50 rounds, but the pilots were told to ram any Zeppelin they might meet rather than to try to shoot it down.

## Armament

A basic problem with tractor aircraft was fear of hitting one's own propeller. A bullet could take off one blade, and the imbalance created would then tear out the engine. Thus, such 'fighters' as the F.E.2 family, the Vickers EFB (which became the F.B.5 Gunbus), and the later D.H.1A and 2 were all pushers. In France Spad built a strange fighter of tractor design but with an extra machine-gun cockpit held by struts in front of the propeller! Other designers fixed pairs of guns on struts to fire past the propeller, or on inclined mountings overhead or diagonally at the side; both methods made aiming a bit of a lottery. It was eventually realized that the surest way to hit the target was to fix a machine gun to fire directly ahead and aim it by aiming the whole aircraft. So a method of protecting the propeller had to be devised, and five groups of inventors in America, Britain, Switzerland, France, and Russia proposed mechanisms to 'synchronize' a machine gun to a rotating propeller so that the bullets missed the blades.

The French inventor was Raymond Saulnier, of the Morane-Saulnier aircraft company, and when he abandoned his mechanism in the face of official indifference his chief pilot, Roland Garros, decided to adopt a cruder solution and fit steel bullet deflectors to the propeller blades.

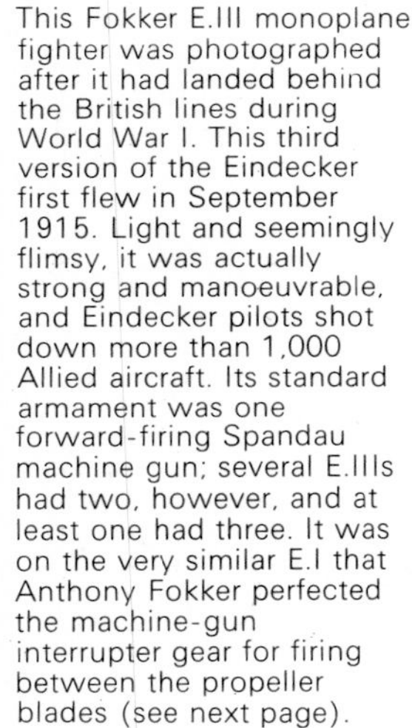

This Fokker E.III monoplane fighter was photographed after it had landed behind the British lines during World War I. This third version of the Eindecker first flew in September 1915. Light and seemingly flimsy, it was actually strong and manoeuvrable, and Eindecker pilots shot down more than 1,000 Allied aircraft. Its standard armament was one forward-firing Spandau machine gun; several E.IIIs had two, however, and at least one had three. It was on the very similar E.I that Anthony Fokker perfected the machine-gun interrupter gear for firing between the propeller blades (see next page).

**Left** Based on a 1912 design, the Vickers F.B.5 Gunbus entered service in France in 1915. Its rear-mounted engine with pusher propeller allowed the front-mounted gunner, with one (occasionally two) machine guns, an excellent field of fire. It proved too sluggish, however, to match the lively Fokker scouts (fighters) with their synchronized guns.

Even this took prolonged research and testing; but in March 1915 Garros, who had joined the Aviation Militaire, returned to his squadron in a Type L monoplane fitted with thoroughly proven deflectors.

By this time several aeroplanes and airships had been forced down in aerial combat. The first victory went to Lt Harvey-Kelly of the RFC who forced down a Taube monoplane on 25 August 1914 by sheer harassment. On 5 October 1914 the observer of a French Voisin, Sgt Louis Quénault, scored the first aerial victory in actual combat, shooting down an Aviatik with his Hotchkiss machine gun. A sprinkling of other aircraft had gone down, but 'kills' were still rare. Imagine the sensation, then, when Garros shot down five Germans in the first 18 days of April 1915.

Garros became a star figure, the first pilot to be called an ace. Then things went stupidly wrong. On 19 April he was forced to land in German-occupied territory, and was captured before he could burn his machine. The Dutch designer Anthony Fokker, who built aircraft for the Germans, was told to copy the deflector scheme. Instead he resurrected the idea of an interrupter or synchronization gear, claiming he had invented it, and produced the Fokker E.I (the 'E' stood for *Eindecker*, monoplane). Although only a little 80 hp machine of 1913 vintage, it was the first aeroplane to mount a gun whose bullets unfailingly passed between the two revolving blades of a tractor propeller – usually at the rate of one bullet every two revolutions of the propeller. The result was death for Allied airmen. The nimble little monoplanes hacked down the unwieldy, poorly armed British and French machines in droves. Air combat had arrived – and it was to prove one-sided for many months.

In spite of the E.I's success, the classic World War I fighters were mostly biplanes of 180 to 400 hp, with two machine guns mounted above the engine, firing ahead with a synchronization gear. The gear was eventually of a hydraulic type, linking the engine and guns by pipes filled with oil under pressure. This was versatile, and could easily be adjusted to match different engines, guns, and aircraft types. The top-scoring fighter of World War I, with 1,294 confirmed victories, was the British Sopwith Camel, so named because of its hump-backed appearance. The close-coupled Camel was one of the most agile aircraft ever made; but it was tricky, and killed many novices who lacked the skill to control what one pilot described as 'a highly strung, lop-sided racehorse'.

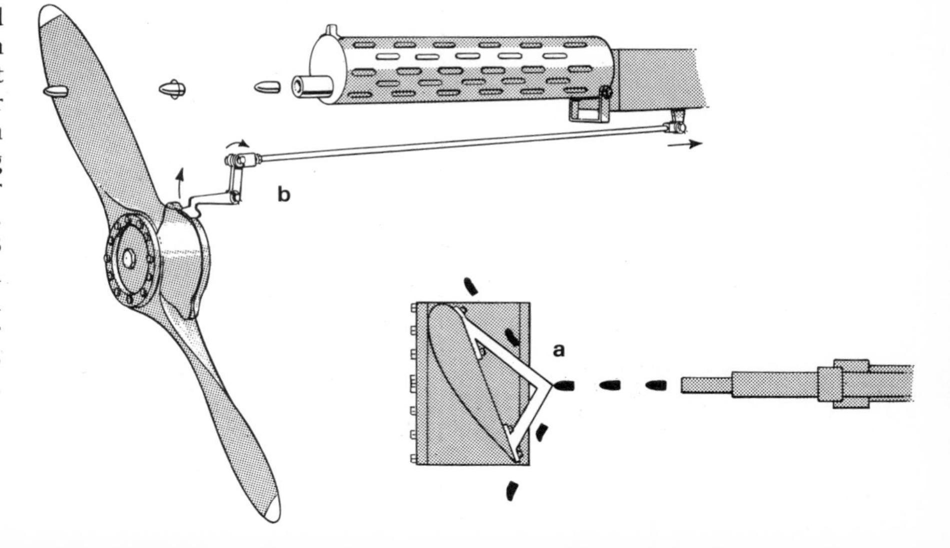


**Below** The Garros bullet deflector (a), fitted to both propeller blades, allowed a forward-firing machine gun to be mounted immediately in front of the cockpit on single-engined, tractor-driven fighters. Fokker's interrupter gear (b), which synchronized the gun-firing mechanism and the rotation of the propeller, ensured that the bullets passed between the propeller blades. Much less wasteful of bullets than the Garros deflector, the synchronization gear gave the Fokker scouts decisive superiority over Allied fighters for many months.

**Left** Dogfight: a German war artist's impression of a battle in 1918 between two-seat Albatros C.XII reconnaissance aircraft and Martinsyde G100 'Elephants' – single-seat scouts that served in the RFC only in small numbers. Another Martinsyde aircraft, the very fast Buzzard, was delivered too late to see wartime action.

Other great World War I fighters included the French Nieuports, at 450 kg (1,000 lb) among the lightest and smallest ever built; the tough and powerful Spads (some of which had a cannon firing shells through the hollow shaft of the propeller); the British S.E.5a, with one synchronized Vickers and one Lewis firing over the propeller from above the upper wing; the German Albatros D.III and D.V. and the Fokker D.VII; and the Sopwith and Fokker triplanes, which used three slender wings for tighter turns. Bristol's F.2B was a superb two-seat fighter, with guns at front and rear. Various companies in Italy, Austria, and Germany made fighter seaplanes and flying boats, and the Royal Navy pioneered sea-going air power with aircraft carriers. Effective fighters, incidentally, meant the demise of the military airship – except possibly for ocean surveillance.

## Inter-war Progress

In the two decades between the World Wars fighter tactics and techniques changed hardly at all, but there was a revolution in technology. While engines consolidated at around 500 hp – about double the average in World War I – with very greatly improved reliability, airframes were made in new ways needing less labour and reduced maintenance. One of the favoured new structures was welded steel tube, which eliminated the need for internal cables and wires. Most air forces gradually changed from wood to metal, although at first fighters merely used traditional structures with the frames made of steel or aluminium tubing instead of wood; the outer covering was almost always fabric. But by the 1930s the light-alloy stressed-skin structure (*see* Chapter 8) was revolutionizing aircraft design – and it was especially advantageous to fighters because it opened the way to almost perfect streamlining.

By 1935 the open-cockpit biplane of some 500 hp, with two machine guns, fabric covering, and fixed landing gear, had given way (on the drawing boards, if not yet in the squadrons) to a stressed-skin monoplane of 1,000 hp, with retractable landing gear, flaps, enclosed cockpit, variable-pitch propeller, and harder-hitting armament. But this decisive advance was preceded and accompanied by many curious or backward-looking developments in fighter philosophy.

Ever since experiments with Camels during World War I, for instance, there had been attempts to hang fighters under airships for the latters' protection. Eventually this idea was abandoned, as were Soviet trials using fighters launched from the wings of bombers. Fighters for naval use from aircraft carriers were generally conventional, but were fitted with large-area wings that could fold for below-decks stowage and an arrester hook and catapult attachment for launch and recovery aboard ship. Japan and France continued the tradition of fighter seaplanes and flying boats, while Italy persisted with the traditional two-gun biplane. A few air forces and manufacturers built 'light fighters' in an attempt to compete with less power while gaining in manoeuvrability and low cost; while from 1935 an increasing number went to the opposite extreme and built long-range heavy fighters, escorts, and *Zerstörer* (destroyers) with two engines and devastating armament.

Armament was in the melting pot. France's Hispano-Suiza company led the development of engines with geared propeller drives able to mount a cannon firing through the propeller hub. In 1928 Britain built twin-engined fighters fitted with a pair of gun cockpits from which large cannon could be aimed, and also fitted monster 37 mm ($1\frac{1}{2}$ in) guns firing up at a steep angle from small single-seaters. The British Air Ministry favoured fixed forward-firing machine guns, but the number was doubled from

**Below** The Boeing P-26, which entered service with the US Army Air Corps in 1934, was the first American monoplane fighter, and its top speed of 378 km/h (235 mph) made it one of the fastest in the world at that time. Powered by a 600 hp Pratt & Whitney Wasp nine-cylinder, direct-injection radial engine, the 'Peashooter', as it was nicknamed, saw service in China in the 1930s and in the Philippines in 1941–2.

Developments in fighter armaments from the 1930s: 1 Dewoitine D 500 series (1932), with 20 mm Hispano *moteur-canon* firing through propeller hub and two wing-mounted 7.5 or 7.7 mm machine guns; 2 Curtiss P-36 (1937), with four machine guns of various calibres; 3 Hawker Hurricane I (1937), with eight 0.303 in Browning machine guns in wings; 4 Northrop P-61 Black Widow (1942), with four 20 mm M-2 cannon in belly and four 0.5 in machine guns in remotely controlled dorsal turret; 5 Messerschmitt Me 262A-1a (1944), with four 30 mm MK 108 cannon in nose; 6 Northrop F-89D Scorpion (1952), with 104 Mighty Mouse 2.75 in spin-stabilized rockets in wing-tip pods; 7 Grumman F-14A Tomcat (1970), with one 20 mm M61 multi-barrel cannon in fuselage and varied missile armament, including up to six Phoenix missiles with 160 km (100 miles) range.

two to four in 1933–5 and doubled again to eight in 1936. The United States stuck to the 7.5 mm (0.3 in) and 12.7 mm (0.5 in) guns, but the Soviet Union produced not only a world-beating machine gun, the 1,800 rds/min ShKAS, but also a series of large recoilless cannon of calibres up to 100 mm (4 in), and fitted them under the wings or in the tail booms of various fighters. Britain became side-tracked by the success of power-driven turrets for the defence of bombers, and from 1935 until 1941 mistakenly devoted great energy to developing turreted fighters.

## Day and Night Fighters

In the first year of World War II the chief fighters were all monoplanes in the 1,000 hp class (860 hp in France), which could often be outmanoeuvred by the old biplanes but had far higher all-round performance and much greater firepower. The Luftwaffe's ubiquitous Messerschmitt Bf 109E was a sleek, stressed-skin machine with one to three cannon and two machine guns, and in spite of its faults it had few equals. The British Hawker Hurricane was larger and slower, and still had fabric covering; but it was tough, manoeuvrable, and carried eight machine guns which were deadly at close range. The Supermarine Spitfire was a stressed-skin machine slightly larger but even faster than the 109, and was at least a match for it. But the much-vaunted Messerschmitt Bf 110 twin-engined escort proved incapable of holding its own against the best modern single-engined fighters, and after the Battle of Britain it spent the war either in safer skies or working in tactical attack and night-fighting roles.

Until the war night fighters were just like day fighters, apart from trivial details of equipment or guns that did not blind the pilot when fired. But in 1939 the RAF fitted some Bristol Blenheim twin-engined fighters with the world's first AI (airborne-interception) radar. Although it did not work well, the airborne radar enabled a fighter to stalk an enemy aircraft on the darkest night, gradually closing on it from astern and finally to spot it visually and shoot it down. In September 1940, after very rapid development, Bristol Beaufighters entered service. These were soon fitted with a better AI radar, and with their unprecedented armament of four cannon and six machine guns they

**Above** One of the few truly authentic examples remaining of the classic Messerschmitt Bf 109E fighter of World War II. This particular aircraft, a Bf 109E-3, made a forced landing at RAF Manston on 27 November 1940. After flying as Air Ministry DG200 in RAF markings, it had a varied career before being restored by RAF St Athan.

**Below** The F.XVIII was one of several late-wartime models of the peerless Supermarine Spitfire. It was powered by a two-stage Griffon engine (with a five-blade propeller turning the opposite way to that on the Merlin) and was quite unlike earlier Spitfires to fly. No fighter ever developed more dramatically than the Spitfire; its engine power, for instance, increased from 990 hp in the Mark I to no less than 2,300 hp in F.XVIII and other versions.

gradually became masters of the night sky. In 1942 the much faster de Havilland Mosquito, originally designed as a reconnaissance-bomber, emerged as a radar-equipped night fighter, and went on to dominate the sky over all Europe. The chief Luftwaffe night fighters were the Bf 110G and various versions of the excellent Junkers Ju 88, with heavy armament and devices for homing on the tell-tale electronic emissions of RAF heavy bombers.

Nobody disputed that night fighters should be large and twin-engined, but for day dogfighting there was no simple answer. Most of the 150,000-odd fighters built during World War II weighed about 3,200 kg (7,000 lb) and had one engine of about 1,500 hp. A few makers flew light fighters of half this power, but these remained prototypes: the trend was unmistakably towards bigger and more powerful machines. In 1940 the US industry, keenly watching the events in Europe, scrapped several proposed fighters with Wright Cyclone or Pratt & Whitney Twin Wasp engines (1,200 hp) and replaced them with P & W Double Wasp engines of 2,000 hp. These air-cooled radials were built in enormous numbers for large fighters such as the Vought F4U Corsair, Grumman F6F Hellcat, and Republic P-47 Thunderbolt, each weighing 5,450–8,200 kg (12,000–18,000 lb). Nobody except the German and Japanese pilots regretted this decision. These massive aircraft consistently outfought the German and Japanese fighters and were also devastating with bombs and rockets against land and sea targets.

There is, however, another side to this argument. The Russian designer Alexander Yakovlev asserted: 'Before the war there was a clear-cut tendency in fighter aviation towards an increase in gross weight. . . . The war proved this tendency to be wrong. . . . The test of war proved that Soviet designers were right when they resolutely took the path of developing the light fighters which dominated the air throughout the second half of the war.' The Russian machines in question, the liquid-cooled Yaks and the radial-engined Lavochkins, began with wooden structures which progressively gave way to light alloy. They were quite small, and often had only one cannon and two machine guns, but their performance and serviceability in harsh conditions were excellent. Which was right, 3,200 kg (7,000 lb) or 6,800 kg (15,000 lb)? Clearly, both types of fighters did well and each helped to win the war.

## Jet Power

Jet propulsion was ideal for fighters, although at first it reduced range and endurance and often increased the take-off run. The German Messerschmitt Me 262 and the British Gloster Meteor twin-jets saw action in 1944, together with the tailless Me 163 rocket interceptor which sacrificed range and endurance for astounding climb and speed in defending local areas against heavy bombers. Germany was far in front of other countries in another factor too: armament. A range of 30 mm ($1\frac{1}{8}$ in) cannon, radically new high-speed cannon with multiple-revolver chambers, very large recoilless guns, spin-stabilized air-to-air rockets fired in salvoes, and wire-guided air-to-air missiles were all under test before the Luftwaffe's defeat. They gradually inspired similar developments in other countries: one German gun, the Mauser MG 213, led to the American Pontiac M-39, the

French DEFA, the Russian NR-30, the Swiss Oerlikon KCA, and the British Aden, all of which are still in use.

Many early jet fighters were fitted into more or less conventional airframes. The fighter often considered the ultimate achievement of the piston era, the long-range North American P-51 Mustang (which could accompany bombers from Britain to Berlin or Prague and then defeat the best fighters in the Luftwaffe before starting the long flight home), appeared both in a twinned double-fuselage form and, with few changes, as a US Navy jet. But the US Air Force decided to wait a year until its makers could sweep back the wings and tail at 35°, which German research had shown could lead to higher speed. The result was the F-86 Sabre, which in 1948 set a speed record at 1,080 km/h (671 mph) and outflew all other fighters. Later versions carried radar and rockets and reached 1,150 km/h (715 mph). During the Korean War (1950–3) the F-86 met a previously unknown machine built in the Soviet Union, the somewhat lighter and simpler MiG-15, and although the MiG could climb higher and had heavy cannon, the Sabre's skilled pilots and better equipment gave it the edge in combat.

North American's next fighter was the F-100 Super Sabre, which exceeded the speed of sound in level flight. The MiG bureau built the twin-jet MiG-19, which was even faster, and is still in wide use. The US Air Force ordered various all-weather interceptors with largely automatic radar and flight-control systems so that, with guided missiles, they could intercept and destroy enemy aircraft without the pilot ever seeing them. The British ordered a jet-fighter flying-boat, but discovered that this way of doing without airfields yielded an inferior fighter. The Americans suffered similar problems with a 'hydroski' fighter, which could dive faster than sound but took off and landed on retractable water skis. Two even stranger fighters were designed around powerful turboprop engines and, standing on their tails, screwed themselves vertically into the air (they were intended to operate from the confined decks of warships or merchant vessels). Britain built high-altitude supersonic fighters with 'mixed power' from a turbojet and a

**Above** The Grumman F6F Hellcat, the big and tough carrier-based fighter that swept the Japanese from Pacific skies, was in service little more than 6 months after the first flight of the prototype in June 1942.

**Below** The North American P-51 Mustang (the version here is a restored P-51D) made its name as a long-range escort fighter from 1942. The F-86D Sabre **(inset)**, which first flew in 1949, was the most sophisticated jet-powered fighter of its day.

rocket. In 1957 the British Minister of Defence suggested there would soon be no more manned fighters at all, only missiles. The Americans stuck to fighters, but made them very large and armed them with missiles but no gun.

Today the wheel has turned full circle. In the past 10 years there has been a powerful wish to get back to the 'eyeball-to-eyeball' type of confrontation of the man in the Sopwith Camel. The pre-eminent Western fighter, the McDonnell Douglas F-4 Phantom, was rebuilt with an internal gun – a rapid-fire 20 mm (0.79 in) cannon with six barrels firing up to 6,000 rds/min – and a slatted wing to pull tighter turns in combat. New small fighters have appeared, such as the General Dynamics F-16, which, although bigger and heavier than any single-engined fighters of World War II, are nevertheless small and light by comparison with such impressive machines as the Grumman F-14 Tomcat, McDonnell Douglas F-15 Eagle, and MiG-25 Foxbat. The RAF's next interceptor, the ADV (Air-Defence Version) of the Panavia Tornado, is a careful midway compromise, smaller than the three monsters just listed but with two engines, long range, powerful radar, and extremely effective Skyflash missiles.

Modern interceptors defend vast blocks of airspace up to 160 km (100 miles) in radius, with powerful radar able to look down at the surrounding land and water and spot low-flying intruders trying to slip through the defences unnoticed. Their task is eased by the presence of special surveillance, early-warning, and AWACS (Airborne Warning and Control System) aircraft, with enormous radars and sophisticated command and control systems to manage all a nation's defences in the most efficient way.

**Above** The Mikoyan MiG-15, best-known of the early Russian jet-powered fighters, first flew in 1947 and had a top speed of about 1,080 km/h (670 mph).

**Left** The McDonnell Douglas F-4E Phantom of 1967 onward had a 20 mm multi-barrel rapid-fire gun (ahead of the open nose-gear doors) and a slatted wing. The white missile is a Maverick, an air-to-ground weapon with pinpoint guidance by various methods including television, infra-red, and laser beam.

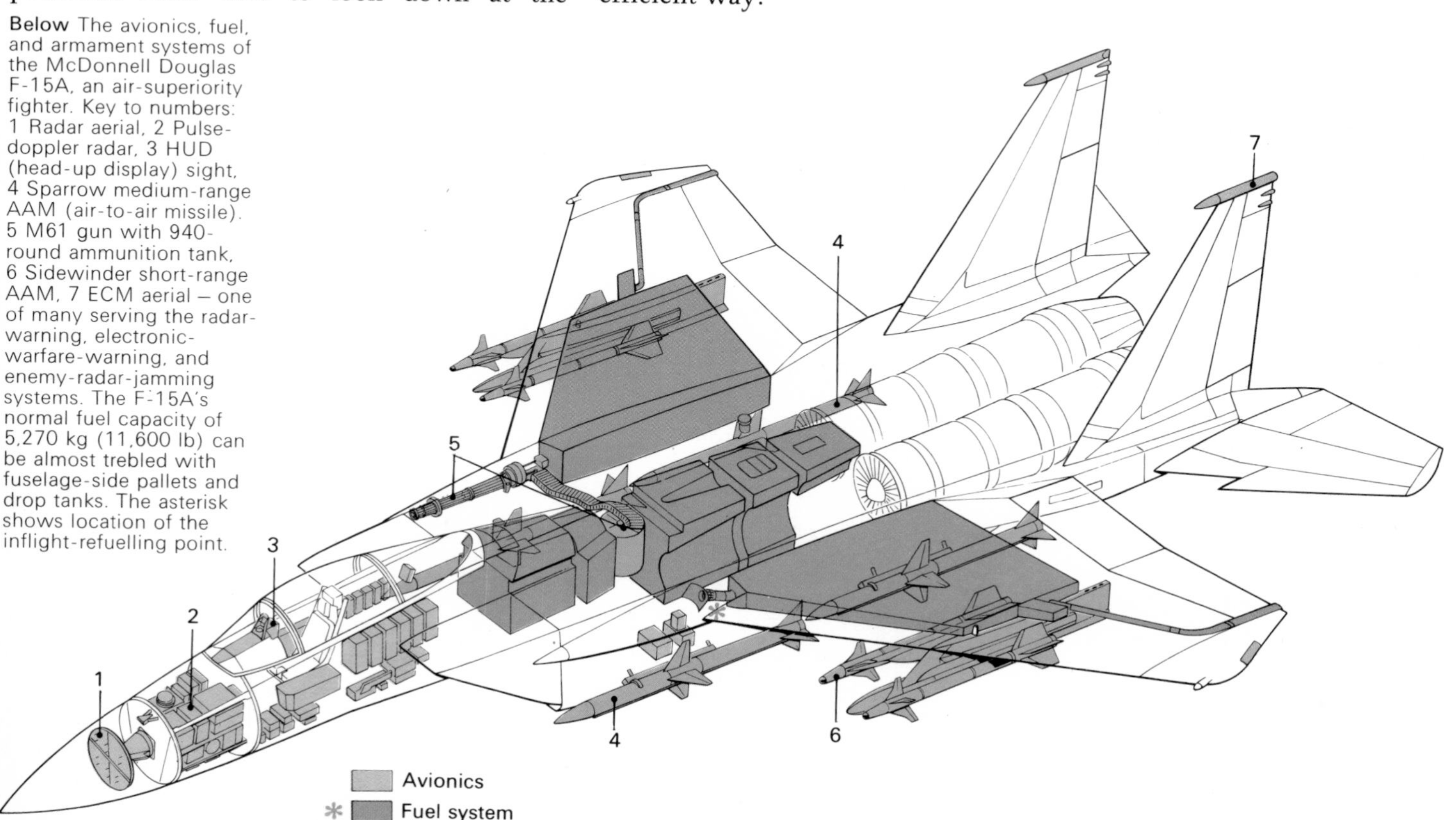


**Below** The avionics, fuel, and armament systems of the McDonnell Douglas F-15A, an air-superiority fighter. Key to numbers: 1 Radar aerial, 2 Pulse-doppler radar, 3 HUD (head-up display) sight, 4 Sparrow medium-range AAM (air-to-air missile), 5 M61 gun with 940-round ammunition tank, 6 Sidewinder short-range AAM, 7 ECM aerial – one of many serving the radar-warning, electronic-warfare-warning, and enemy-radar-jamming systems. The F-15A's normal fuel capacity of 5,270 kg (11,600 lb) can be almost trebled with fuselage-side pallets and drop tanks. The asterisk shows location of the inflight-refuelling point.

# 14 Anti-submarine Warfare

Although at the start of World War I the submarine (or U-boat, as the German sub was called) was a severely limited device, small in size and hardly capable of doing any damage except to vessels that blundered into its narrow orbit, a U-boat handled with skill and daring could be lethal. On 22 September 1914 the U9 managed to sink a British cruiser. Another cruiser came to help the survivors, and U9 sank that. Then up came a third, and the tiny U-boat sent it after the others. The Royal Navy lost 3,600 men to a vessel with a crew of 28 packed like sardines in a cramped hull. Such incidents spurred frantic work on the new science of anti-submarine warfare (ASW), but there were only three practical weapons. One was the depth charge, a canister of explosive dropped into the sea and detonated by a hydrostatic pistol when it had fallen to what was thought to be the correct depth. A second was the stick bomb, a smaller charge attached to a strong stick so that it could be thrown to a particular place in the sea. And to help decide where that place might be, the hydrophone was a kind of loudspeaker in reverse, lowered into the water on a cable (from a ship with engines stopped) to listen for the noise of a submarine.

From the start, aviation did what it could to help. The most useful machine was the airship. Only this had the long endurance needed to watch over friendly ships for hours or days, or to search for U-boats by day and night. On 28 February 1915 the Royal Navy ordered a fleet of quite small airships to help fight the growing U-boat menace. Called the SS series (Submarine Scout), these were followed by 24 of the C (Coastal) series and, in 1917, by a much better airship called the NS (North Sea) type. Unlike the Zeppelins, all were non-rigid, their flexible gasbags supporting a car containing the crew, engines, and equipment. To be useful or even safe they had to have quite powerful engines; earlier dirigibles had been unable to make any headway against even a light breeze, but the anti-submarine airships had to cope with gales over the open ocean. The NS, of which 200 were made, were substantial craft with a crew of 10 and the ability to fly for longer than 60 hours. They carried a variety of AS (anti-submarine) bombs, and had at least one machine gun. But it was difficult to spot submerged submarines in the grey and almost opaque waters around north-western Europe; and not until after the war did the records of the U-boat commanders reveal how remarkably effective the eyes in the sky had been in protecting Allied convoys and in occasionally sinking submarines.

Aeroplanes were at first too fragile and feeble to be much use in ASW, but they improved fast. One of the important builders was Fairey, which built the Campania and the III/IIIA/IIIB series, all quite large seaplanes with folding biplane wings, able to carry a crew of two, bombs, and a gun. Short Brothers built a series of even larger single-engined seaplanes, some of which could carry a torpedo. One, the Type B, had a monster Davis recoilless gun, firing 2.27 kg (5 lb) shells, for use against submarines or airships. Curtiss in the United States, the Felixstowe seaplane station, and a group of industrial firms in Britain concentrated upon large flying-boats which included ASW in their duties; one had a Davis gun firing 2.72 kg (6 lb) shells. On the whole, however, aircraft had to develop further before they could prove a real menace to submarines.

Development between the wars was remark-

This historic photograph was taken on 4 August 1917 as Commander E. H. Dunning's Sopwith Pup went over the side of the aircraft-carrier HMS *Furious*, in spite of attempts by the deck party to grab it; Dunning was drowned. Only two days before, he had become the first pilot to land on a carrier, although aircraft had been taking off from ships since 1910. The *Furious*, a converted light battle cruiser, was later fitted with a longer landing deck.

ably slow. Although the aircraft themselves became faster and more capable, little was done to find better ways of seeking out submarines from the sky. But submarines had grown much more formidable during the latter part of World War I, and in 1939–42 the increasing numbers of German U-boats wrought terrible havoc among the vital merchant convoys that kept Britain (and to some degree the Soviet Union) alive, often out of range of Allied aircraft entirely. The US Navy built a fleet of blimps, very like the non-rigid airships of World War I, but although these had long endurance they were too slow to get far from land. It needed the Consolidated B-24 Liberator and similar extra-long-range aircraft to close the gap in mid-Atlantic. Also needed were new search equipment (sensors) and better weapons.

## Electronic Aids

By far the most important of the new sensors was the *sonobuoy*. The old-fashioned hydrophone had been developed in the 1920s into Asdic, a much better listening device, and this in turn was developed into sonar. Sonar can be made in two forms. Active sonar sends out intense pulses of sound, which travel through the water, bounce off solid objects such as a submerged submarine, and send back echoes

**Above** The Felixstowe F.2A flying-boat, which flew in 1917, was used by the RNAS for patrol and anti-submarine duties.

**Left** The Martin T4M-1 seaplane was a large but slow torpedo bomber serving with the US Navy in the late 1920s and early 1930s. The somewhat similar Fairey Swordfish, which first flew in 1934, served the Fleet Air Arm throughout World War II.

which can be analysed to give the submarine's apparent position. Passive sonar merely listens to the sound of the enemy submarine. Passive sonar cannot so easily detect a submarine lying on the sea-bed with its engines switched off, but on the other hand active sonar can be deceptive by bouncing submarine-like echoes off, for instance, sunken wrecks and even schools of whales. At first sonars were large installations in ships, but by 1943 they were being miniaturized and packaged into bomb-like sonobuoys which could be dropped by aircraft. By 1945 a Liberator or a Consolidated PBY Catalina or Short Sunderland flying-boat could release a string of sonobuoys, pinpoint a submerged U-boat, and kill it with depth charges. They were so deadly that the U-boat – from 1943 bristling with as many as eight rapid-fire anti-aircraft guns – often chose to fight it out on the surface.

The late-World War II aircraft also had other aids to ASW. One was the Leigh Light, a multi-million-candlepower searchlight that could illuminate a U-boat from miles away on the darkest night when submarines thought it safe to surface and recharge their batteries. Another was radar, which by 1943 was being installed in ocean-patrol aircraft in forms able to spot a surfaced submarine at night from many miles away, and under favourable conditions even the tip of a periscope or a snorkel breathing pipe. Some aircraft, including fast fighter-bombers such as the de Havilland Mosquito XVIII, carried large-calibre cannon able to pierce a submarine before it could dive out of range. Depth charges increased in size and destructive power, but the real advance was the sensors that told the crew of the ASW aircraft where to release the charges.

In the 1950s two major advances increased the lethality of the submarine, and two quite different ones helped the ASW aircraft. The submarine was transformed from a sluggish short-ranged craft that occasionally went under the surface into a nuclear-powered projectile with astonishing power and endurance, able to outrun almost any surface ship and stay down for months at a time. The second advance was its dramatic change in role, from a ship-destroyer to a city-destroyer armed with intercontinental ballistic missiles, such as Polaris, fitted with nuclear warheads.

**Above** The Fairey Barracuda II saw action in 1943 as a carrier-based dive bomber and torpedo bomber, and was fitted with a .53 m (21 in) torpedo with air-launching tail controls and a primitive ASV (air-to-surface-vessel) radar.

**Below** Final touchdown by a Short Sunderland flying boat of RAF Coastal Command (No 201 Sqn) in 1957, after the famed 'Flying Porcupine' had seen 20 years of hectic service (a few served for two further years from Singapore). The only countries still operating military flying-boats are the Soviet Union and Japan, both of which use turboprop-powered aircraft.

**Right** A Westland Lynx attacking a ship with Sea Skua missiles. After falling free the missile lights up and flies at high subsonic speed just above the waves, finally homing on its target. (The sketch greatly foreshortens the attacking range. In fact, the Sea Skua would be launched at sufficient distance from its target to give the Lynx 'stand-off' protection against the ship's anti-aircraft armament.)

**Left** Unlike the US Navy's Sikorsky SH-3 from which it was derived, the Royal Navy's Westland Sea King contains a tactical compartment for independent ASW operations, using its own sensors and weapons. Thus it could, in theory, operate from a merchant vessel as well as from a warship. Here a Sea King HAS.1 of 706 Sqn dunks its sonar to listen for a submarine target.

## Helicopters

On the other side of the coin came the ASW helicopter with MAD (magnetic anomaly detector). MAD equipment measures the extremely small local disturbance that a submerged submarine causes to the Earth's magnetic field. To do so at all is very difficult, especially with modern submarines that can dive to great depths. The MAD-carrying aircraft has to fly as low as possible. It also has to carry the MAD equipment as far away as possible from its own magnetic parts. On large fixed-wing ASW aircraft the usual place is on the end of a long boom projecting behind the tail. Helicopters package the MAD gear into a 'bird' towed on a long cable so that it searches just above the surface of the waves.

Helicopters are useful for ASW because they can match the speed and agility of the nuclear submarine and track it even more accurately than the big patrol aircraft, which must fly overhead at too high a speed and then turn and come back. While the patrol aircraft has to release perhaps dozens of costly sonobuoys, often using a retro-launcher exactly to cancel the aircraft's own speed so that the buoys fall vertically, the ASW helicopter can 'dunk' a single buoy in the water. After listening to the submerged buoy it can fish it up, move to a better spot, and dunk it again. Using only one buoy is more economical, leaves more weight available for weapons, and also enables a much more powerful than usual sonobuoy to be used. Most aircraft sonobuoys are of the active type, and they do more than detect the presence of a submarine: they fix its position in direction and distance and also, by the Doppler effect, determine its speed. Dunked buoys are permanently connected to the helicopter, whereas the usual dropped kind have to radio their information up to the aircraft.

It has not been easy to choose the ideal kind of aircraft for ASW operations. Until 1960 many were water-based, either amphibians or flying-boats, but today the only aircraft in that class are the Japanese Shin Meiwa PS-1 and the Soviet Beriev M-12. The PS-1 is one of the aircraft still using an American sonar of the 1950s, Jezebel/Julie, in which explosive charges are used to generate the sound waves. Canada chose a very large ASW patrol aircraft based on the Bristol Britannia civil airliner but with compound piston engines giving very long endurance. Most countries use large turboprop aircraft, some of which are intended to shut down two engines in extended search operations. Britain has probably the most advanced ASW aircraft, the Hawker Siddeley Nimrod MR.2, having four turbofan engines giving a speed of nearly 600 mph and outstanding all-round performance. Most of the large ASW machines contain a tactical compartment, similar to those aboard ASW ships, where comprehensive displays fed by radars and managed by computers give the navigators and battle controllers all the information they need to take decisions.

Some helicopters, notably the British Westland Sea King, also have tactical compartments, making them capable of self-contained action. Most, however, are limited to acting as a short-range extension to the eyes, ears, and punch of the surface ship on which they are based. They radio back the information from their sensors and the decisions are taken aboard the ship. In fact, in the early 1960s the US Navy even got as far as buying 300 DASH (Drone Anti-Submarine Helicopter), small pilotless machines which could take off from an AS frigate and drop two torpedoes at a place discovered by the frigate's own sensors, with radio control from a 'pilot' on the ship. Although it is theoretically possible, the idea was found much too complicated and potentially hazardous, and was abandoned.

## Maritime Policing

Today many ocean-patrol aircraft are needed for law-enforcement. Almost every available kind of short-haul turboprop or turbofan airliner has been developed in a maritime-patrol form, usually with few sensors other than a good over-water radar (difficulties with reflec-

**Right** Artist's impression of the British Aerospace Sea Harrier, which was undergoing flight trials in autumn 1978. Powered by a Pegasus 104 vectored-thrust engine similar in output to that of the 103 in other Harriers, it has a largely new airframe with radar, raised cockpit, and different sensors and weapons for maritime air combat, anti-ship, and surface-attack missions. External armament will normally include two Aden guns and two Sidewinder missiles, as shown here.

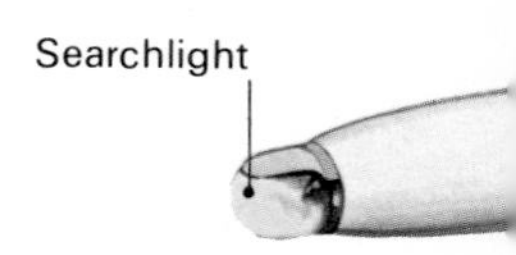


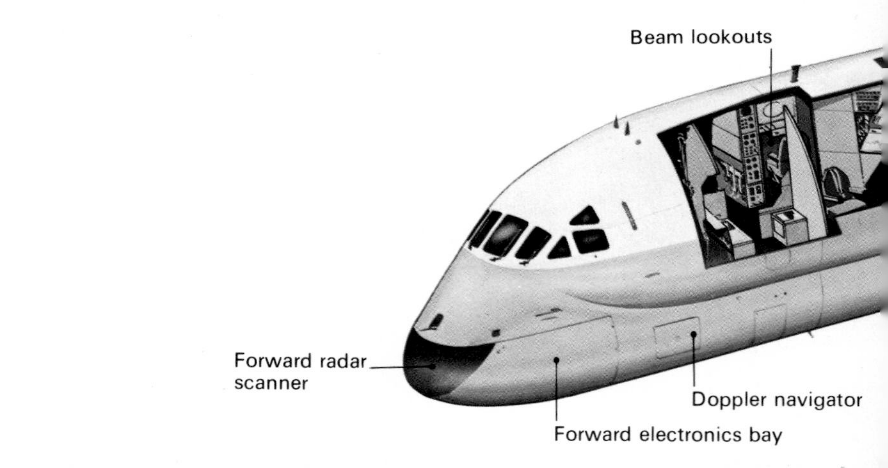

tions off the waves are one of many problems that call for a special kind of radar in ASW or any over-water mission). Together with exceptionally comprehensive and precise navigation aids, this enables a sea-surveillance aircraft to prevent smuggling or illegal immigration, arrest trawlers violating new fisheries laws, patrol oilfield installations, monitor sea pollution of all kinds, and fly search/rescue missions. Although nearly all these aircraft are landplanes, they can help rescue operations by dropping dinghies; for example, the British Hawker Siddeley 748 Coastguarder can drop twelve 30-man dinghies to a stricken ship or ditched airliner. Perhaps surprisingly, the US Coast Guard chose a small jet for this work, a special version of the Fanjet Falcon with new ATF-3 turbofan engines, of which 41 were ordered in one contract in 1977.

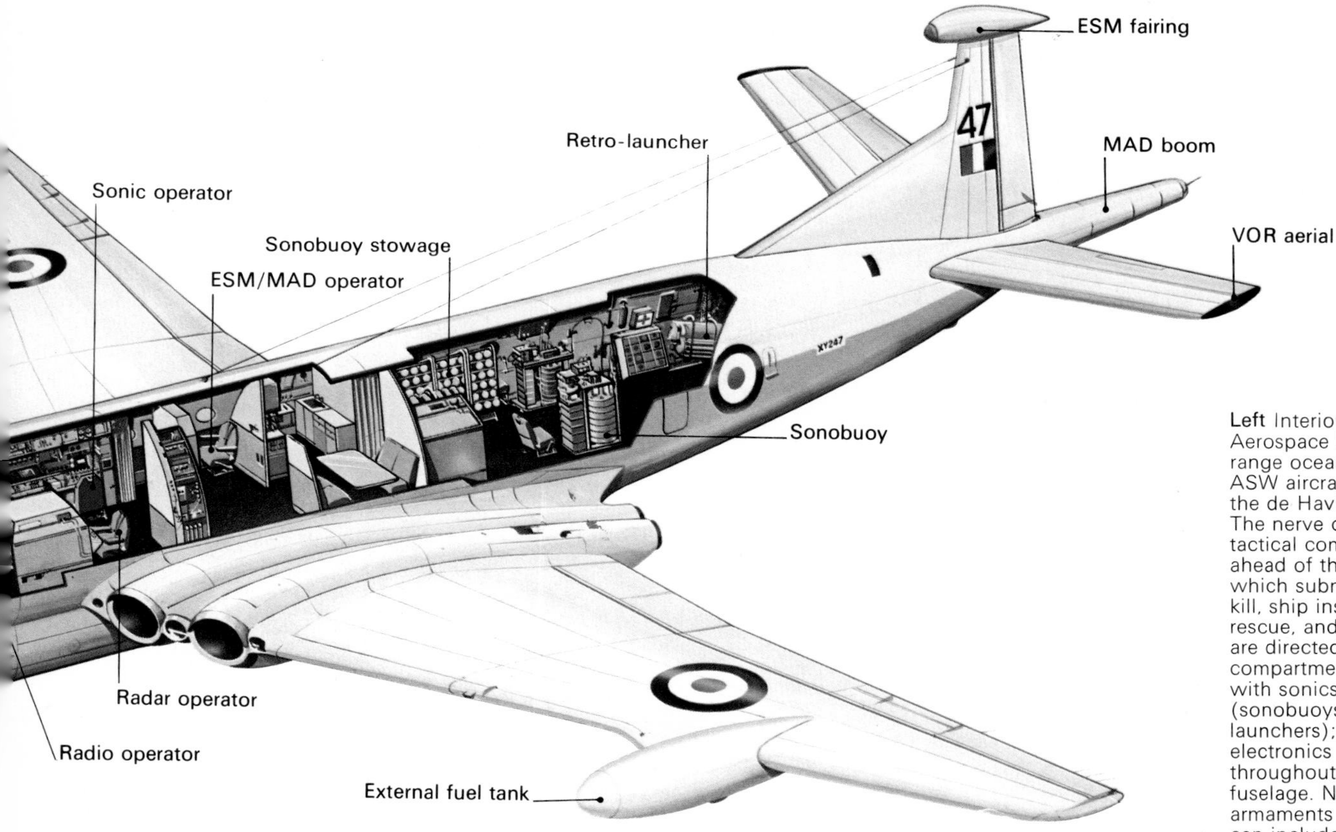


**Left** Interior of the British Aerospace Nimrod, a long-range ocean-patrol and ASW aircraft derived from the de Havilland Comet. The nerve centre is the tactical compartment, just ahead of the wing, from which submarine search or kill, ship inspection, sea rescue, and other missions are directed. The large compartment aft is filled with sonics sensors (sonobuoys and their launchers); mission electronics fill large racks throughout the pressurized fuselage. Nimrod's armaments (not shown) can include torpedoes, mines, depth bombs, air-to-surface missiles, and nuclear weapons.

**Left** The US Navy's Lockheed P-3 Orion, powered by four 4,910 hp turboprops, has been steadily improved since 1959, and in the P-3C version shown here it has reached an impressive level of sensing, data-handling, and submarine-killing capacity.

# 15 Aircraft in Land Warfare

Previous chapters have shown that from 1911 until about 1916 there was hardly any such thing as 'air warfare', but plentiful use of aeroplanes, airships, and balloons in land warfare. The primitive use of hand (or foot-thrown) bombs, and the firing of rifles or the occasional machine gun, did not significantly influence land battles but did make vast numbers of front-line troops edgy about all flying machines. Aircraft recognition had not emerged as an instructible subject. Even although by October 1914 it had become part of official policy to paint national markings on all aircraft, trigger-happy soldiers soon got into the habit of firing at every aircraft that came within range. The French tried to minimize self-inflicted casualties of this kind by concentrating on pusher aircraft. It was said (essentially correctly) that the Germans did not use any pushers, so such machines should be left in peace by Allied troops. But as most machines were of the tractor type they were still fired on indiscriminately by both sides.

This made it hazardous to use aircraft in direct support of one's own front-line troops. With almost incredible optimism the Imperial German Army decided that airships could play a direct close-support role in land battles, mainly by dropping bombs. Although there were no established anti-aircraft defences anywhere and field guns could not be elevated to a high angle, the airship losses were almost crippling. On both the Western Front and over Russia the early Zeppelins and Schütte-Lanz ships went down like ninepins. Soon the survivors were all sent to the slightly less dangerous Russian theatre, with not much success, and the Army Airship Service was finally disbanded on 1 August 1917.

Aeroplanes, however, stood a better chance. In the early months of World War I the official policy of most combatants was that the aeroplane had no role in war except, possibly, reconnaissance. But the men in the front-line squadrons itched to participate directly in the fighting, as Italian pilots had done in 1911. Two of the first pilots in the RFC in 1912 had been Capt Penn-Gaskell and 2nd Lt Strange, and both had experimented in that year with firing rifles from the air. On 22 October 1914 the inevitable happened. Strictly in defiance of authority, these two young pilots of No 5 Sqn fitted an Avro with a mounting (made from the tail boom of a wrecked Farman) for a Lewis gun. Strange took the Avro to Perenchies, where he found a train and German troops, and Penn-

**Below** The Sopwith Salamander was one of the first specialized ground-attack aircraft, about 40 of them reaching the RAF late in World War I. Based on the airframe of the Snipe fighter of 1917, it had some 295 kg (650 lb) of armour to protect the pilot and fuel tanks. Although the Salamander's armament was confined to the Snipe's two 0.303 in Vickers machine guns, other ground-attack machines of 1918 had as many as six guns, and many carried light bombs.

The Russian Ilyushin Il-2 Stormovik, with its heavy armour, was one of the most durable fighting aircraft of World War II. Designed in 1935 as a specialized close-support attack bomber and tank killer it developed through several versions in 1942–5, with harder-hitting cannon, ground-attack rockets and, as seen here, a rear gun to discourage fighters. Often operating at zero altitude, the Il-2 was consistently successful against heavily armoured German tanks on the Eastern Front.

Gaskell in the rear cockpit raked them with fire.

By 1917 such ground-strafing had become commonplace. On 11 May of that year the Battle of Arras was marked by the use of two officially organized squadrons, RFC No 11 (F.E.2b) and No 60 (Nieuport XI and XVI), to clear a path near Roeux for the held-up British 3rd Army. They were effective, but suffered heavy casualties. Later, at the Battle of Cambrai in November–December 1917 the RFC used hundreds of single- and two-seat scouts in determined attacks on the German defences. Among the aircraft were the still-new Sopwith Camels as well as the D.H.5. Both were especially well suited to ground attack because they offered a superb forward view and were strong and highly manoeuvrable. The D.H.5 pilot looked ahead over a single Vickers gun and the Camel pilot over a pair. Both could also carry four 11.4 kg (25 lb) bombs under the belly, but neither was fitted with any armour. According to the official history *The War in the Air*, 'The casualties to the low-flying aircraft were high, averaging . . . 30 per cent for each day. . . . That is to say, a squadron . . . had to be replaced every four days'.

Such losses were intolerable, and by February 1918 Sopwith had produced an armoured Camel called T.F.1 (Trench Fighter 1). In April there followed the T.F.2 Salamander, the world's first production ground-attack aircraft. Derived from the 230 hp Snipe fighter, it carried the usual two guns and four bombs as well as comprehensive armour all around the pilot and other protective features. Large-scale production was in hand by the summer of 1918 and the RAF had several dozen at the Armistice. Another Sopwith, the triplane Snark, had no fewer than six machine guns, but this armament was as much for aerial combat as for ground-strafing. The Royal Aircraft Factory at Farnborough also produced a ground-attack machine, the A.E.3 (Armoured Experimental), known as the Ram. A substantial pusher biplane, it had an armoured nacelle for a front gunner able to fire two Lewis to the front and a third to the side or rear, the third gun mainly for defence. Germany's Junkers J.1, a large biplane used mainly for trench strafing, was made entirely of metal.

## Arms and Armour

After 1918 development stagnated, but the US Army Air Service, after fruitless attempts to produce heavily armed and armoured groundstraffing machines in 1918–22, decided in 1926 to introduce a new category of aircraft called 'attack'. The first to go into service was the Curtiss A-3, a conventional tandem-seat biplane almost identical to the widely used O-1 Falcon observation machine but equipped with guns and light bombs. Curtiss later supplied the Army Air Corps (as it was from 1926) with monoplane and then twin-engined attack machines, and in 1938 Douglas flew the prototype of what became one of the most important attack bombers in history, the 7B, later developed into A-20 Havocs, Bostons, P-70s, and F-3s for almost every tactical purpose. From this stemmed the A-26 (later B-26) Invader, which played a major role in World War II, Korea, and Vietnam owing to its splendid performance, weapon load, and long endurance.

It was natural that air weaponry in the Soviet Union, from the time when that vast country emerged from bloody civil war, should have been polarized around land warfare. In 1928 Nikolai Polikarpov, leader of the most prominent design collective, produced the first R-5 biplane, a classic and outstanding tandem-seat aircraft destined to serve in immense numbers up to the middle of World War II. One of its many versions was the R-5Sh 'assault' or close-support model with seven guns (four of them fixed firing ahead) and a 500 kg (1,102 lb) bomb load. There followed a profusion of other Russian attack aircraft, of which by far the most important was the Ilyushin Il-2 Stormovik. It is odd that, although the specification of the Il-2 very closely paralleled that of the Fairey Battle, the British machine proved a disaster while the Stormovik was described by Stalin as 'as necessary to our troops as air and bread'. It went on to sustain the biggest production run of any aircraft in history, a total of some 35,000, or 41,400 including the later Il-10. Its chief features were very thick armour forming a primary structure around the crew, engine, and fuel tanks, and heavy armament including tank-piercing cannon and rocket projectiles. The first were single-seaters, but from October 1942 the standard model had a manually aimed rear gun for defence.

In the fierce fighting on the Eastern Front the Il-2 was probably the single most important weapon, and its technology in guns and anti-tank bombs kept pace with German tank design. Its counterpart in the Luftwaffe was a rather unimpressive machine, the Henschel Hs 129, with two French engines of 690 hp each, which had various guns of up to 75 mm (2.9 in) calibre and numerous arrangements of bombs or rockets (or even a battery of downward-firing recoilless cannon fired automatically as the aircraft flew over a tank). Many of the Luftwaffe's ground-attack and anti-armour aircraft were mere lash-ups. The Ju 88 and even the big He 177 were pressed into service with various guns of 37 mm (1.5 in), 50 mm (2 in), or 75 mm (2.9 in) calibre, and even larger guns were under development in 1945. The Luftwaffe also pioneered the use of air-to-ground guided missiles, mostly intended for strategic targets but in practice misused in attempts to demolish key bridges and similar tactical battlefield objectives. Among these missiles were complete pilotless aircraft, chiefly modified Ju 88s in the Mistel (Mistletoe) composite scheme in which the missile was aimed and released by a piloted fighter riding on top of it.

Three important developments that came much to the fore in World War II were the dive bomber, the light attacker, and the fighter-bomber. Dive bombers, briefly discussed in Chapter 10, were (in the shape of the Ju 87 Stuka) the dominant aerial influence on land battles in 1939–42. Thereafter the Ju 87, in many versions, was used mainly in close-support, night attack, and anti-tank work, trying to avoid the daylight sky if possible. Equally stealthy were the light biplanes, many of them designed years before as trainers, which from 1942 blossomed forth as effective front-line operational machines. Among them was the Soviet U-2 (later restyled Po-2 in honour of designer Polikarpov), of which some 20,000 were built from 1927. Every night it kept the Germans from sleep, fired guns and dropped light bombs, and, according to a German prisoner, 'even looked over the window sills to see if any of us were inside'. The Luftwaffe retaliated with the Arado Ar 66, Go 145, and several other light or obsolescent machines. Years later similar aircraft were to become important in 'limited' or 'brush-fire' war situations calling for close-support machines able to operate from short front-line airstrips and carry out light attack, psy-war (psychological warfare, with loudspeakers and leaflets), forward air control, and casevac (casualty evacuation) missions.

Fighter-bombers began in 1935 in the Soviet Union and the USA (especially in the US Navy) as fighters fitted with light bombs. By 1941 aircraft such as the Messerschmitt Bf 109 and Hawker Hurricane were filling in time with close-support bombing, and such indifferent fighters as the Bell P-39 and Curtiss P-40 spent most of the war in attack duties, where they were extremely valuable. This led to a confusion of nomenclature. A 'fighter' is designed for air combat, and most close-support or attack aircraft try to avoid hostile fighters. Yet many attack machines since 1943 have been called 'fighters'. Some, such as the Hawker Typhoon, which in 1944 carried air-to-ground rockets and was a major factor in defeating the German armies in France, could give a good account of themselves in air combat – at least at low level. More recent examples of such machines include the Republic F-84 series, North American F-100, Lockheed F-104, Northrop F-5, SEPECAT Jaguar, Dassault Mirage 5, Sukhoi Su-7 and MiG-27. But these are hardly fighters; none (except special versions such as the F-104S) carries air-to-air radar or weapons, and even the basic Panavia Tornado is an attack aircraft rather than a fighter. The crowning example of mis-designation is the General Dynamics F-111, nearly all of which have no air-to-air capability or mission but are outstanding all-weather-attack delivery systems.

## Helicopters

A special class of aircraft called by various names has seldom carried armament but has played a major role in land war. In World War I they spotted for the guns and reconnoitred the battle area, either throwing down written messages or using cumbersome army radio. In

**Right** The Russian Mil Mi-24 (NATO code-named Hind) is larger than most battlefield helicopters and combines the duties of troop carrying, logistic supply, and anti-tank attack. Another version, code-named Hind D, has a two-seat nose with special sights and weapons for attacking armour or defensive bunkers.

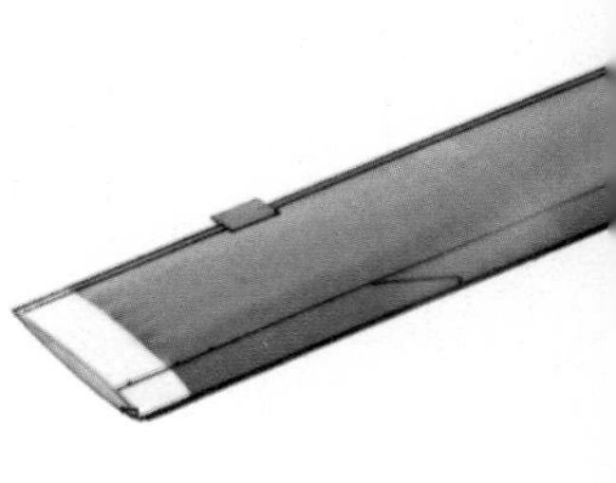

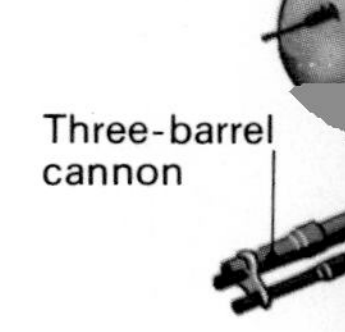


**Above** The AH-1J Sea-Cobra is a US Marine Corps version of Bell's 'Huey' ground-attack helicopters. Based on the Model 212, unlike the HueyCobra it has twin turboshaft engines (Pratt & Whitney T400s of 1,100 shp each). The AH-1 Cobra family has several hundred armament variations based on 25 different types of weapon. Although they are vulnerable to surface-to-air missiles, ground-attack helicopters rely on agility to keep out of trouble.

World War II observation or liaison machines tended to have short-take-off-and-landing (STOL) qualities, important types being the German Fieseler Storch, British Auster, and American L-4 Grasshopper, all very slow and short-ranged but endowed with excellent all-round visibility and the ability to land safely almost anywhere. Today many of the numerous front-line duties of such machines have been taken over by helicopters, which began to find roles in land warfare during the Korean War (1950–3). The first of such helicopters could

**Right** Apart from 'tank busting', tactical helicopters play a vital role airlifting men and equipment to and from the front line. Here an RAF Aérospatiale/Westland Puma, powered by two 1,328 shp turboshaft engines, brings up a 105 mm Light Gun specially designed for airlifts.

lift only seven men, but they had the previously unavailable asset of being able to rescue downed aircrew or troops from the jaws of the enemy. Gradually the US Army, the French (in Indo-China and, especially, in Algeria), and the Soviet Union experimented with helicopters fitted with armament. At first the weapons were simply air- or land-warfare machine guns, cannon, or rocket launchers, but by the mid-1960s the attack helicopter had become one of the most important and sophisticated aerial-weapon platforms. The US Army launched a development programme for an Advanced Aerial Fire Support System which eventually flew as the Lockheed AH-56A Cheyenne, one of the most complex and costly tactical aircraft ever built (it was said to have electronics surpassing even those fitted to a B-52). The AH-56A was eventually dropped in favour of the simpler Bell AH-1 HueyCobra, derived from the UH-1 Iroquois or 'Huey' family, the world's most mass-produced military aircraft of recent years. Various Cobra versions carry remotely aimed turrets for machine guns, cannon, and grenade projectors, numerous rockets and fixed guns, and guided anti-tank missiles such as the tube-launched TOW.

The Cobra family were the first helicopters to go into service (in June 1967) that could play a positive offensive role on any battlefield. Although retaining similar power plant and rotor systems as the UH-1 family, the AH-1 series have a much narrower body, seating a gunner in the nose and a pilot above and behind. Both have a superb view and are surrounded by armour and equipment that bring the unladen weight up to more than that of the larger cabin Hueys. Among the equipment are all-weather and night-sighting systems, comprehensive communications, and various optional radar warning or radar infrared (IR) counter-measures. Combat helicopters of this kind, backed up by the simpler but well-armed Hughes and Bell LOH (Light Observation Helicopter), developed helicopter battlefield techniques to a fine art in Vietnam and are today deployed in large numbers as a main defence against massed attacks by armour. Undoubtedly the biggest user of tactical helicopters is the Soviet Union, whose large and powerful Mi-24 is used in various troop-carrying, attack, and anti-armour versions.

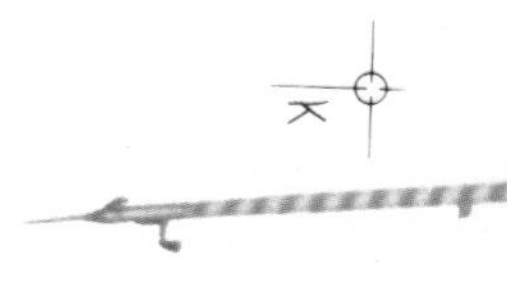

Obviously a helicopter is vulnerable to ground fire. As far as possible it stays out of sight behind natural cover or buildings. It can sometimes land a squad of troops with their own anti-armour missiles, or it can briefly rise into full view, sight on a tank, and attempt to kill it before being seen by the enemy. Clearly the requirements are a sight system that can look at the enemy without exposing the helicopter, and a fast-flying missile that either hits its target within a second or two or does not need to be steered by a human operator all the way to the target, having what is called 'fire-and-forget' guidance. Great emphasis is still being placed on the development of long-range magnifying sight systems that can see clearly at night or in bad weather. Increasingly these battlefield sighting duties are being taken over by simple, and potentially expendable, RPVs (remotely piloted vehicles) fitted with miniaturized optical or IR sights and a laser designator to point out targets for homing missiles.

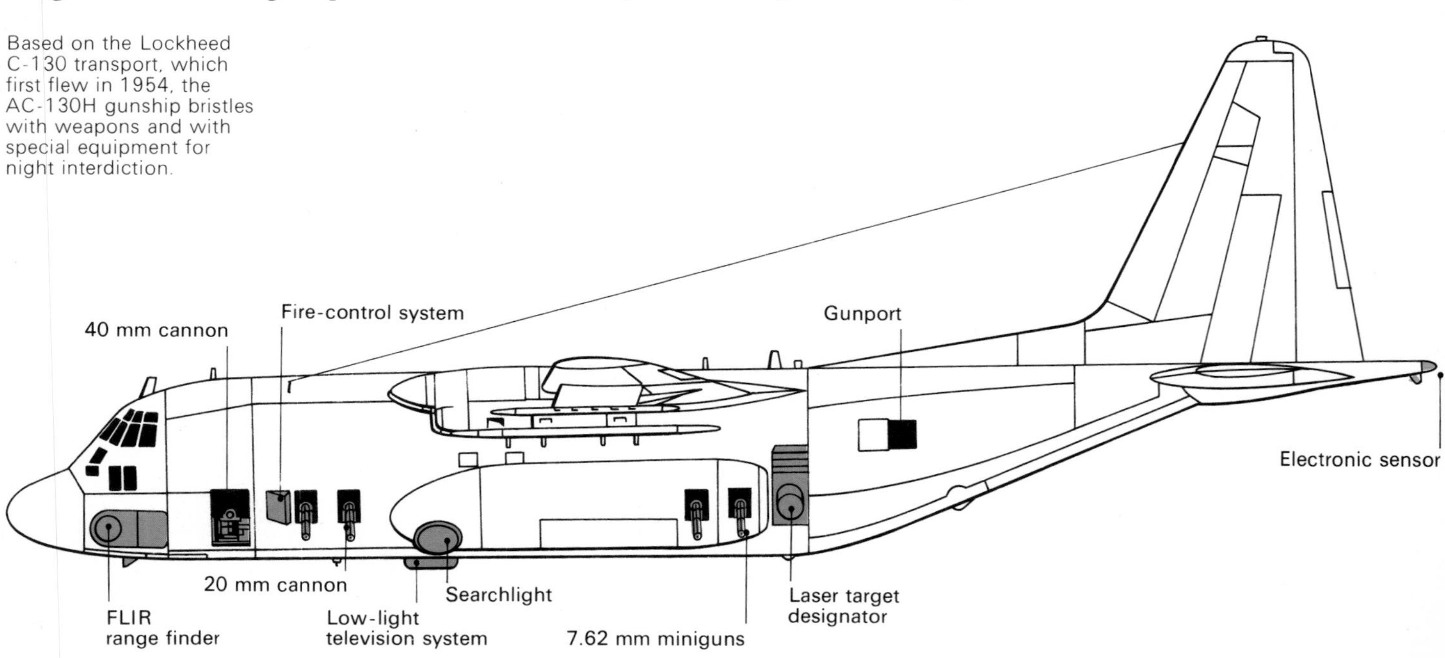


Based on the Lockheed C-130 transport, which first flew in 1954, the AC-130H gunship bristles with weapons and with special equipment for night interdiction.

The Fairchild A-10A is the most completely specialized ground-attack weapon platform flying today. Although slow, with a top speed of about 612 km/h (380 mph), it is designed to survive hits by most kinds of bullets and shells. Its armaments include rockets, bombs, missiles, and a 30 mm GAU-8/A high velocity cannon – the most powerful rapid-fire gun ever fitted to an aircraft. The A-10A shown here carries free-fall bombs on its external pylons.

Today aircraft would play a major, and often dominant, role in any land war. Central to their role would be laser-guided weapons, which are especially suited to aerial use because aircraft have an overall view of the battle area and can easily carry the required laser. The laser emits a pencil-thin light beam, with precise characteristics that may be hard for an enemy to interfere with. Once the target has been spotted, it can be designated by aiming a laser at it. The laser may be in the aircraft carrying the weapons, in an accompanying aircraft (or small RPV), or even aimed by a soldier on the ground. Today missiles can be made to fly towards the light diffused or scattered from the target, and do so with unfailing reliability and perfect accuracy. What this means is that, taken in conjunction with modern sight systems, aircraft can so dominate a land battle that it becomes virtually an air battle. According to the Soviet Union's Frontal Aviation, by far the world's largest tactical air force, 'Anything we can see, we can hit. Anything we can hit, we can destroy – usually with one weapon.' Night, snow, mist, rain, or camouflage offer no refuge against such armaments.

There remains, as always, the problem of aircraft vulnerability. Until the 1970s most tactical aircraft could usually be brought down even by small arms (rifle-calibre or pistol) if strikes were taken in a vital area. Of course, the pilot has been armoured since the 1930s, and fuel cells are protected or self-sealing and purged with nitrogen or other inert gas, but the basic aircraft has been so penetrable by bullets and so complex that even primitive troops could be dangerous to aircraft flying really low, while fast-firing cannon and small SAMs (surface-to-air missiles) were potentially lethal at ranges of a mile or more. Much thought therefore went into making tactical aircraft 'survivable'. Today's front-line combat types are supposed to be able to survive hits by cannon shells of 37 mm (1.5 in) or even 57 mm (2.2 in) calibre, and by most shoulder-fired SAMs used by infantry. Aeroplanes and helicopters are heavily armoured, made with 'two of everything', and cunningly built so that even severe damage will not stop them flying.

Two of the important battlefield aircraft are the British Hawker Siddeley Harrier and the American Fairchild A-10A. The former is a jet V/STOL (*see* Chapter 23) that can be based on any small clearing or reasonably level patch of ground and is thus often able to reach front-line troops within seconds of a call for help. It carries all normal weapons and a laser designator, and has the advantage of presenting an extremely small, agile, and elusive target to the enemy. The A-10A is much bigger, and is a conventional straight-wing aeroplane with a take-off run of 900 m (3,000 ft), or rather more at full load. It was designed around a monster tank-killing gun, bigger than any other currently in use, and it can also carry about 8,150 kg (8 tons) of weapons of many kinds.

# 16 Military Transports

Until after World War I aircraft were not used for military transport duties except in isolated emergencies. There were plenty of examples of cut-off troops being supplied from the air by rather hastily devised methods, but there was no such thing as a supply container, such as was later developed to enable urgently needed items to be dropped, with or without parachute. Ammunition, food, water, and medical supplies were usually wrapped in blankets, canvas (such as was used for folding camp-beds), and similar coverings, and dropped from low level. Results were usually poor, both in accuracy and payload survival. Very occasionally troops were flown by large aircraft such as the Russian Sikorsky *Ilya Mourometz* heavy bombers, but there is no record of aircraft being specially equipped for this role by either the Allies or the Central Powers.

After 1918 the scene changed dramatically, especially in the case of Britain and France, which had large colonial empires, and increasingly by Italy. It was suddenly realized that the aeroplane – and, it was often suggested, the airship – could play a central role in policing enormous areas. The problem was how to deliver a well-armed force quickly to any point of major disturbance. To keep large forces permanently stationed throughout the colonies would have been prohibitively expensive in men and equipment. The alternative of hurrying troops over distances of hundreds or even thousands of miles, which was what happened in the Indian Mutiny of 1857 and various late-19th-century wars in southern Africa, was equally unattractive. Trains or motor trucks were slow-moving in difficult terrain and were vulnerable to sabotage or attack. Air transport, by contrast, offered a means of shuttling adequate forces from place to place in a matter of hours, and the colonial powers were quick to see its advantages.

## Early Transports

Probably the first purpose-designed military transport was the British Vickers Vernon, which entered service with the RAF in 1922. At this time the grand design of imperial communications was as important as colonial policing, and a handful of RAF squadrons accomplished an immense task in surveying and opening up reliable air routes in various parts of the empire. Some of the earliest services were flown by 58 and 216 Sqns in Egypt with Vickers Vimys originally built as bombers. Another Vimy unit, 45 Sqn, was first to receive the Vernon, which had a fat fuselage seating 12 and an open cockpit for the crew of two sitting side-by-side. A tractor ploughed a furrow across the featureless desert of Jordan and Iraq, so that pilots would not get lost on the tough 1,300 km (810 miles) route from Cairo to Baghdad. Every 150 miles or so a landing ground was prepared, and if possible stocked with petrol, oil, and water, which had to be kept under guard. Gradually the routes were built up to India, while in North Africa the French used aircraft extensively in the Rif wars in Morocco and gradually opened up aerial links to Senegal (Dakar) and, eventually, the Camerouns in the course of establishing their airline service to South America (*see* Chapter 18).

Week by week the burden on the infant air forces grew. Before 1922 was out a contingent of British troops in remote northern Iraq was

The Bristol Braemar typified the approach to large aircraft in 1918. Originally intended as a bomber (as here), it was only later considered for a transport role before being scrapped. Its designer built a transport version, called the Tramp (by analogy with the term 'tramp steamer'), with similar wings and tail but with four 230 hp Puma engines in the fuselage with long shafts to the propellers. It never flew.

smitten with disease, and within a few hours Vernons had flown all the men to hospital in Baghdad. The following year war between Kurds and Arabs in Kurdistan was nipped in the bud by the first troop airlift. Policing grew in efficiency and technique. Instead of the costly punitive column of troops, all the RAF needed to do was drop leaflets telling the trouble-makers – who were invariably fighting each other rather than the colonial power – that if they did not desist their villages would be bombed at a specified time on the morrow. Once or twice the bombing had to be carried out, after the last villager had left, and it was soon judged that tribal wars were not worth while. The notable exception was the rioting in Afghanistan in 1928. By this time the Vernon had largely been replaced by the 22-seat Victoria, and with other aircraft eight Victorias of 70 Sqn in two months evacuated 586 civilians (including the king) and 11,000 kg (24,200 lb) of baggage from Kabul, the capital. This was the first large-scale airlift and it went like clock-work despite chaos around the city. Aircraft also played a major role in restoring order in the spring of 1929 and in bringing the civilians back to the capital.

In the United States the first military transport was the US Army's T-2 bought from Fokker in 1921. Powered by a 400 hp Liberty, this efficient monoplane could lift 4,925 kg (10,850 lb), and if loaded to above this weight with nothing but fuel it could fly for more than 24 hours. On 2 May 1923 a T-2 took off from New York in the hands of Lieutenants John A. Macready and Oakley G. Kelly and flew the 4,060 km (2,520 miles) to San Diego in 26 hr 50 min, the first American non-stop coast-to-coast flight in history. By 1925 Fokker was building F.VII/3m-type trimotors in the United States, and the US Army's C-2 versions made many notable flights. On 9 May 1926 one took off from skis on Spitzbergen, flown by US Navy Lieutenant-Commander Richard Byrd and Floyd Bennett, and made the first flight over the North Pole. In 1928 a C-2, ably crewed by Army Lieutenants Lester Maitland and Albert Hegenberger, made the first flight from California to Hawaii; and in January 1929 a C-2A named *Question Mark* was skilfully kept topped up with petrol through a primitive air-to-air hose system to set a flight endurance record of 150 hours.

## World War II

Many of the RAF's aircraft in the 1930s were so-called bomber-transports. They could do both jobs, but they were not as good at either as single-purpose machines. Until well into World War II there were hardly any really useful military transports because those in use were just like civil airliners in having small side doors that could not admit anything larger than troops, stretcher (litter) casualties, drums of fuel, and perhaps a motorcycle. Some bomber-transports, such as the Handley Page Harrow

Loading a stretcher (litter) casualty in through the nose hatch of a Vickers Victoria, the RAF's most important transport in the decade 1926–36. Powered by two Jupiter air-cooled engines, the Victoria was much more reliable than any previous aircraft in RAF service, and several of them lasted until World War II. It is a reflection of the relative simplicity of these and other large aircraft of the time that their laden weight was less than a quarter of that of modern fighters such as the Grumman F-14.

(originally purely a bomber) and the Italian Savoia-Marchetti S.M.82 Canguru, could carry bulky loads slung inside an internal bomb bay, but this was highly inconvenient. It was left to the Germans to build the obvious type of machine with a level cargo floor close to the ground, with front or rear doors giving access over the whole cross-section of the fuselage.

The first such aircraft was the Messerschmitt Me 321 Gigant, an enormous glider. Its cavernous interior was opened by left- and right-hinged nose sections through which an 88 mm gun, a light tank, or most German wheeled vehicles or half-tracks could be loaded, up to a total cargo load of 22 metric tons (48,500 lb). It was intended to carry heavy equipment on a planned invasion of Britain in 1941, the use of small assault gliders having been proved in extremely successful operations in the invasion of the Low Countries on 10 May 1940. Having the doors in the nose prohibited the dropping of heavy loads in flight, but the giant glider was so useful that it was soon developed into a powered version, the Messerschmitt Me 323, with six 1,140 hp engines. The Blohm und Voss 222 Wiking (Viking), a flying boat of similar size and power, also spent most of its war service as a strategic transport. Near the end of the war the Luftwaffe used the Junkers Ju 290 and 352, which had large rear doors which could be hinged downward to form a loading ramp or to allow heavy items to be dropped in flight. The Arado Ar 232 was an even better design, with a high wing, multiple wheels for soft airstrips, and an unobstructed cargo hold with a level floor at truck-bed height.

There were many Allied transports in World War II, by far the most numerous being the Douglas C-47 and other versions of the ubiquitous DC-3 – which, like the main Luftwaffe machine, the Ju 52/3m, had a small interior with sloping floor and constricted side door. The first really good military transport appeared only in August 1954, when Lockheed flew the C-130 Hercules. In this all the logical features were combined: unobstructed level floor, full-section rear doors for bulky loading or heavy dropping, high-lift wing, four powerful turboprop engines, pressurization and air-conditioning, full all-weather equipment and navigation aids, and rough-field landing gear retracting into bulges on the sides of the streamlined fuselage. After 24 years the C-130 is still in production, with over 1,500 built. Like the wartime German transports the C-130 owed nothing to any prior civil design, yet in its turn it has sold well in civil versions, usually for carrying varied kinds of cargo on standard-size pallets which are loaded on the ground, then winched along conveyers and into the aircraft in a matter of minutes.

**Above** Derived from the huge Me 321 Gigant glider, the six-engined Messerschmitt Me 323 was probably the first specialized military transport in history, with vast capacity and full-section nose doors. Its rows of landing wheels with low-pressure tyres enabled the Me 323 to operate from rough front-line airstrips.

**Below** Although the Germans pioneered airborne assault with gliders, the Allies used gliders in far greater numbers in World War II. Apart from the tank-carrying Hamilcar, the biggest type to see action was this British Airspeed Horsa, made of wood. More than 3,650 of these 25-seaters were built.

## Refuelling in Mid-air

Flight refuelling, the key to the endurance record of the *Question Mark*, was intensively experimented with just before World War II but played virtually no part in it. By 1948, however, the new USAF Strategic Air Command (SAC) was eager to extend its global striking power and it examined two methods of aerial refuelling to accomplish this. One was the British-developed probe/drogue technique in which the tanker trails from one to three hoses from power-driven reels, each connected to its transferable fuel supply. The receiver thrusts a fixed or retractable probe – a forward-facing pipe connected to its fuel system – into a stabilizing drogue on the end of the trailing hose. The probe finally makes a fuel-tight connection, and fuel is pumped through until the receiver is full, disconnects by throttling back, and allows the self-sealing valves in the probe and drogue to slam shut. The other method, which was chosen for SAC, was developed by Boeing. An operator lying in the tail of the tanker pays out a long pivoted boom by means of two control surfaces near its end. When he judges it is pointed correctly at a special receptacle somewhere in the upper surface of the receiver, he fires the telescopic boom so that it quickly extends and makes a fuel-tight connection. The two aircraft ride together while fuel is transferred until the boom operator pulls the telescopic boom back, its self-sealing valves shutting off the fuel. It will be noted that in the SAC flying-boom method the onus of connecting up is on the boom operator, while in the probe/drogue scheme it lies with the receiver pilot.

Later the probe/drogue method was widely adopted by the RAF, Royal Navy, US Navy, and USAF Tactical Air Command, as well as by many other forces, and the hose-reels are often carried under the wings of combat aircraft. SAC, however, needed specialized tankers, and after building 888 piston-engined KC-97 versions derived from the B-29, the Boeing company delivered 732 even larger jet aircraft of the basic 707 type (though slightly smaller and with slimmer bodies), designated KC-135. This was the largest fleet of jet transports ever constructed, and most have now been converted to other uses.

## Giant Transports

Today the largest military transports are the USAF Military Airlift Command's Lockheed C-5A Galaxy and the Soviet Union's V-TA (Transport Aviation) Antonov An-22 Antei. Both are high-wing, long-range aircraft with cavernous fuselages able to carry battle tanks – among the heaviest airlift loads at around 61,000 kg (60 tons) – all normal military vehicles, earth-moving equipment, and any other large mobile items needed by military

**Above** The Hawaii Mars was one of a series of enormous transport flying-boats, originally planned as ocean patrol and anti-submarine aircraft, that served the US Navy on both the Atlantic and Pacific immediately after World War II. They were the last reminders of an era in which all really giant aircraft had to take off and alight on water.

**Below** Contemporary with the Mars was the Fairchild XC-120 Pack Plane. Derived from Fairchild's C-82 Packet freighter, it had a detachable freight hold in the form of a giant container, complete with doors, which could be attached at its destination airport and quickly replaced by another that had been loaded before the aircraft's arrival. Although the idea was not adopted (the picture shows the sole XC-120 to be built) it is similar in principle to the 'containerization' system that is today a feature of integrated ship, rail, and road transport.

forces. The C-5A has four TF39 turbofan engines, each developing 18,600 kg (41,000 lb) of thrust, and with 28 wheels it can operate from sand desert or soft earth. The An-22 has four 16,000 hp NK-12 turboprops driving large eight-blade, contra-rotating propellers. The An-22 has enabled the Soviet Union to send vast quantities of modern weapons and other war material over long routes to places such as Angola and Ethiopia, a task that until the early 1970s would have been impossible to undertake. This global logistic capability is something that Britain no longer has, although the RAF still has a small force of shorter-range C-130 Hercules transports.

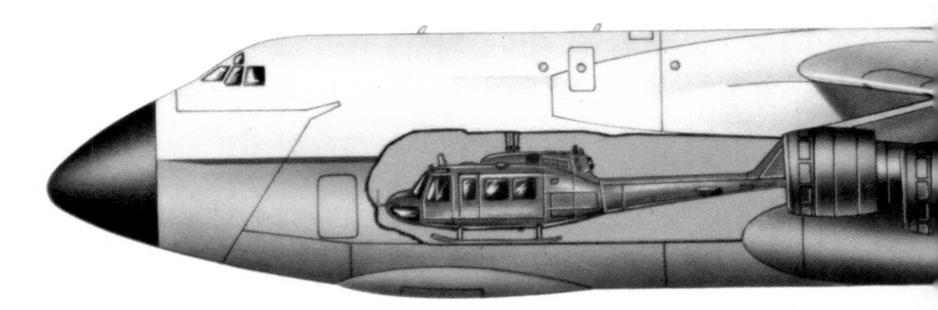

## Helicopter Transports

Transport helicopters are usually more powerful but shorter-ranged than equivalent aeroplanes, but their priceless asset of VTOL often makes them the only possible answer. Early helicopters could not carry more than two or three people in addition to the pilot. The Sikorsky S-55 (US Army H-19) of the Korean War period could easily carry seven troops, and the S-58 (H-34) more than doubled this number, as did the Vertol H-21 Work Horse. Then in 1957 the Soviet Mil bureau knocked everyone for six with the Mi-6, a monster capable of carrying 65 passengers or up to 12,200 kg (12 tons) of cargo. Hundreds are in use, many equipped to airlift assault teams with missiles and launch crews, anti-tank sections, or anti-aircraft troops with armoured vehicles carrying missiles or fast-firing AA guns. There are also small numbers of the Mi-10, a crane version able to carry 15,250 kg (15 tons), including items too big to fit in the Mi-6 fuselage, and the gigantic Mi-12 (V-12) with two rotors and a world-record helicopter payload of over 40,200 kg (40 tons) (*see* Chapter 23).

## The Future

Considerations of sheer cost make it unlikely that larger airlift transports than those now in use will appear in the immediate future, except possibly in the Soviet Union. But technology is constantly striding ahead, and six years ago the USAF asked for ideas from manufacturers on how best it should replace the evergreen C-130. Eventually two proposals were picked and evaluated in a prototype fly-off competition – although, for the present, there is no money in the US defence budget to enable either to be put into production.

Both these contenders, the Boeing YC-14 and McDonnell Douglas YC-15, are bigger than the C-130, have advanced and very powerful turbofan engines, and use extremely advanced high-lift systems with exhaust from the engines directed on to special wing flaps (*see* Chapter 7). Such integrated lift/thrust systems confer STOL performances on these new military transports, and will eventually come to be of great importance in the design of short-field airliners for civil use.

**Above** Progress in transport aircraft (drawings to same scale). The Junkers Ju 52/3m (upper left), slow, strong, and reliable, served the Luftwaffe throughout World War II. The Lockheed C-130 Hercules (upper right), which flew in 1954, had many times greater payload than the Ju 52, was three times as fast, and had four times its range. The gigantic Lockheed C-5A Galaxy (bottom) of 1968, offers global mobility. It is 75.6 m (248 ft) long, its cavernous interior has full-section doors at each end, and it weighs about 400 tons fully laden.

**Left** A Boeing KC-97 L tanker/transport of the 1950s serving the US Strategic Air Command. Its Flying Boom pod and the control fins (projecting behind tail) are clearly visible, as are the jet-booster under-wing pods.

**Right** Douglas transports: the C-74 Globemaster I (1946) had a large belly hoist and side-loading door; the C-124 Globemaster II (1950) had a much deeper fuselage, often with two decks (see broken lines), and clamshell nose doors; the C-133 Cargomaster (1956), with pressurized hold, had access via a rear ramp on most versions.

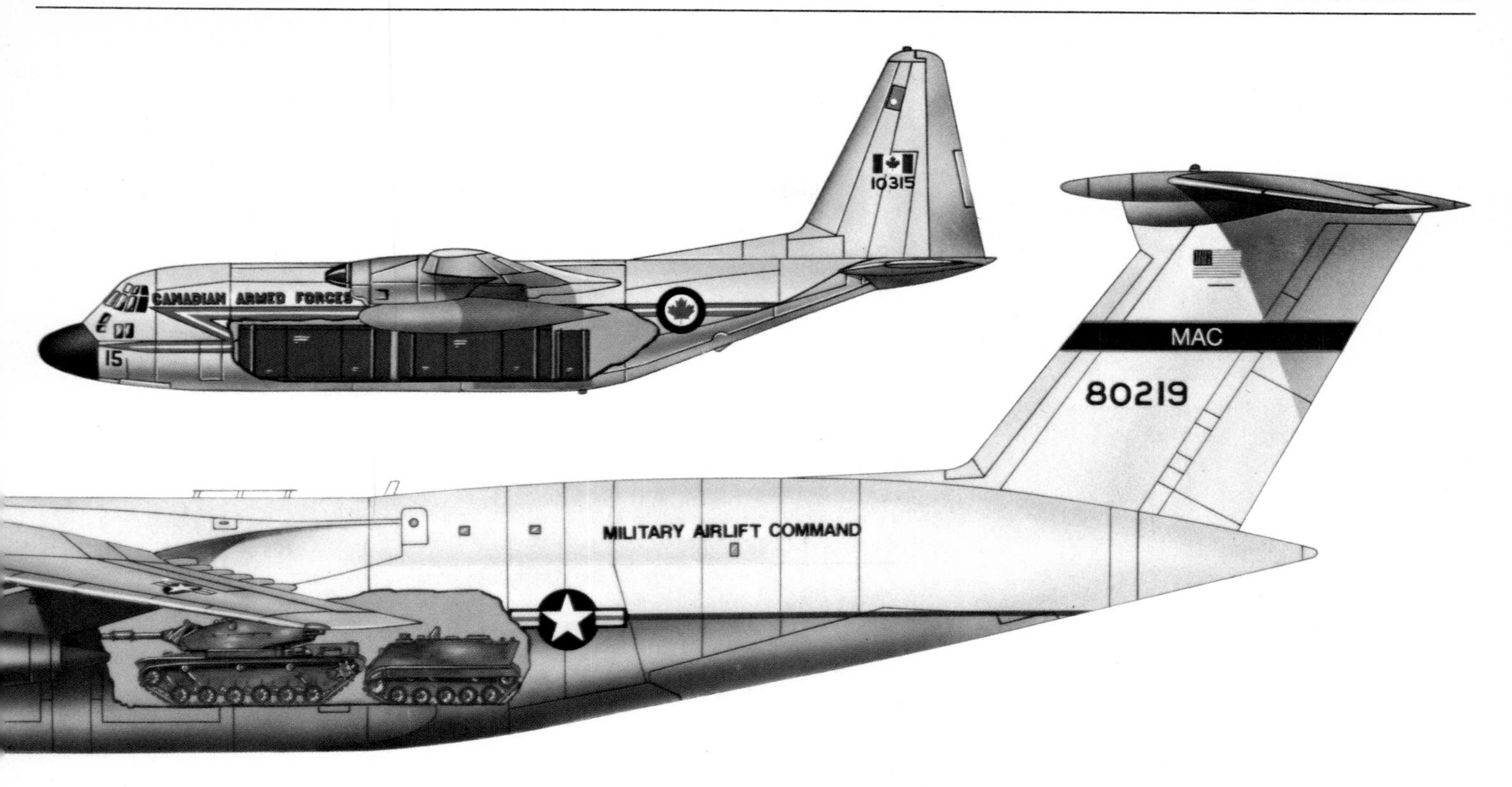
CANADIAN ARMED FORCES
15
10315
MAC
80219
MILITARY AIRLIFT COMMAND

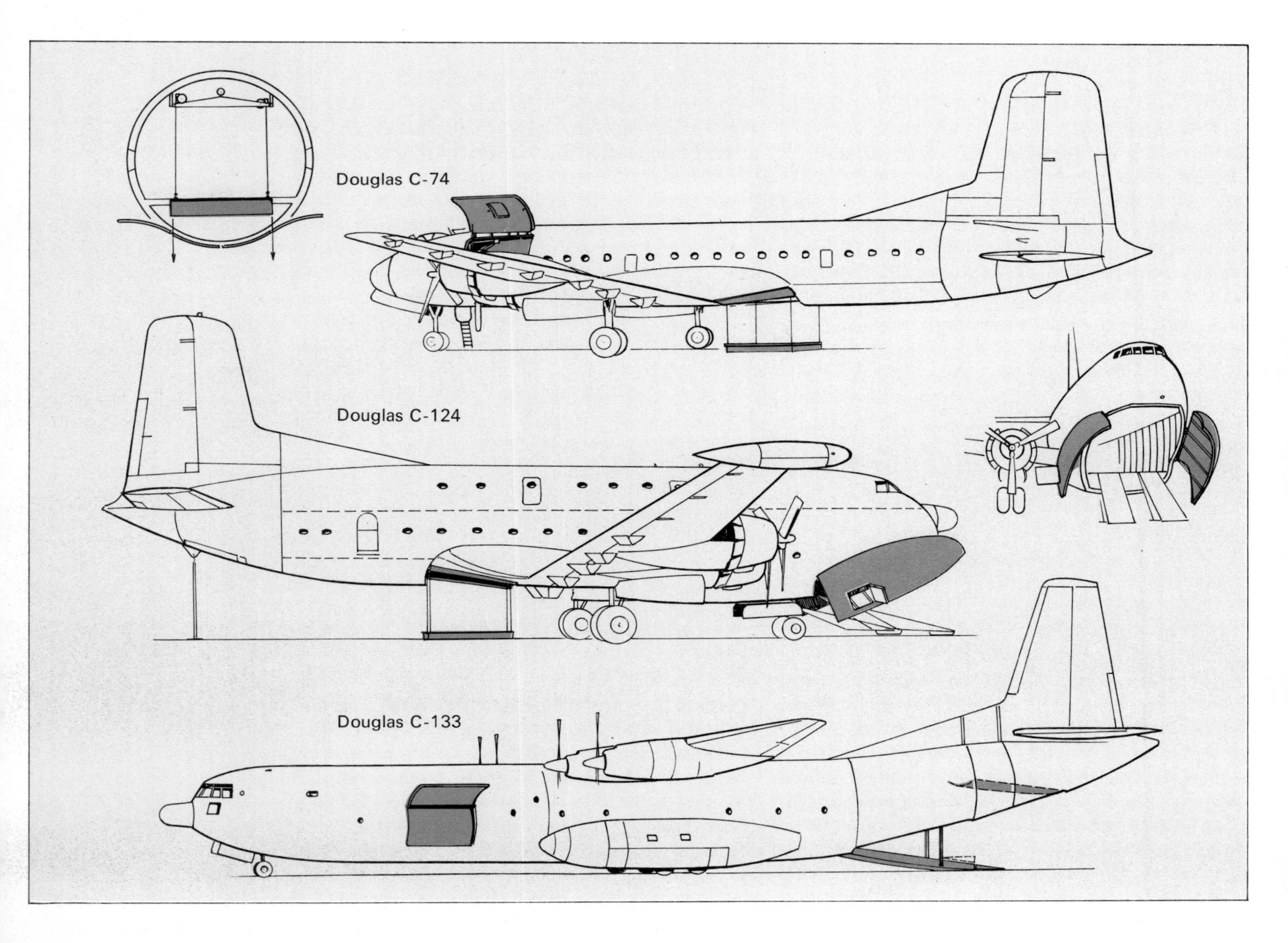
Douglas C-74
Douglas C-124
Douglas C-133

# 17 Trainers

In the early days of aviation a pupil was given a brief course of flying instruction on the ground, after which he climbed aboard the Wright, Voisin, Farman, or whatever, and squeezed in behind the pilot. Reaching round, he attempted to feel the motions of the 'stick' (control column) and to remember how far the pilot moved it, and in which direction, to achieve each desired result. Rudder-pedal movements often could not even be watched. In a few cases dual controls were provided side by side, pupil and instructor sitting on simple seats on the leading edge of the lower wing, and in this ideal arrangement the pupil could 'follow through' with all the controls while taking care not to interfere with them.

Yet pilot training in World War I was gravely deficient. It was, of necessity, skimpy and hurried. It was invariably done by instructor pilots who had been wounded or were being 'rested' from combat, and many of them were dispirited and uninterested in teaching others. There was no detailed syllabus or even a standard technique, each instructor simply having his own way of cajoling, bullying, or otherwise trying to get the mystified and frightened tyro to master a probably dangerous and temperamental flying machine. Casualties were terrible. It was left to one man, Robert Smith-Barry, to change the system. Badly wounded on operations at the start of the war, Smith-Barry complained to General (later Viscount) Trenchard that flying training ought to be better organized. The characteristic reply was, 'Go and do something about it'. So Smith-Barry wrote the first detailed scheme of pilot training, which has been the basis for all such training to this day. He chose specially designed trainer aircraft, instead of whatever happened to be available, with proper dual controls for instructor and pupil and handling qualities that made the pupil fly accurately without exposing himself to excessive risk. He also insisted that instructors were specially chosen and trained, and the RAF Central Flying School has ever since been the premier training unit in the world, qualifying instructors for more than half the world's air forces.

Undoubtedly the most important early trainer, in the pre-Smith-Barry era, was the Maurice Farman 'Longhorn', a large and ponderous pusher biplane that was almost totally at the mercy of the wind. After Smith-Barry the chief trainer was the trim Avro 504, a superb machine which also served as an important bomber, fighter, and anti-Zeppelin aircraft, and remained in production in various forms until 1933. The first purpose-designed trainer to be mass-produced was the 504J, which was the main type at Smith-Barry's School of Special Flying at Gosport, Hampshire, from which came all subsequent training methods. (The flying school lent its name to the Gosport tube, the speaking tube linking instructor and pupil which preceded the modern 'intercom'.) Also called the Mono-Avro because of its 100 hp *monosoupape* (single-valve) Gnome engine, the 504J was one of the most numerous versions of an aircraft built in larger numbers – more than 8,000 – in World War I than any other. It had a simple wooden structure with fabric covering, and the instructor sat in a separate cockpit behind the pilot. After the war the 504K, which could have any of a variety of rotary engines, soon became the most important model, giving way from 1927 onwards to the cleaned-up 504N or Lynx-Avro (from its Lynx static-radial engine), which led to later machines such as the Avro Tutor and the de Havilland Tiger Moth (of which almost 9,000 were built before and during World War II).

## Trainer Types

By the 1930s pilot training had been recognized as ideally calling for two types of trainer, a primary or *ab initio* type, followed by a more advanced machine to give experience of more powerful engines, higher performance, and possibly even gunnery and bombing. Curiously, the US Army Air Corps, while using designations prefixed PT for its primary trainers, chose BT (for basic trainer) to designate the more advanced type, at one time adding BC (for basic combat). Before World War II the basic category had been replaced by the more logical AT for advanced trainer, and production of trainers in the United States in 1941–5 exceeded any previous output of one class of aircraft in history. The total of some 72,000 included large numbers of multi-engined aircraft for training navigators, radio operators, gunners, and bombardiers, and for giving multi-engine instruction to pilots. Twin- or four-engined aircraft need special piloting techniques, especially in case of failure of one engine, which leads to a potentially hazardous asymmetric condition. In World War II this had to be demonstrated in the air; today it can be practised more safely in simulators (see below).

## Jet Trainers

During the decade following World War II military flying training introduced first jet advanced trainers and then jet *ab initio* types. Some jet trainers seat pupil and instructor side by side, where each can watch the other and gain confidence from close teamwork. Others position the instructor to the rear, as in the old days, giving the pupil the feeling he is on his own and realistically on the centreline, as in an

Three historic biplane trainers. **Below** The American Curtiss JN-4 (whence its nickname 'Jenny') was built in enormous numbers during World War I and was later used by postal services and 'barnstormers'; this one is a replica. **Inset** E3404 was first of a batch of 500 Avro 504K trainers built in 1918. Later models of the 504 series continued in production until 1933. **Right** The best-known British successor of the 504 was the de Havilland 82 and 82A Tiger Moth, in which about half of all the Commonwealth pilots of World War II learned to fly. It was finally retired from the RAFVR in 1951.

14

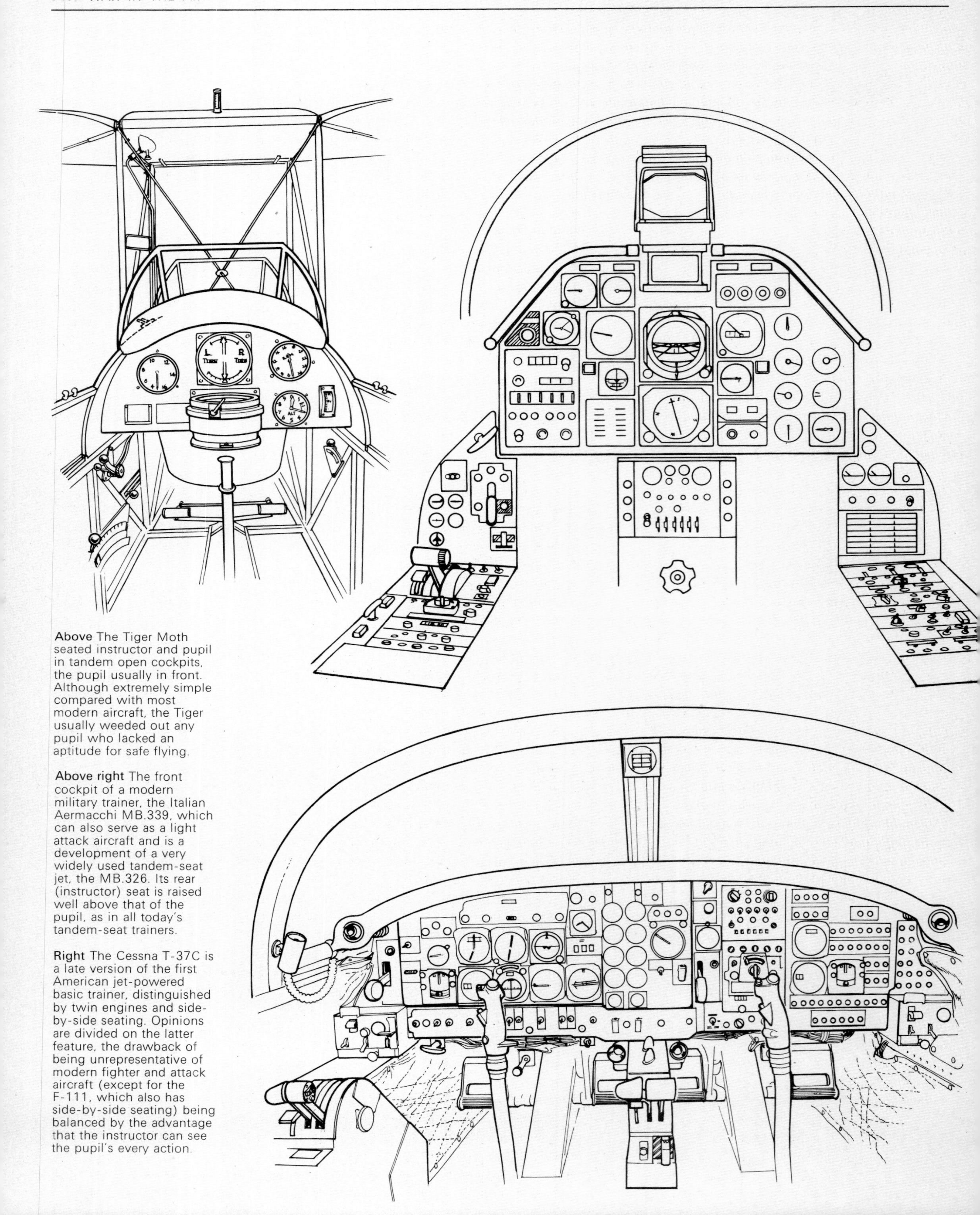


**Above** The Tiger Moth seated instructor and pupil in tandem open cockpits, the pupil usually in front. Although extremely simple compared with most modern aircraft, the Tiger usually weeded out any pupil who lacked an aptitude for safe flying.

**Above right** The front cockpit of a modern military trainer, the Italian Aermacchi MB.339, which can also serve as a light attack aircraft and is a development of a very widely used tandem-seat jet, the MB.326. Its rear (instructor) seat is raised well above that of the pupil, as in all today's tandem-seat trainers.

**Right** The Cessna T-37C is a late version of the first American jet-powered basic trainer, distinguished by twin engines and side-by-side seating. Opinions are divided on the latter feature, the drawback of being unrepresentative of modern fighter and attack aircraft (except for the F-111, which also has side-by-side seating) being balanced by the advantage that the instructor can see the pupil's every action.

operational fighter. The syllabus and techniques have hardly changed, and of course include demonstration of the effects of the controls, all the normal modes of flight, recovery from spinning and other abnormal situations, aerobatics, night flying, and cross-country flying. Possibly half the early flight-time is spent on circuits of the airfield and in achieving safe and consistent landings. The first solo might come after 4 to 10 hours of flight instruction, and qualification at 'wings' standard after 200 hours or more. If anything, modern jets can be worked harder around the clock than the much smaller and less powerful piston-engined machines of the past.

Today's military jet trainers differ from their predecessors in that if the seats are in tandem the instructor is raised at least a foot above the pupil, giving him a better view, especially on the landing approach when the aircraft will fly in a nose-up attitude. Both occupants sit in ejection seats, which need a course of instruction in themselves, and most trainers are equipped with at least a simple sight, a fixed gun, and racks for light bombs or rockets. Many trainers are powerful enough to double as light attack aircraft, and important types such as the USAF Cessna T-37, RAF Hawker Siddeley Hawk, Luftwaffe Alpha Jet, and Swedish Saab 105 are built in special attack versions. One trainer, the Northrop T-38 Talon, is supersonic; so is the tandem-seat SEPECAT Jaguar, but the latter is used not as a mere trainer but as a combat-ready conversion trainer for the single-seat Jaguar. Nearly all single-seat attack and fighter aircraft are today also made in two-seat forms as conversion trainers.

A slight hiccup in the normal routine of trainer aircraft was introduced in the late 1950s by the concept of limited wars or 'brush-fire' conflicts, which appeared to call for combat aircraft of limited power and performance suited to austere airstrips in primitive environments. Possibly the concept itself was faulty, in that today there are no 'primitive' enemies except possibly urban guerrillas, against whom aircraft have had to discover new roles. But 20 years ago there was a rush to equip trainers with guns, rockets, casevac (casualty evacuation) equipment, and such devices as loudspeakers and leaflet dispensers. Favoured limited-war aircraft included the North American T-28, which was the last of the powerful piston-engined trainers and 'son' of the AT-6 Texan (called Harvard by the British), which with over 20,000 built in many versions was the most numerous trainer of all time. Even the old AT-6 itself was used in counter-insurgency and para-military roles by several countries, notably France.

## Flight Simulators

Almost 50 years ago an American, Edwin A. Link, invented a simple device intended to help a pupil learn to fly on the ground. By 1935 it was launched and in production by Link's own company, and many thousands were made in World War II. Then popularly called the 'Blue Box', because of its colour, it consisted of a pilot cockpit mounted on a system of pivoted rods connected to air-operated jacks. Movement of the pilot's stick to the left caused the blue box to bank to the left, pulling back caused the nose to rise, and pushing on the right rudder pedal made the device yaw to the right. All the while the readings on the instruments indicated what might be expected in a real aircraft, governed by aircraft attitude and

The Hunting Provost, developed from the Percival P.56 Provost designed just after World War II, was the RAF's last piston-engined trainer. Powered by an Alvis Leonides engine, it seated instructor and pupil side-by-side.

**Left** Although originally based on the Hunting Provost, the Hunting Jet Provost, powered by a Rolls-Royce Viper turbojet, has evolved into a quite dissimilar aircraft.

the position of a simulated throttle lever looking like the real thing. Even the earliest Link Trainer was a tremendous help because it allowed a pilot to gain the experience needed to enable his reactions to become instantaneous, without exposing him to the hazards of flight.

Before World War II the basic Link had become better and more complex. The computing was still done mechanically, or pneumatically, but optional accessories were available which enabled the installation to be used for teaching or practice in navigation, the Lorenz beam approach (*see* Chapter 26) and later ILS; now most flight operations can be simulated. The first navigation adjunct was a small device called a 'crab', which ran on three tiny wheels across a large glass-covered map table under the eye of an instructor. The driving wheel was governed in speed by the supposed true air speed of the aircraft, taken in conjunction with any desired 'wind' fed into the apparatus by the instructor, and in its direction by the aircraft course or heading (again adjusted for the effect of wind). The crab could be made to carry a fine pen to trace its course, and by 1941 military pilots were sweating in Link rooms trying to draw perfect flight and navigation patterns.

Today the simple Blue Box has matured into the *simulator*, often a million-pound package (in terms of both weight and cost) occupying a large building and needing a large staff, yet still able to show vast savings over the course of its life compared with use of the actual aircraft. Some simulators are general installations to teach multi-engine techniques, radars, navigation, and similar subjects (notably ASW: *see* Chapter 14). Most are precisely representative of particular types of aircraft, and their cockpit or flight deck is indistinguishable from the real thing. If the real aircraft has a crew of five, so does the simulator. Thanks to modern electronics, every one of possibly hundreds of dials or controls works just as in the real aircraft. The whole flight deck may be mounted high above the ground on monster rams so that it can move up and down, tilt or rotate in any direction, move from front to rear or vice versa to simulate acceleration, and comes complete with the exact noises of engines, slipstream, even the squelch of the tyres on a runway.

**Below** Most modern fighter and attack aircraft are also available in two-seat dual-control versions for pilot training. This SEPECAT Jaguar T.2 seats two pilots in tandem but is otherwise basically similar to the RAF GR.1 single-seater.

Outside the windows or canopy there can now be a real scene that exactly simulates the expected world outside. In order to make the scene precisely reproduce the 'aircraft motion' it is created by a television camera which is driven across a gigantic model of the airfield and its surroundings according to the pilot's commands. Usually the model is up-ended and arranged in the vertical plane, and the camera picture is amplified and processed so that it exactly fills the pilot's field of view. It is possible, for example, to practise take-offs and landings, using all the automatic systems and blind-approach equipment both on the airfield and in the aircraft.

**Above** Its odd exterior gives little hint of the complexity of this TriStar simulator built by Redifon. It precisely reproduces the characteristics of the Lockheed airliner, and gives the crew a realistic exterior view.

# Civil Aviation

# 18 The Pioneers

The rigid airship LZ 7 *Deutschland* was the first aircraft designed for the carriage of passengers. Like other Zeppelins it had a light-metal structure and was fabric covered. The forward car, or gondola, contained the controls and a 125 hp Daimler engine which drove two propellers; the rear gondola housed two more engines. The 24-passenger cabin can be seen amidship in the V-shaped keel. The 18 gas cells had 19,300 m³ (681,574 cu ft) capacity, and the gross lift was about 20,865 kg (46,000 lb). Normal cruising speed was 53 km/h (33 mph).

Sustained development of air transport began in 1919, only a few months after the end of World War I, but there had been two earlier attempts at operating passenger air services. In Germany Count Ferdinand von Zeppelin had designed and built a series of rigid airships, and on 16 November 1909 Deutsche Luftschiffahrts AG (Delag) was founded, as the world's first airline, to operate passenger flights with Zeppelins and to train Zeppelin crews.

Delag's first Zeppelin, LZ 7 *Deutschland*, was 148 m (485 ft) long. It made its first flight on 19 June 1910 and was the first powered aircraft to carry passengers. Delag planned to operate a network of airship services in Germany and, later, to other countries. Crews were trained, sheds built, and maps of the routes published, but no services were operated until after the war. Seven Zeppelins, however, were used to carry passengers on limited voyages, and by August 1914 they had made 1,588 flights and carried 33,722 passengers.

The world's first scheduled, fare-paying service was operated in Florida by St Petersburg-Tampa Airboat Line. On 1 January 1914 a single-engined Benoist 75 hp flying-boat flown by Tony Jannus left St Petersburg for the 23-minute crossing of Tampa Bay to Tampa. The service, which was subsidized by the city of St Petersburg, operated until the end of March 1914 and carried 1,204 passengers.

No other civil airline operations took place until after the war, but some military air services were operated by the French, German, and Italian armed forces, and an Austro-Hungarian military air-mail service was opened between Vienna and Kiev (Soviet Union) on 20 March 1918.

## Post-War European Airlines

Britain operated a number of military services and on 10 January 1919 began regular London–Paris flights in connection with the Peace Conference. These continued until that September, with 91 per cent regularity, and 934 passengers were carried, mostly in single-engined de Havilland 4s and twin-engined Handley Page O/400s.

The first British airline was George Holt Thomas's Aircraft Transport & Travel (AT & T). It was founded in 1916 but could not operate commercial services until 1919, when the necessary agreements covering international air transport had been made. On 25 August 1919 AT & T opened regular London–Paris services when Major Cyril Patteson flew four passengers in a single-engined D.H.16 biplane from Hounslow (near the present Heathrow) to Le Bourget in 2 hr 25 min. On the same day Handley Page Transport (HPT) made a special flight over the route but did not begin regular service until 2 September. HPT used a fleet of O/400-type converted bombers, with enclosed cabins and seating for two passengers in an open cockpit in the nose. AT & T and HPT opened several cross-Channel services, and AT & T flew the first London–Amsterdam service in collaboration with KLM Royal Dutch Airlines on 17 May 1920.

The third British airline was the Instone Air Line, which began private services in 1919 and its first public London–Paris service in February 1920. This airline extended to Brussels and Cologne, using the famous Vickers Vimy Commercial *City of London*, a passenger adaptation of the type which had made the first non-stop crossing of the North Atlantic.

AT & T ceased operation at the end of 1920 but a successor, Daimler Airway, began London–Paris services on 2 April 1922, and later flew as far as Berlin. Daimler introduced the eight-passenger, single-engined D.H.34 which made its first flight only *one week* before entering regular service.

Before Daimler appeared the British airlines had ceased operation for nearly three weeks owing to lack of finance and competition from French airlines, and a temporary scheme of Government subsidies had to be introduced. Further difficulties enforced a new scheme whereby Handley Page, Instone, and Daimler were subsidized to operate specific routes; and the subsidy applied also to British Marine Air Navigation when it opened a Southampton–

**Previous two pages** BAC/Aérospatiale Concorde on its take-off run.

Guernsey service with Supermarine Sea Eagle flying-boats in September 1923. But the continuing financial problems were ultimately resolved only on 31 March 1924, when the British Government established Imperial Airways to take over the four existing airlines.

In those early days conditions were quite primitive. Airports were grass fields of modest size and frequently muddy, most flying was confined to daylight, and few 'airliners' were equipped with radio. Engines were water-cooled and unreliable – on one occasion a London–Paris aircraft made more than a dozen forced landings en route – and the low speed

**Above** The British airline Aircraft Transport & Travel inaugurated the first cross Channel air services on 25 August 1919 with a fleet of two-passenger de Havilland 4A and four passenger de Havilland 16 single-engined biplanes. One of the D.H. 4As is seen here at Hounslow, London's terminal airport until March 1920.

**Left** The interior of a Farman Goliath of the 1920s seen from the back of the rear cabin, which normally had eight passenger seats. The pilot's seat, with open cockpit, was beyond the bulkhead on the left. To its right was the four-seat nose cabin, which provided a superb view for its passengers. The diagonal fuselage bracing is clearly visible.

**Below** A Grands Express Goliath taxi-ing across Plough Lane at Croydon airport in the early 1920s.

and short range of the aircraft meant that weather could be a major obstacle. Nevertheless, regularity and safety were of quite a high standard, although passenger comfort left much to be desired. The number of passengers was small: in the financial year 1921–2 just over 11,000 passengers crossed the Channel by air, and cargo loads were equally modest.

The most active nation in those formative years was France. Numerous airlines were launched, a daily Paris–Lille service was begun as early as 1 May 1919, and Paris–London services began on 16 September that year. Single-engined Breguet 14s were among the most widely used aircraft, but twin-engined 12-passenger Farman Goliaths soon appeared on the more important routes. Considerable effort went into establishing trans-Mediterranean flying-boat services linking France with its North African possessions. But France's great ambition was to open air services to South America and the Far East.

Germany opened Europe's first daily civil passenger air service when Deutsche Luft-Reederei (DLR) began flying between Berlin and Weimar on 5 February 1919. Using mainly L.V.G. C VI single-engined biplanes with room for two passengers in the open rear cockpit, DLR built up a large network of routes. Of many German airlines founded about this time the two most important were Deutscher Aero Lloyd and Junkers-Luftverkehr (an offshoot of the aircraft manufacturer), and in 1926 these merged to form Deutsche Luft Hansa.

**Below** The Junkers-F 13 played an important part in the development of civil aviation as the first all-metal cantilever monoplane transport. Its basic features, including corrugated skin, were continued by Junkers up to the Ju 52/3m. Luft Hansa's *Nebelkrähe* (Hooded Crow) is seen here fitted with skis.

**Left** *Hannibal* was one of eight large Handley Page 42/45 biplanes built for Imperial Airways in 1930–1. Powered by four 490/555 hp Bristol Jupiter engines, these aircraft had accommodation for 38 passengers on European routes and 24 on the trunk routes. Their well-appointed cabins and inflight meals service set new standards in air travel. *Hannibal* is seen here at Khartoum (Sudan) on the England–South Africa route, when the type worked the Cairo–Kisumu (Kenya) sector.

Many of the early German services were flown by four-passenger all-metal Junkers-F 13 single-engined monoplanes, and this type played a major part in establishing air transport in much of Europe, South America, and elsewhere. The F 13 was the first all-metal passenger aircraft; it first flew in June 1919, more than 300 were built, and the famous Ju 52/3m of the 1930s was a direct descendant.

Elsewhere in western Europe famous names were beginning to make their mark: KLM Royal Dutch Airlines was founded in October 1919; DDL (Danish Air Lines, one of the constituents of the present SAS) opened its first service, from Copenhagen to Warnemünde, in August 1920; and in Belgium SNETA paved the way for the formation of the national airline Sabena in 1923. Lack of aerodromes and difficult terrain forced many of the early airlines to use seaplanes or flying-boats. The original Swedish, Finnish, Norwegian, and Italian air services were operated by marine aircraft, and seaplanes also worked some Hungarian services along the Danube river.

## The Trunk Routes from Europe

The outstanding achievements of air transport between the wars were the establishment of the great trunk air routes and of the US transcontinental air services. First to operate in the tropics was SNETA, which began flying-boat services in the Belgian Congo (now the Republic of Zaïre) in July 1920. That was the first step towards a comprehensive network of Congo air services; Sabena's link between Belgium and the Belgian Congo was not established until 1935, although the first flight had been made 10 years earlier.

Imperial Airways' main task was development of services linking the United Kingdom with its territories overseas. Routes were surveyed and aircraft designed and built with the initial aim of a London–India service. As early as June 1921 the RAF had opened the Desert Air Mail route between Cairo and Baghdad, using mostly twin-engined Vickers Vernons, and Imperial Airways took over this route and extended it south-eastwards to Basra. A fleet of

This sectional drawing of the Imperial Airways Short S.23 C-class flying-boat *Canopus* shows the original interior layout. As mail loads increased, passenger accommodation was reduced and the forward smoking cabin became a mail hold. The flying-boats were not used as sleeper aircraft, but sometimes a lower berth was made up in the midship cabin for the convenience of any passenger who was unwell. The windows in the port side of the promenade cabin were higher than those in the starboard side so that passengers could stand and watch the view.

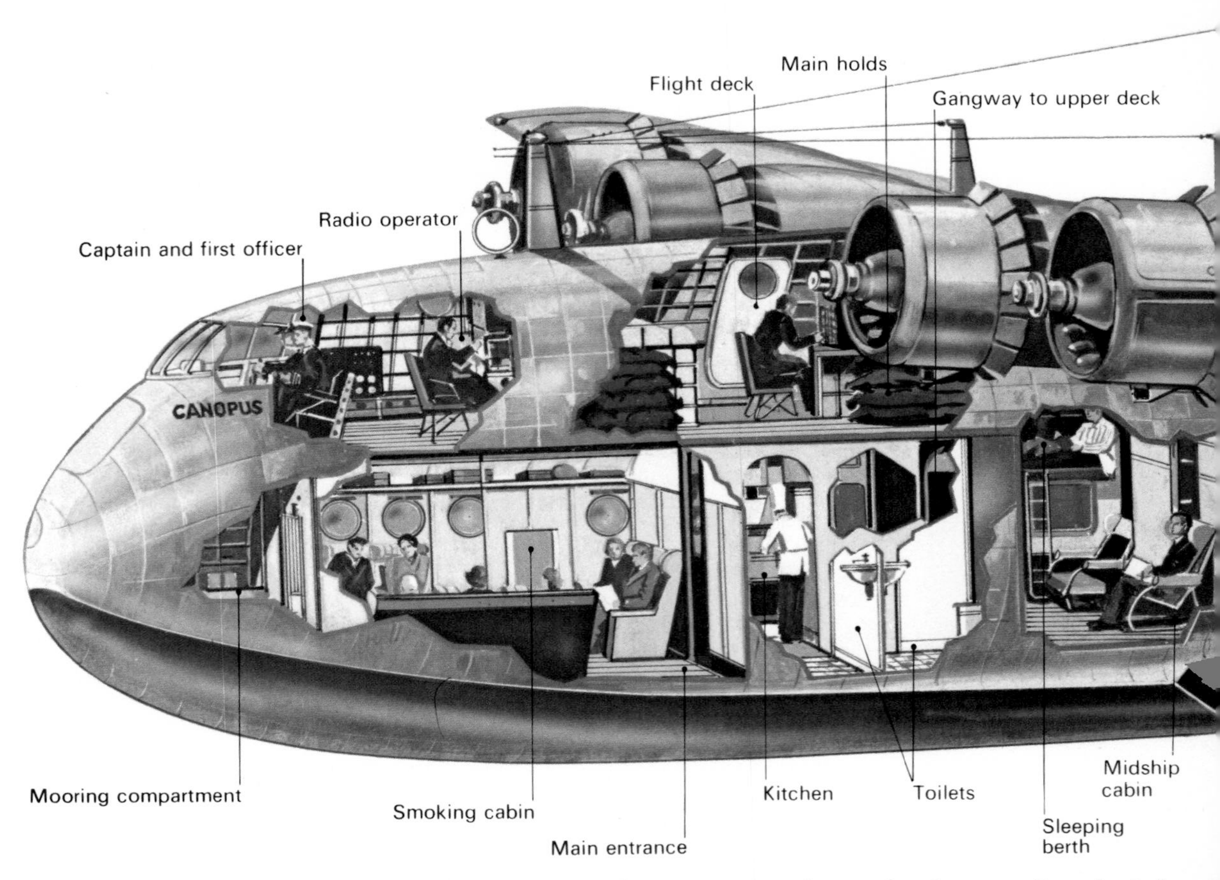


three-engined de Havilland Hercules biplanes was built for this task, and the first service left Basra for Cairo on 7 January 1927.

Political and technical difficulties delayed the extension to India and the link between England and Egypt, but on 30 March 1929 the Armstrong Whitworth Argosy *City of Glasgow* left London to fly the first sector, via Paris and Basle, of the inaugural through mail to India – although mail and passengers had to go by train between Basle and Genoa. Short Calcutta three-engined flying-boats worked the trans-Mediterranean sector, and the Hercules flew from Cairo to Karachi (in what is now Pakistan). The entire journey took seven days. The service was extended to Delhi that December, to Calcutta in July 1933, and to Singapore in December 1933 – four-engined Armstrong Whitworth Atalanta monoplanes working east of Karachi. The four-engined Handley Page *Hannibal*-class biplanes had by then taken over the Egypt–India sector. Finally, in December 1934, the mail route was extended to Australia, with passengers being carried from April 1935; the Australian Qantas Empire Airways was responsible for operating the Singapore–Brisbane section. The England–Australia schedule was 12½ days.

The second British trunk route was to Africa. The nature of the terrain and high altitude of landing grounds made this difficult for the aircraft of the period, but the first stage from Egypt to Mwanza, on Lake Victoria, was opened with the departure from London on 28 February 1931. Passengers were initially carried only as far as Khartoum (Sudan) where Calcutta flying-boats took over from the Argosies to fly the mail southwards. The entire journey took 10 days. In January 1932 the mail service was extended to Cape Town, with passengers being carried from April.

West Africa was reached in February 1936 with the opening of a branch service from Khartoum to Kano (northern Nigeria), later extended to Lagos and Accra; and in March 1936 Hong Kong was linked by a branch from the Australia route – first from Penang (Malaysia) and later from Bangkok (Thailand). These two extensions to the trunk routes were operated by D.H.86 *Diana*-class biplanes, and Qantas used similar aircraft on the Brisbane–Singapore services.

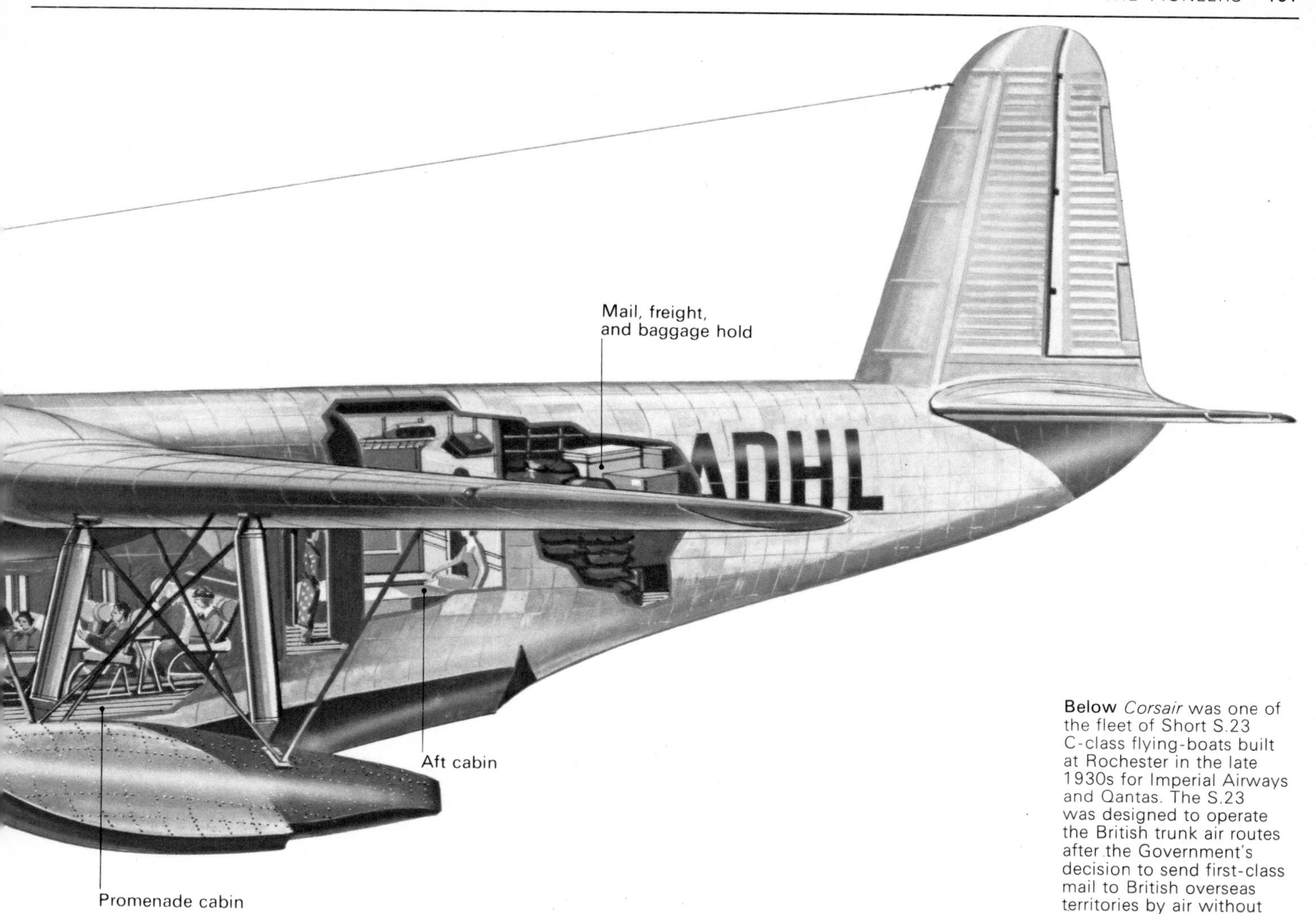


**Below** *Corsair* was one of the fleet of Short S.23 C-class flying-boats built at Rochester in the late 1930s for Imperial Airways and Qantas. The S.23 was designed to operate the British trunk air routes after the Government's decision to send first-class mail to British overseas territories by air without surcharge. The last service operated by C-class 'boats was in December 1947.

In 1934 the British Government decided to send all possible first-class mail to British territories by air without surcharge, and a fleet of Short C-class flying-boats was built for Imperial Airways and Qantas. Beginning operation late in 1936, they flew right through from Southampton to Durban from June 1937 and to Sydney from June 1938. One of the C-class 'boats, the *Cavalier*, began working a Bermuda–New York service in June 1937; and at the end of that year another, the *Centaurus*, was used on the first survey flight to New Zealand.

Although KLM established a network of European services, its finest early achievement was to develop a trunk route to Batavia (now Jakarta) in Java – an achievement made possible by another great Dutch enterprise, Fokker. This outstanding aircraft manufacturer built a series of single-, twin-, three-, and four-engined monoplanes with cantilever all-wooden wings that were soon to be used by many of the world's airlines.

The first KLM flight to the Far East took place in 1924 with a single-engined Fokker F.VII; another carried a paying passenger from Amsterdam to Batavia and back in 1927, and

Berlin's Tempelhof Airport in the 1930s. The Junkers-Ju 52/3m (several are visible in the centre of the picture) first flew in April 1932 and soon became one of the most widely used European transport aeroplanes, for many years forming the main fleet of Lufthansa. A number of earlier single-engined Junkers-F 24s and W 34s can also be seen, and in the foreground is Lufthansa's less successful Junkers-Ju 86 *Brocken*, which was powered by heavy-oil engines.

proving flights followed in 1928 and 1929. Then on 1 October 1931 a three-engined F.XII left Amsterdam to open regular service. It had luxurious seats for four passengers and worked to a 10-day schedule, with 81 hours of flying on what was then the world's longest air route. In 1934 came the first of the new breed of modern, all-metal low-wing monoplanes to be used by KLM. This was the Douglas DC-2, which began by taking second place in the Mac. Robertson England–Australia race, carrying three passengers and flying 19,680 km (12,300 miles) to Melbourne in just over 90 hours total elapsed time. It operated to Batavia from June 1935 and cut the schedule to six days.

France was concerned with establishing a route to the Orient at a very early date, and founded Cie Franco-Roumaine de Navigation Aérienne (later re-named CIDNA) on 23 April 1920. On 20 September that year the airline opened a Paris–Strasbourg service, extending it to Prague that October and to Warsaw in the summer of 1921. By a series of further extensions the route was opened to Constantinople (Istanbul) by October 1922. This was a very difficult route for the early aircraft, involving flight over high mountains and through some of Europe's worst weather. Much of the flying was done with single-engined Potez biplanes, but after the introduction of three-engined Caudron C.61s night flying began on some sectors from September 1923. Plans were made to extend the route as far as Baghdad; this extension was in fact operated by Air Orient over a different route and the entire Eastern route was opened to Saigon (in modern Vietnam) in January 1931.

France's great pioneering effort to establish an air service to South America was vividly described in the aviator-novelist Antoine de Saint-Exupéry's *Vol de nuit* (1931; Night Flight). The first survey flight, between Toulouse and Barcelona, was made in December 1918, and by April 1920 a mail service had been opened between Toulouse and Casablanca (Morocco). Services were worked under the title Lignes Aériennes Latécoère; the name was changed to Cie Générale d'Entreprises Aéronautiques in 1921 and to Cie Générale Aéropostale in 1927; but to many it was known simply as La Ligne (The Line).

Extension of the route from Casablanca to Dakar (Sénégal) involved a long desert crossing and a harsh climate, and hostile tribesmen were a constant danger to crews whose Breguet 14s were forced to land in the desert. In spite of such hardships La Ligne was opened to Dakar in June 1925.

Aéropostale and its associates in South America prepared aerodromes and installed radio stations along the east coast of South America, and in November 1927 opened the route between Natal (northern Brazil) and Buenos Aires. The dream of a mail service linking France and South America became reality on 1 March 1928, when the entire route was opened with an eight-day schedule, although the Dakar–Natal sector still had to be worked by fast ships. In July 1929 the route across the sub-continent to Santiago (Chile) was pioneered by brave men struggling over the towering Andes in all weathers in single-engined, open-cockpit biplanes.

Jean Mermoz made the first flight across the South Atlantic for Aéropostale on 12–3 May 1930 in a Latécoère 28 single-engined seaplane, taking 21 hours. But the Laté 28, although widely used by then as a landplane, was unsuitable for transoceanic work, and a whole series of landplanes and flying-boats was built for the South Atlantic. Regular, though infrequent air crossings began in May 1934 with the three-engined Couzinet *Arc-en-Ciel* (Rainbow). By that time Air France had been founded as the national airline and had taken over the pioneer lines, including Aéropostale. The most famous of the South Atlantic aircraft was the Latécoère 300 four-engined flying-boat *Croix du Sud* (Southern Cross), which made its first ocean crossing on 3 January 1934. In December 1936 the flying-boat, Mermoz its commander, and the crew were lost without trace.

Germany meanwhile was catering to its many interests in South America and had used two methods of establishing air services over the route. After proving flights the LZ 127 *Graf Zeppelin* airship began regular passenger services between Germany and Brazil on 20 March 1932 – the world's first transoceanic passenger air service. Operations continued with great regularity until May 1937, when all Zeppelin services ceased following the loss of the LZ 129 *Hindenburg* at Lakehurst, New Jersey, at the start of its second season of North Atlantic crossings.

The second German method was to catapult seaplanes and flying-boats from depot ships at sea. Germany had used this method for speed-

**Below** The Boeing 247 was the first of the modern all-metal, smooth-skinned, cantilever monoplane transports with retractable undercarriage, entering service in 1933. This preserved example has United Air Lines' livery of the period.

**Inset, left** Most early United States' air transport was concerned with mail carriage. Here mail is unloaded and checked – with the usual security – from a National Air Transport Douglas M-2 at Hadley Airport, New Jersey, in 1927.

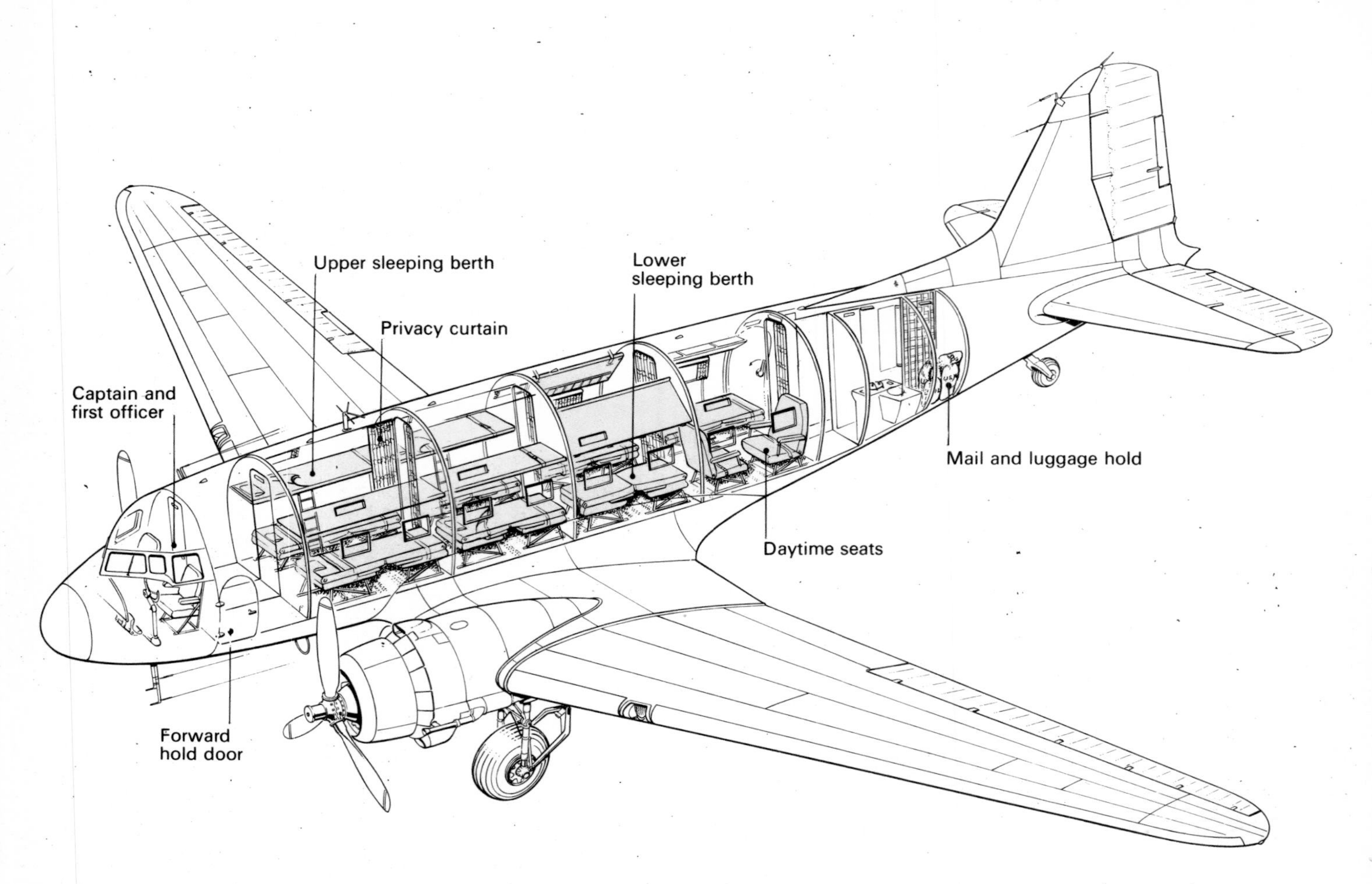


The Douglas DC-3, of which 10,655 were built in the United States, was originally conceived as the Douglas Sleeper Transport (DST), with 14 sleeping berths, as in the drawing above, or 21 daytime seats. Forty DSTs were built for American Airlines and United Air Lines, and the first US trans-continental sleeper service with DSTs began in September 1936. The DSTs and most early US-operated DC-3s had the passenger door on the starboard side.

ing mail deliveries between Germany and New York, catapulting seaplanes from the liners *Bremen* and *Europa*, and for the South American service she used twin-engined Dornier Wal flying-boats from depot ships off West Africa and South America. Regular service began in 1934, the overall Berlin–Buenos Aires schedule being four to five days. Hundreds of ocean crossings were made by Wals, Do 18s, Do 26s and Blohm und Voss Ha 139 seaplanes before war brought the service to an end.

Outside Europe the Italian airline Ala Littoria concentrated on services to Italian territories in North and East Africa, but also made proving flights to South America. Another Italian airline, LATI, inaugurated a Rome–Rio de Janeiro service in December 1939 with three-engined Savoia Marchetti S.M.83s. The service was withdrawn two years later.

## US Transcontinental Services

The development of air services in the United States differed from that in Europe. Concentration was on mail services, and no regular sustained passenger services were operated until 1925.

The task of building up a nationwide air mail service began with an Army service between Washington and New York on 15 May 1918. This was taken over by the US Post Office that August, and late that year the USPO acquired a large fleet of war-surplus aeroplanes, including more than 100 US-built D.H.4 single-engined biplanes. The principal target was a coast-to-coast service, and on 15 May 1919 the Chicago–Cleveland sector was opened as the first stage. The entire route between New York and San Francisco was in operation by 8 September 1920, and there was a number of branch routes. Flights were confined to daylight, but it was soon realized that to take full advantage of the aeroplane it was essential to fly by night as well. Aerodromes were illuminated and beacons installed every three miles along the entire route, and by the end of 1925 the night service was ready for operation.

The Kelly Act of 1925 and the Air Commerce Act of 1926 led to private companies running the services and to the regulated operation of civil air transport. On 7 October 1925 the first five air-mail contracts were awarded, and in 1926–7 12 airlines began feeder services to and from the transcontinental route. In 1927 Boeing Air Transport took over the San Francisco–Chicago sector and National Air Transport the Chicago–New York sector. Boeing encouraged passenger traffic from the start, and from that time passenger and mail services grew rapidly. There were numerous airlines and a wide variety of aircraft – the all-metal Ford Tri-

Motor making a major contribution to the development of passenger traffic.

The 'Big Four' airlines finally emerged – American, Eastern, TWA, and United – of which three were concerned with the transcontinental route. Provision of aircraft for this route was of major importance and this led in 1933 to the introduction of the Boeing 247, the first of the modern all-metal monoplanes with retractable undercarriages. The 247 was followed by the Douglas DC-2 and then in 1936 by the DC-3 – one of the most outstanding transport aeroplanes, of which well over 10,000 civil and military examples were built.

United States overseas operations were mainly in the hands of Pan American Airways, which flew its first service, between Key West (off the southern tip of Florida) and Havana (Cuba), in October 1927. Pan American opened air routes throughout the Caribbean and (with associates and subsidiaries) Latin America before setting about the enormous task of opening trans-Pacific services. This involved very long over-water stages, and specially built Martin M-130 flying-boats were acquired. The first, *China Clipper*, inaugurated the mail service when on 22 November 1935, it left San Francisco for Honolulu, Wake Island, Guam and Manila (Philippines) – flying the 12,525 km (7,783 miles) in 59 hr 48 min. Passengers were carried from October 1936.

Following joint proving flights with Imperial Airways from 1937, Pan American became the first airline to operate heavier-than-air services across the North Atlantic. The large Boeing 314 flying-boat *Yankee Clipper* began mail services between New York and Lisbon and Marseilles on 20 May 1939 and between New York and Southampton over the northern route on 24 June. Passenger services began over these routes on 28 June and 8 July respectively.

The first significant Russian participation in air transport was the opening in 1922 of the joint German–Russian Deruluft air service between Königsberg (now Kaliningrad) and Moscow, using single-engined Fokker F.III monoplanes. In 1923 Dobrolet, Ukrvozdukhput, and Zakavia were founded and they opened air routes in various parts of the country using mainly German aircraft. In 1925 Zakavia was taken over by Ukrvozdukhput, and in 1930 the latter was merged with Dobrolet and reorganized as Dobroflot. Further reorganization in 1932 brought Aeroflot into being.

Considerable development took place, mainly with small aircraft, and by 1935 Aeroflot was carrying more than 100,000 passengers a year. Most of the aircraft were single-engined, but there were small numbers of two- and three-engined Tupolev ANT-9s, twin-engined ANT-35s and a few Douglas DC-3s, the last two types being used when Aeroflot opened a Moscow–Riga–Stockholm service in July 1937. By the end of 1940 the airline had a route network of some 146,300 km (90,905 miles) and was carrying nearly 359,000 passengers and about 45,000 tonnes of mail and freight a year.

In other parts of the world considerable development took place. Australia established a first-class system of air services, as did New Zealand; Tata Air Lines (later Air-India) and Indian National Airways linked most of the main cities on the sub-continent; numerous airlines were operating a wide range of services in Canada, and by 1938 Trans-Canada Air Lines (now Air Canada) had established transcontinental services; most South American countries had air services; and Japan had an air-route network. In China airlines were developed with German and United States assistance. Misrair was operating from Egypt. Kenya had Wilson Airways, Rhodesia RANA, and the biggest African network of all was that of South African Airways.

Pan American Airways' Martin M-130 flying-boat *China Clipper*, the first of the type, inaugurated the American trans-Pacific services between San Francisco and the Philippines. It carried mail from November 1935 and passengers from October the following year

# 19 Mature Growth

The unpressurized Douglas DC-4 flew in February 1942 and was put into war service as the C-54 Skymaster. It became available to the airlines in 1945, when American Overseas Airlines introduced DC-4s between New York and the United Kingdom. A total of 1,242 military and civil examples were built and many saw airline service, initially mostly with 44 seats but later some with as many as 86. The Airlines of New South Wales DC-4 seen here is at Sydney/ Kingsford Smith Airport.

World War II halted the operations of many European airlines, and most of those that did continue were fully engaged in essential war operations. BOAC (British Overseas Airways), founded on 1 April 1940 as successor to Imperial Airways and the private British Airways, maintained communications with British overseas territories and also began North Atlantic flights. Qantas, too, did outstanding war work, and Tasman Empire Airways began flying-boat services linking New Zealand and Australia – for most of the war this was the only link between the two countries. United States operators found themselves working on a global scale on behalf of their government and this prepared the way for worldwide airline operations by the United States after the war.

The war also brought into being a worldwide system of airports with paved runways, new navigational systems, and radar – all vital to future air-transport development. And it was responsible for the production in the United States of thousands of military transport aircraft which, when the war ended, became available to commercial airlines and charter companies.

After the war the original IATA (International Air Traffic Association), founded in 1919, was reconstituted as the International Air Transport Association and became widely representative of world airlines, whereas previously membership had been confined mainly to Europe. Today the IATA regulates fares, conditions of carriage and service, and many other aspects of the airline business, and contributes to safety. ICAO (International Civil Aviation Organization), a United Nations agency, deals with standards and practices concerned chiefly with safe and reliable operation.

In spite of the technical advances made during the war, post-war air transport was at first austere and primitive. Many airlines, particularly in Europe, had to reconstruct their entire operations; their airports and maintenance facilities had been destroyed or badly damaged, and for many airlines there was an acute shortage of aircraft and an even more acute shortage of finance with which to purchase them. Nevertheless, dramatic recovery took place; many of the pre-war airlines got back into operation – with even larger networks and longer routes – and countries such as India, Pakistan, Ceylon (now Sri Lanka), New Zealand, and Canada gradually began to operate long-distance intercontinental commercial services.

In Europe BOAC had a variety of aircraft, including pre-war types, flying-boats, and converted bombers; Air France restored some of its pre-war fleet; and Italy had some wartime Savoia-Marchetti and Fiat transports. Germany and Japan were forbidden to own or operate aircraft for several years. In the United States the airlines were still equipped mostly with Douglas DC-3s, and throughout the world airlines acquired the few new examples of this aircraft, some second-hand ones, and many of the surplus military variants which became known generally as Dakotas. New and interim types were put into production in Britain and France, numerous bombers were given austere passenger cabins, and many French-built Junkers Ju 52/3ms were put into service.

Before the United States became involved in the war the major American airlines had ordered two important types of four-engined transport aircraft, the unpressurized Douglas DC-4 and the much sleeker, pressurized Lockheed Constellation; but when they were ready these aircraft were taken for military services, while airline crews operated many on government work.

When peace returned the airlines were able to acquire these advanced aeroplanes, which were also purchased by many non-American airlines as their first long-distance aircraft. DC-4s were adopted by Air France, Sabena, KLM, Swissair, the Scandinavian airlines (later to merge as SAS), and by numerous airlines elsewhere including Qantas and other Austra-

**Right** Much of the success enjoyed by the Douglas company as a manufacturer of transport aircraft was due to its production of families of aircraft. From top to bottom this illustration shows how Douglas developed the DC-4 into the pressurized DC-6 and the further improved DC-6B, with development ending in the very-long-range DC-7C. Later, Douglas (now McDonnell Douglas) developed its DC-8, DC-9, and DC-10 jet transports into further families, and this tradition is continuing.

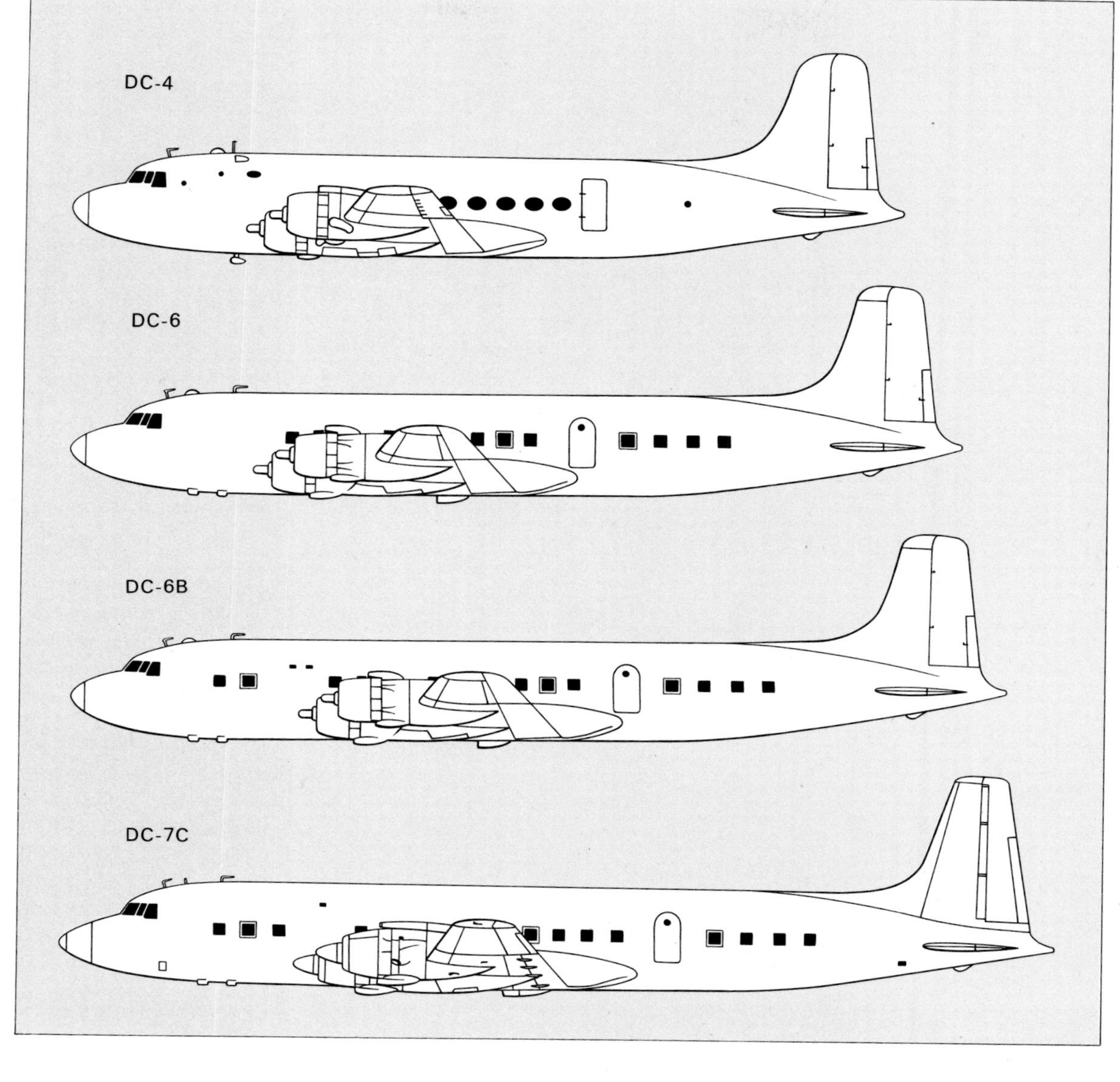


**Below** The Vickers-Armstrongs Viking was one of the first post-war British transport aircraft. It entered service with British European Airways on September 1946, and for several years formed the backbone of the airline's fleet. Vickers built 163 Vikings and 263 of the Valetta military version. Many were exported, among them Suidair International Airways' *Rex*, seen here with South African registration.

lian airlines, India, and some Central and South American airlines. Many airlines operated their first North Atlantic services with DC-4s.

Constellations were mostly newly built for the airlines. Pan American introduced them on round-the-world services, TWA had a large fleet, Qantas used them for its Australia–United Kingdom services, and Air-India International used them to inaugurate its Bombay–London service.

The Constellation provided superior passenger comfort and was much faster than the DC-4. This led Douglas to develop the pressurized DC-6, which in turn led to the slightly bigger DC-6B – one of the most economical and reliable piston-engined airliners ever built. For many years the Constellations, DC-4s, and DC-6s dominated medium- to long-haul routes and many of the busier short-haul routes.

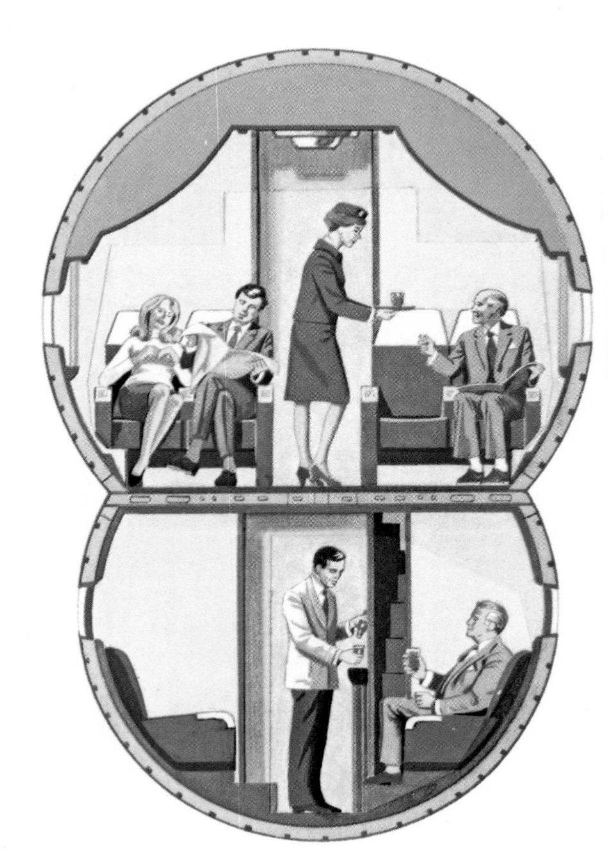

**Right** The Boeing Stratocruiser, which entered airline service in September 1948, had a pressurized double-deck fuselage. The main accommodation was on the upper deck, but a section of the lower deck was arranged as a lounge and bar, the two decks being connected by a stairway.

In Britain, in 1946, air transport was nationalized and three corporations were set up. BOAC was responsible for long-distance routes except to South America, BEA (British European Airways) took over internal and European services, and BSAA (British South American Airways) operated Latin American routes. BSAA used the Avro Lancastrian (the civil version of the Lancaster bomber), the Avro York, and the unsuccessful pressurized Avro Tudor. BEA used a fleet of Dakotas and D.H. Dragon Rapide biplanes and introduced the new Vickers Viking. BOAC used Short flying-boats, Dakotas, Lancastrians, and Yorks and ordered Tudors and the new Handley Page Hermes 4, but the Tudors were never used. Prototypes of the large Bristol Brabazon transatlantic landplanes and Saunders-Roe Princess flying-boats were built and flown but never put into service.

France, too, built several prototypes that were never used, but it did produce 100 four-engined Sud-Est Languedocs which gave good service to Air France and some other airlines, mostly in Spain, North Africa, and Egypt. Sweden built the twin-engined Saab Scandia, which saw service with SAS and in Brazil.

## Dominance of US Aircraft

It was the American airliners that dominated world markets and had the best performance and economics. The first British commercial North Atlantic service, on 1 July 1946, was operated by Constellations, and BOAC was destined to buy many more US aircraft, although BEA managed for many years to operate mainly British aircraft, apart from its Dakotas and some captured Ju 52/3ms. BSAA was merged with BOAC after the failure of the Tudor – two of these aircraft having disappeared without trace.

Canada decided to produce its own versions of the DC-4 and DC-6 with Rolls-Royce Merlin liquid-cooled engines in place of the usual air-cooled Pratt & Whitneys. Known as Canadair Fours they were operated by Trans-Canada Air Lines, under the title of North Star, on Canadian transcontinental and transatlantic services, by Canadian Pacific Air Lines over the Pacific, and by BOAC as Argonauts.

Traffic was growing and so was the need for long-range airliners, and Boeing developed the double-deck Stratocruiser from its B-29/B-50 bombers. This was introduced on trans-Pacific routes by Pan American, United Air Lines, and Northwest Orient Airlines, and on the North Atlantic by BOAC. These large aircraft could normally fly non-stop from New York to London, but against the usual west winds on the return (east-west) leg had to make refuelling

**Below** One of the most successful short-range piston-engined airliners in the 1940s and 1950s was the Convair-Liner. The 56-seat CV-440 Metropolitan seen here was operated on Swedish domestic services by Linjeflyg.

**Bottom** An Ilyushin Il-14, operated by the East German airline Interflug, seen at Helsinki Airport. The Il-14, introduced in 1954, was an improved version of the Il-12.

stops in Iceland, Greenland, or Newfoundland.

Modern, pressurized, relatively long-range airliners were by now established; but in the early post-war years there was a great need for equally modern high-performance short-stage aircraft. The Viking, Britain's contribution to this market, was scarcely more than a faster DC-3, and it was the US series of Convair-Liners which achieved most success. First, in 1948, came the CV-240 with 40 seats, followed by the improved CV-340 and CV-440 Metropolitan. Large fleets were used by many American airlines and others were used in Latin America, Europe, Asia, and Australia. More than 1,000 were built for civil and military customers; and many were later re-engined with propeller-turbines and are still in service.

During the early post-war years the airlines of the more developed countries gave assistance to those of the less advanced. TWA and Pan American were much involved in providing technical assistance, as were Air France, BEA, and BOAC. In Eastern Europe the Russians at first set up jointly controlled services with the other communist countries, Bulgaria, Hungary, and Yugoslavia combining their national airline operations with those of Aeroflot; later, however, the Russian interest was withdrawn and the airlines became nationally owned. Most of them were supplied with Li-2s (Russian-built DC-3s). The Czechoslovak ČSA and Polish LOT, however, had some Ilyushin Il-12s, the first Soviet post-war transport aircraft, and many adopted its improved successor, the Il-14. For many years Aeroflot, recovering from the devastation of war, operated large fleets of Li-2s, Il-12s and Il-14s, and used single-engined aircraft widely in remote areas. The Soviet Union also began to increase its international operations, although these did not become extensive until the turbine era got under way.

The rebuilding of passenger services after the war was vital, but cargo, too, had to be flown in increasing quantity. Much was carried in freighter versions of DC-3s, DC-4s, and Avro Yorks, but one of Britain's first post-war civil aircraft was the Bristol Freighter – a twin-engined high-wing monoplane with non-retractable undercarriage, box-like fuselage, and large nose-loading doors. This found favour in various parts of the world with airlines and air forces, notably in Australasia, but it is best remembered in Britain as the Silver City Airways vehicle ferry, which opened a cross-channel service in 1948.

## The Long-haul Airliners

By 1950 large numbers of reliable and relatively economical aeroplanes were serving the world's air routes, but there was a growing demand for increased performance, particularly range. The two main needs were for non-stop coast-to-coast services across the United States and non-stop operation in both directions over the

**Above** The pressurized Lockheed Constellation was designed as a civil aircraft but first went into service as a military transport. Civil operation began in April 1946, and the Constellation underwent continuous development, with the much longer L.1049 Super Constellation going into service at the end of 1951. The TWA version seen here, the L.1049G, entered service in January 1955.

**Below** The DC-7C, last of the Douglas piston-engined transport aircraft, was the long-range transoceanic development of the DC-7 and DC-7B. Its useful life as a mainline transport was curtailed by the introduction of jet airliners and only 121 were built. Because of its transoceanic capability, the DC-7C was known as the Seven Seas.

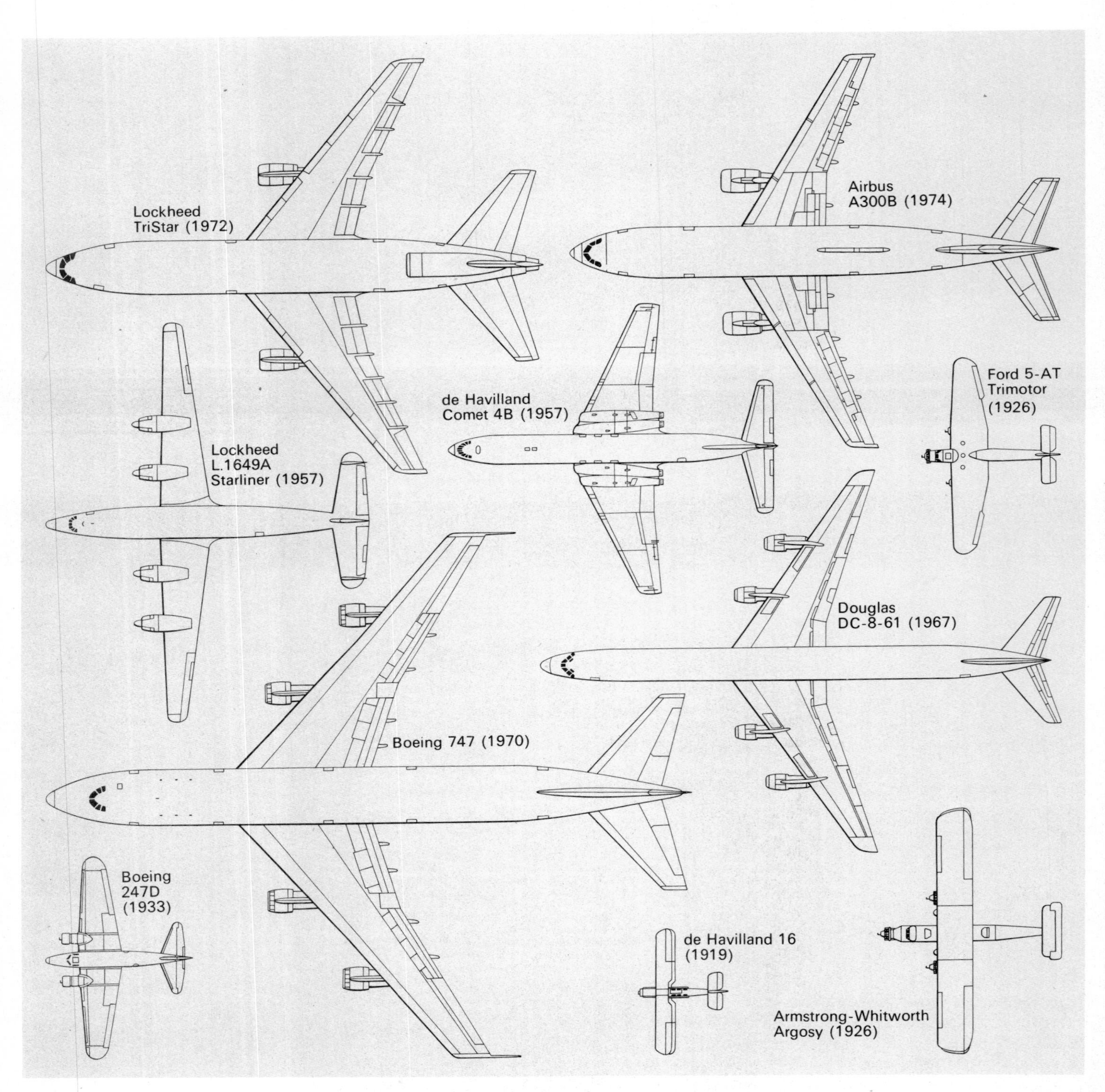


North Atlantic. There was also a need for longer-range aircraft on the increasingly important Pacific routes. There was competition between the US transcontinental airlines to produce the fastest schedules and between Douglas and Lockheed to secure the orders.

Lockheed stretched the Constellation into the L.1049 Super Constellation. The coming of the 3,400 hp Wright Turbo-Compound engine made possible further improvement: Lockheed built the L.1049C and Douglas developed the DC-6 series into the DC-7. TWA began the first sustained non-stop transcontinental service with the L.1049C in October 1953, and American Airlines followed a month later with the DC-7. Regular non-stop transcontinental services had become a reality with a schedule of about eight hours.

**Left** This illustration (with aircraft drawn to the same scale) shows the growth of the transport aeroplane, from the single-engined de Havilland 16 of 1919 to the very large wide-bodied jet transports of today. The aptness of the term 'wide body' is clearly shown in the drawings of the Lockheed TriStar, Airbus A300B, and, biggest of all commercial aeroplanes, the Boeing 747. The term 'long body' is illustrated by the Douglas DC-8-61.

The DC-7 was further stretched as the DC-7C to give Pan American non-stop North Atlantic flights in both directions, and it went into service between New York and London in June 1956. Lockheed, in order to keep up, designed a completely new wing for the Super Constellation and produced the L.1649; it had even greater range than the DC-7, and was introduced over the North Atlantic routes by TWA in June 1957.

With such long-range aircraft available a new route between Europe and North America and the Orient became an attractive objective: the so-called polar route. The first was SAS's operation from Copenhagen to Los Angeles with DC-6Bs, but it involved refuelling stops and was not strictly polar. However, after acquiring DC-7Cs SAS opened a route linking Scandinavia and Tokyo with an intermediate stop at Anchorage (Alaska) and, given favourable winds, this route traversed the north polar regions. Air France followed in 1958 using the long-range L.1649As on a Paris–Anchorage–Tokyo service.

The DC-7C and Lockheed L.1649 were the last of the long-range piston-engined aircraft. Both had been stretched to their limit, and the jet era was fast approaching. Indeed demand for the L.1649 was shortlived and only 43 were built. Douglas and Lockheed had catered for the cargo market with their DC-6A and L.1049H. Neither was ordered in quantity, although some of the late Super Constellation orders were for the H because, while it could be used as a passenger aircraft, it had a better resale value as a freighter.

During the post-war era the operating speed of the piston-engined airliners had risen from about 260 km/h (162 mph) to around 485 km/h (301 mph); their range had increased from a few hundred miles to transcontinental and transoceanic; their comfort had been improved by higher cruising levels made possible by cabin pressurization; and their reliability and regularity had been improved by increased use of radar, better navigational aids, and better airport lighting. Safety had also improved after a fairly poor record during the first post-war winters. But the great increase in traffic growth was brought about by the introduction of cheaper coach-class fares in the United States from 1948 and by transatlantic tourist-class fares in 1952 and later of economy class.

These lower rates were adopted generally and, as the airlines' capacity increased, the fare structures became extremely complex, with a whole range of cut-rate promotional fares for night journeys, students, varying-period excursions, advanced-booking charters, and many other categories – all designed to solve the problem of too many empty seats. They increased traffic but produced a much lower yield per passenger. At the same time, standard fares continued to rise, so that there was no fixed relationship between the distance flown and the fare paid.

Outside the Soviet Union and China, world air passengers increased from 21 million in 1947

**Below** The double-deck Breguet 763 Deux-Ponts airliner, which went into service with Air France on trans-Mediterranean routes in March 1953 as the Provence. The upper deck had 59 tourist-class seats; the lower deck could have 48 second-class seats or be used for cargo. Only 12 of these aircraft were built, and six of them were converted to Universal freighters; the last was withdrawn from airline service in March 1971.

Used in very large numbers throughout the world are two very successful British STOL transports powered by piston engines: the twin-engined Britten-Norman Islander (see page 175) and this three-engined Trislander. Well over 700 Islanders and a much smaller number of Trislanders are serving third-level routes.

to 88 million in 1958, while overall revenue measured in tonne-km went up from 2,110 million to 9,630 million. In 1958, for the first time, the airlines carried more transatlantic passengers (1,292,000) than did ships. Since that time aircraft have completely replaced the passenger liner on routes all over the world.

## Local Services

By the mid-1950s, then, air transport had become the prime form of mass public transport over long-distance and many short-to-medium-length routes. In the United States, where air travel was developing fastest, services were structured on three levels: trunk-route airlines (first level), local-service airlines (second level), and commuter airlines (third level) – the last growing very rapidly from the early 1960s. The local-service airlines had small beginnings and their title was descriptive. Many began with DC-3s or even smaller aircraft, but expansion was rapid and bigger aircraft had to be obtained. Some of these airlines failed. But others amalgamated to form units that were much bigger than some of the trunk airlines in the early post-war years. For example, Allegheny Airlines in 1976 was the sixth biggest US airline in terms of passengers carried (11,031,143) and outside the USA was beaten only by Aeroflot and British Airways. Allegheny began as All American Aviation – operating a small-scale mail service, with the aircraft snatching mail bags from lines stretched between upright posts; by mid-1977 Allegheny was operating 80 jet aircraft, with more on order.

The third-level airlines are now numerous in the United States, where their services radiate from the hub airports to serve large numbers of smaller communities. This type of operation has spread to Australia, South America, Canada, and Europe. A typical example is Loganair, in Scotland, which uses a fleet of Britten-Norman

Islander twin-engined, eight-passenger and Trislander three-engined, 16-passenger monoplanes to serve the Edinburgh–Inverness route as well as routes in the Hebrides and in Orkney and Shetland. Some of these routes are so short that the flight time is little over a minute. The Islander, particularly, has a very short take-off and landing run and operates from some very short, rough strips. More than 700 Islanders have been delivered and they are employed in the operation of third-level and other services all over the world.

In the Soviet Union there is an enormous number of local services, particularly in the more remote regions, and large numbers of Antonov An-2 single-engined biplanes are used, although they are now giving way to more modern types. Designed as an agricultural aircraft, the An-2 first flew in 1947; it is believed to be still in production, and has been built in far greater numbers than any other biplane in history.

## Helicopter Airlines

Before turning to the turbine era it is necessary to deal with one other form of air transport that emerged after World War II: the helicopter airline. In 1946 Los Angeles Airways began experimental helicopter mail services, and in the following year started regular mail services to and from the roof of the city's main post-office. Later, passenger services linked Los Angeles International Airport with a number of outlying communities.

BEA began mail services in East Anglia in June 1948 and in June 1950 operated the world's first helicopter passenger service when it opened a Liverpool–Cardiff route. In 1950 Sabena began helicopter mail services within Belgium and in 1953 started the first international passenger services. It was also in 1953 that New York Airways began passenger helicopter services after an initial period with mail and freight only.

Aeroflot began helicopter passenger services in the Crimea in 1958. The Italian airline, Elivie, began operating from Naples to Capri and Ischia in July 1959. In Australia at the end of 1960 Ansett and Trans-Australia Airlines opened helicopter services between Melbourne's international airport and a city-centre heliport across the Yarra river.

Small single-engined Bell 47s and three-passenger Sikorsky S-51s were used for most of the mail services and mainly S-51s, 8/10-seat S-55s, and 16/18-seat S-58s were used for passengers, although New York Airways and Sabena for a time used the 15-seat twin-rotor Vertol 44B. Initially the Soviet Union employed the Mil Mi-4P, with up to 18 seats. Later some of these airlines were to introduce the twin-turbine 26/30 seat Sikorsky S-61, and it was this type that operated Pakistan International Airlines' helicopter services in East Pakistan (now Bangladesh) for some time from 1963. British Airways uses the S-61 to maintain its Penzance–Scilly Isles service, and has announced a new service between London's Heathrow and Gatwick airports. In Greenland, Grønlandsfly employs S-61s on coastal services.

Although the helicopter has done magnificent work all over the world, its high operating costs have severely limited its use in commercial scheduled service.

The Sikorsky S-61, a twin-turbine helicopter with accommodation for up to 30 passengers, is one of the most widely used airline rotorcraft. This one, an S-61L, is operated by Los Angeles Airways.

# 20 The Turbine Era

**Right** The de Havilland Comet was the world's first turbojet transport. It entered service with BOAC in May 1952, cutting flying times by about half. After a successful introduction, three aircraft were lost owing to inflight structural failure, but a strengthened version, the Comet 4, was produced and three different models entered service. The example shown was one of BEA's fleet of 89-seat Comet 4Bs.

World War II forced the rapid development of the gas-turbine aircraft engine, and before its end both Britain and Germany had turbojet aircraft in active service. Two forms of turbine were developed initially, the turbojet or 'straight' jet, and the propeller-turbine or turboprop; the turbofan came much later.

While the war was still in progress the Brabazon Committee in the United Kingdom had recommended the types of transport aircraft which should be designed and built for post-war operation. These included the turbojet de Havilland Comet (first flown on 27 July 1949) and the propeller-turbine Armstrong Whitworth Apollo (which did not go into production) and Vickers-Armstrongs Viscount. No such early development took place in the United States; but in Canada a prototype jet transport, the Avro C-102 Jetliner, flew two weeks after the first flight of the Comet, although it did not go into production.

The Comet 1 with 36 seats was a very clean, low-wing monoplane powered by four de Havilland Ghost turbojets buried in the wing roots. Its pressurized fuselage enabled it to fly at up to 12,000 m (40,000 ft); it could cruise at 790 km/h (490 mph) and had a range of 2,820 km (1,750 miles). The fuel consumption of the turbojet at that time was high, which ruled out its use on short stages (for economic reasons) and also on transoceanic flights (because of inadequate range). The Comet flew the world's first jet-powered passenger service on 2 May 1952, when BOAC introduced the type on the London–Johannesburg route to a schedule of less than 24 hours including intermediate stops.

Although the Comet was the first turbine-powered airliner to enter service, the short-range propeller-turbine Viscount had flown earlier than the Comet, on 16 July 1948. The Viscount was a conventional low-wing monoplane with a pressurized cabin for 32 passengers,

**Below** The Vickers-Armstrongs Viscount, powered by four Rolls-Royce Dart propeller-turbines, was the first aircraft to enter passenger service with this type of powerplant. The Viscount was one of the best-selling British transport aeroplanes, with many sold in the United States and Canada. BEA's Viscount V.806, the *George Stephenson*, is seen here flying over the south coast of England.

The first United States jet transport was the Boeing 707, which went into service in October 1958. Qantas purchased short-bodied 707s specially suited to its routes but later acquired 707-320 Intercontinentals, one of which is shown here. When it first appeared the 707 was regarded as a very big aeroplane, but it has now been dwarfed by its successor, the 747.

but it was powered by four reliable Rolls-Royce Darts and it was these which made the aircraft faster, higher-flying, quieter and smoother than piston-engined rivals.

The prototype Viscount V.630 operated some of BEA's London–Paris services for a short period from 29 July 1950 and was then transferred to the London–Edinburgh route for a few weeks; but its future was far from assured until Vickers increased its capacity and Rolls-Royce produced more powerful Dart engines. The larger Viscount V.701 entered service with BEA on 18 April 1953 on the London–Rome–Athens–Nicosia (Cyprus) route, and Viscount 700s and 800s became widely adopted by airlines in many parts of the world, a total of 445 being built.

The Comet and Viscount brought absolutely new standards to air travel. The Comet almost halved flight times and operated at almost double the previous cruising levels. Both types had vibration and noise levels well below those of piston-engined aircraft, and the Viscount's very large windows had a special passenger appeal. In many places, such as India, the introduction of Viscounts doubled the volume of traffic. When it appeared the Viscount had no comparable competitor and, as a result, it broke into the United States market – an almost unbelievable achievement.

BOAC introduced Comets on services to Pakistan, India, Ceylon, Singapore, and Tokyo. The French airlines Air France and UAT put Comets into service on long routes, and several other airlines, including some in North America, ordered the type; but three Comets disintegrated in flight, and after the last disaster the Comet was withdrawn from service. Four years later a completely redesigned and larger version appeared as the successful Comet 4.

For more than two years the Viscounts were the only turbine-powered passenger aircraft in service, but on 15 September 1956 the Russian twin-jet Tupolev Tu-104 was introduced by Aeroflot on the Moscow–Omsk–Irkutsk route. There is no record of the Tu-104 ever having been withdrawn from service, and as some are still operating it is the only jet transport to have remained in continuous operation for 22 years.

Until 1956 Aeroflot had had no aircraft more modern than the piston-engined Ilyushins, and the Tu-104 was the first result of a major modernization programme. To speed its development it employed the wings, tail, undercarriage, and engine installation of a Tupolev bomber, and it made its first flight on 17 June 1955. The Tu-104 had high performance, but with only 50 seats it was uneconomic. It was therefore followed by the 70-seat Tu-104A and 100-seat Tu-104B, about 200 of which served Aeroflot on domestic and international routes, with a few more being used by the Czechoslovak airline ČSA.

In early 1947 BOAC issued a specification for a large Medium Range Empire transport, known as the MRE. This was envisaged as a 32/36-passenger piston-engined aeroplane, but when it first flew on 16 August 1952 it had gone through many design changes to become the 90-passenger Bristol Britannia 100 with four 3,780 hp Bristol Proteus propeller-turbines. It was the first large propeller-turbine airliner. Unfortunately, its trials were prolonged and it went into service, on the London–Johannesburg route, only on 1 February 1957. From the Britannia 100 was developed the long-range

Britannia 312, and this operated the first turbine-powered transatlantic service when BOAC introduced it between London and New York on 19 December 1957. A number of airlines bought Britannias, but on the North Atlantic it had less than a year before the jets arrived and only 85 civil and military Britannias were built. Later Canadair produced a number of Britannia derivatives including the CL-44D swing-tail freighter and the 189-passenger CL-44J for the Icelandic airline Loftleidir.

## The Big American Jets

First of the United States jet transports was the Boeing 707, which first flew on 15 July 1954 as the prototype 367–80. This was much bigger and heavier than the Comet, had swept wings and tail, and its turbojets were suspended below the wing in pods. The first production model was the 707–120, with six-abreast seating for 179 passengers, and 5,670 kg (12,500 lb) thrust Pratt & Whitney JT3C-6 turbojets. It entered service with Pan American World Airways on the New York–Paris route on 26 October 1958.

The Boeing 707 specification surpassed any existing airliner's and set the pattern for the modern air transport at what might be regarded as the 'real' start of the jet era. Subsequently the 707 was developed into a whole family which included the bigger –320 and –420 Intercontinental and the smaller, high-performance 720 and turbofan-powered 720B. By November 1977 the total of 707s and 720s ordered had reached 924 (almost all delivered) and they had flown 23,540 mn km (14,620 mn miles) and carried more than 555 mn passengers.

Before the failure of the Comet 1, de Havilland had begun producing the Rolls-Royce-Avon-powered Comet 2 for BOAC. These never went into airline service but gave perfect service to the RAF for almost 20 years. The redesigned Avon-powered Comet 4 was ordered by BOAC, and on 4 October 1958 Comet 4s left London and New York to operate the first west- and eastbound jet services – beating the Boeing 707 by three weeks – although the Comet had not been designed for the route. The Comet 4 had accommodation for 60–81 passengers and from it were developed the long-fuselage Comet 4C and long-fuselage short-span Comet 4B for shorter ranges, the latter initially serving BEA and Olympic Airways. A total of 74 Comet 4 series aircraft was built. The last operator was the British independent airline Dan-Air, whose mixed fleet of 16 Comet 4 series were all acquired from other operators in various parts of the world, Comet 4s having found favour in the Middle East, East Africa, Mexico, and the Argentine.

The second of the big jets was the Douglas DC-8. In spite of its long-enjoyed position as the world's main supplier of transport aircraft, Douglas took some time to decide to risk building a jet airliner and the first DC-8 flew only on 30 May 1958. The aircraft was similar to the Boeing 707 but had slightly less wing sweep. It appeared first as a Pratt & Whitney-powered US domestic aeroplane and this version, the DC-8-10, entered service with Delta Air Lines and United Air Lines on 18 September 1959, nearly a year behind the Boeing. The long-range DC-8-30 and Rolls-Royce-Conway-powered DC-8-40 followed, and there was further development in the -50 series of passenger, cargo, and mixed-configuration aircraft with Pratt & Whitney JT3D turbofans.

DC-8s did not achieve quite the same sales success as the Boeing 707, but Douglas injected new life into the type with the -60 ('Dash 60') series. The DC-8-60 was produced in three main versions. First was the DC-8-61, which retained the -30 wing but had its fuselage 'stretched' by more than 11.25 m (37 ft) to provide seats for up to 259 passengers.

The -61 began service in February 1967 on United Air Lines' Los Angeles–Honolulu route. Next came the DC-8-62 with a 1.8 m (6 ft) increase in span but with a fuselage stretch of only about 2 m (6½ ft) greater than the original DC-8. It had seats for up to 189, more efficient wings and engines, and very long range; it entered service with SAS in May 1967. The third version, the DC-8-63, combined the long fuselage of the -61 with the improved wing and engine pods of the -62, and was introduced by KLM in July 1967. DC-8 production ended in 1972 with delivery of the 556th aircraft.

One other United States manufacturer built large jet aircraft for the airlines. This was Convair (now General Dynamics), which built the CV-880 and CV-990. These were of similar layout to the Boeing 707 and Douglas DC-8 but were smaller and faster and carried less fuel for shorter ranges. The CV-880 had numerous problems during its trials, and initially the CV-990 did not achieve its (extremely high) specified performance.

As a consequence of these setbacks only just over 100 Convair jet transports were built and the financial loss was one of the biggest suffered on any product. The Convairs were powered by General Electric engines, and had five-abreast seating in tourist class with a maximum of 120 seats in the CV-880 and 158 in the CV-990. The CV-880 was introduced by Delta Air Lines on the New York–New Orleans route on 15 May 1960, and the CV-990 went into service in the spring of 1962 with American Airlines (for whom it was designed) and

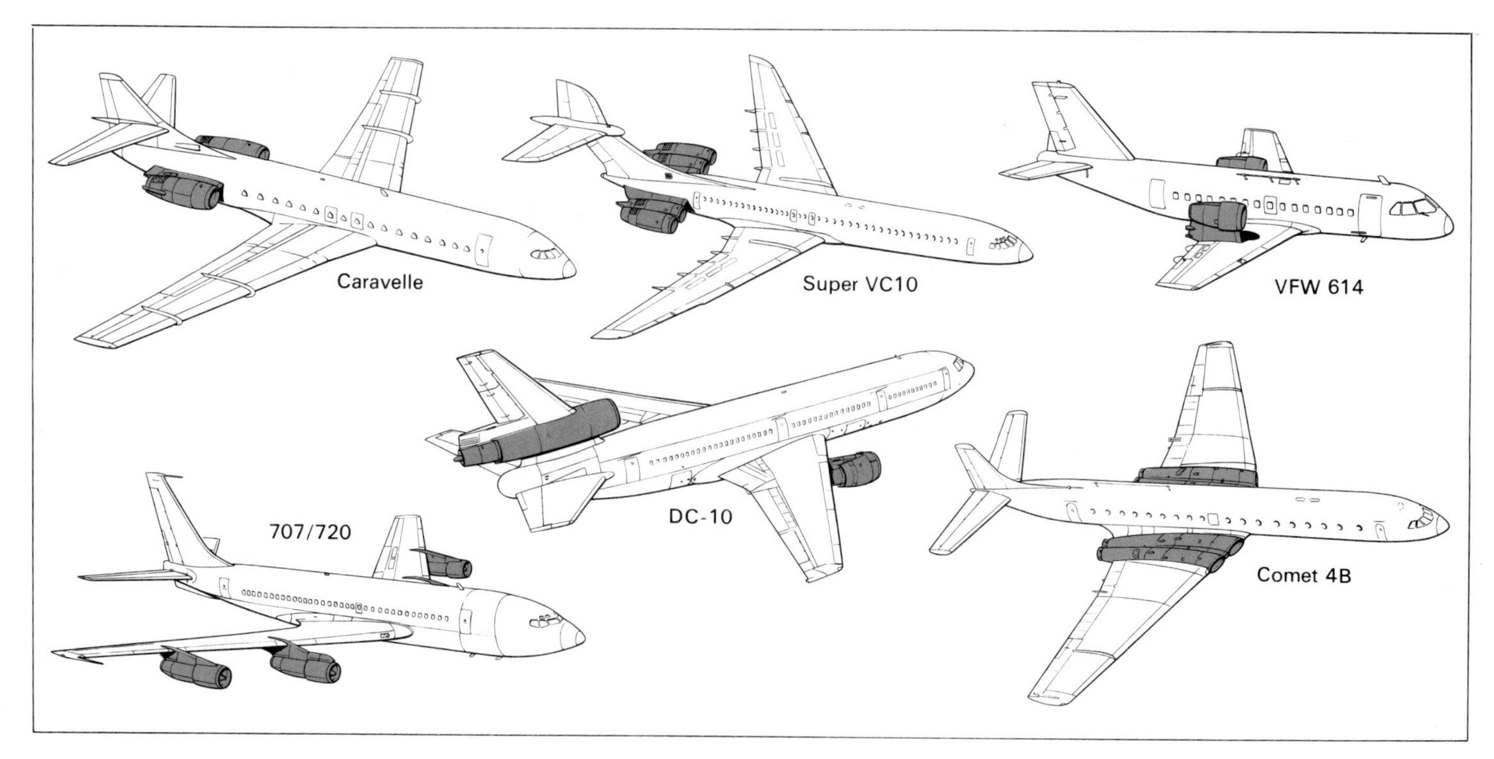


Swissair, who gave it the name Coronado. Some are still flying – the Spanish airline Spantax, for instance, had twelve CV-990s at the beginning of 1978.

Britain also produced a large jet transport. This was the Vickers-Armstrongs VC10, designed mainly for BOAC's African routes where good performance is required at high airfield elevations and temperatures. Unlike the American jets, the VC10 had its four Rolls-Royce Conway engines mounted in pairs on each side of the rear fuselage and its tailplane positioned at the top of the fin. Maximum seating was 151 and the VC10 entered service on BOAC's London–Lagos route on 29 April 1964. From the VC10 was developed the lengthened Super VC10 with maximum seating for 174. BOAC put the Super VC10 into service between London and New York on 1 April 1965. It immediately became very popular with passengers and crews, but most airlines already had Boeing 707s and Douglas DC-8s and total orders for both versions of the VC10, including 14 for the RAF, came to only 54.

An aeroplane which outwardly resembled the VC10 was the Soviet Ilyushin Il-62 which entered service with Aeroflot on 10 March 1967 on the Moscow–Khabarovsk and Moscow–Novosibirsk routes. This aircraft had a number of shortcomings and improved versions have now been built. It is used by Aeroflot on domestic trunk routes and on some of its longer international routes, and it has been sold in small numbers to several communist countries.

## Propeller-Turbines

The period 1958–61 saw the development of several propeller-turbine airliners, mostly for short-to-medium routes. Two were very similar in size and layout, both high-wing monoplanes powered by two Rolls-Royce Darts and having accommodation for up to 56 passengers. One was the Handley Page Herald, which had first flown with four piston engines, and the other was the Dutch Fokker F.27 Friendship, which was also produced in the United States as the Fairchild F-27. The F-27 was the first to enter service, with West Coast Airlines on 27 September 1958; Aer Lingus introduced the Dutch-built F.27 on the Dublin–Glasgow route on 15 December that year. The Herald began operations with Jersey Airlines on 19 May 1961. Only 48 Heralds were built, but the F.27 is still

**Above** Engine location in jet airliners. The de Havilland Comet 4 (like the Russian Tupolev Tu-104 and Tu-124) had turbojets buried in the wing roots. The Boeing 707 (like many later Boeings and others) had engines in pods below and forward of the wing. The Aérospatiale Caravelle (and later the BAC One-Eleven, Douglas DC-9, and Fokker F.28) had engines on the sides of the rear fuselage. The Vickers-Armstrongs Super VC10 (and later the Ilyushin Il-62) had rear-mounted engines in pairs. The McDonnell Douglas DC-10 was fitted with two wing-mounted engines and a third at the base of the fin. The VFW 614 had turbofans above the wing.

**Left** Fokker's first post-war transport, the F.27 Friendship, first flew in November 1955. After more than 20 years in production the Friendship is still finding new customers and new uses and is likely to remain in service for many years. The Friendship seen here was one of the Aer Lingus fleet, the first in Europe.

The Tupolev Tu-134, which replaced the Tu-124, is in service in large numbers as a short-haul jet transport. Its bogie undercarriage retracts backwards into the streamlined fairings visible in this picture. In the background at left is an Ilyushin Il-62; at right is the tail of a Yakovlev Yak-40 trijet.

in production with more than 660 sold to airlines, air forces, and private operators all over the world – the only European airliner to rival the Americans. Very similar to the F.27 is the Soviet Antonov An-24, which began work on Aeroflot's Kiev–Kherson route on 31 October 1962; several hundred have been built.

Although the turbojet and turbofan aircraft were destined to dominate the world's air routes, several quite large propeller-turbine transports followed the Britannia in Britain, the United States, the Soviet Union, and Japan. First to enter service was the Lockheed 188 Electra, powered by four 3,750 hp Allison 501 engines and with accommodation for up to 99 passengers. Over 170 of these fine aircraft were ordered, with Eastern Air Lines and American Airlines introducing them in January 1959. Many other important airlines bought it but two crashed in the United States as a result of gross structural failure. This was not, however, the end of the Electra: it was modified and some examples continue in service.

On 20 April 1959 Aeroflot introduced the Ilyushin Il-18 on its Moscow–Alma Ata and Moscow–Adler/Sochi routes. The Il-18 was powered by four 4,000 hp Ivchenko AI-20 engines and in its various versions had accommodation for 80–122 passengers. More than 600 were built, of which about a fifth were exported, and for many years the Il-18 formed the backbone of the Aeroflot passenger fleet.

A second Soviet propeller-turbine transport was introduced by Aeroflot on 22 July 1959, when the Antonov An-10 began flying between Moscow and Simferopol. The An-10 was about the same size as the Il-18 and had the same engines, but it was designed to operate from unpaved runways and, unlike the Il-18, was a high-wing monoplane. The An-10 was followed by the improved An-10A and by the An-12 freighter with rear cargo-loading doors. On 18 May 1972, however, an An-10 was involved in a serious accident and the civil passenger versions were withdrawn from service.

The last of the big propeller-turbine airliners to go into passenger service with Aeroflot was unique. It was the Tupolev Tu-114 which had a wing span of more than 51 m (almost 168 ft), a length of 54.1 m (177 ft 6 in), and a maximum weight of 175,000 kg (385,800 lb). It could carry up to 220 passengers, was powered by four 12,000 hp Kuznetsov engines, and had a top speed of at least 870 km/h (540 mph), making it the fastest propeller-driven airliner; its cruising speed was only slightly less than that of the Comet, and its maximum range was 10,000 km (6,200 miles). The Tu-114 went into service between Moscow and Khabarovsk on 24 April 1961, later operated the Moscow–Delhi and Moscow–Tokyo routes, and flew Aeroflot's first North Atlantic services. Only about 30 were built and the last was retired in October 1976.

In Britain Vickers-Armstrongs built the large Vanguard with four 4,985/5,545 hp Rolls-Royce Tyne engines and seats for up to 139. Its only customers were BEA and Air Canada, for whom 43 were built. The first Vanguard service was between London and Paris on 17 December 1960, and many of these aircraft are still in service as freighters.

Later came the Japanese YS-11. This had 46–60 seats, was powered by two 3,060 hp Rolls-Royce Darts, and entered service in Japan in 1965. More than 120 were built and YS-11s were exported to USA, Canada, South America, Greece, and elsewhere. It was the first export success for a Japanese airliner.

Three other propeller-turbine transports of

**Near right** The BAC One-Eleven twin-jet was designed to provide economical jet transport over shorter routes. A number of versions have been produced, more than 220 have been sold, and the type is still in production. Shown here is one of British Caledonian Airways' One-Elevens just after take-off.

**Far right** The McDonnell Douglas DC-9, best-selling of all the twinjet transports, has already appeared in five major versions. Still to appear is the DC-9-80 which will be the biggest of the DC-9 series. Seen here is a -30 model operated by the Spanish airline Aviaco.

the 1960s were built in Europe. The Armstrong Whitworth Argosy twin-boom freighter with four Darts was used by BEA, some US carriers, and the RAF, and the Avro 748 (now the Hawker Siddeley 748) is a twin-Dart low-wing monoplane which has been in production in a number of versions for more than 17 years, with about 320 sold. The third aircraft is the French Nord 262, with two Bastan engines and up to 29 seats. It entered service with Air Inter in the spring of 1964, and some in the United States are being re-engined as the Mohawk 298.

## Shorter-range Jets

Following worldwide acceptance of the medium-to-long range big jet transports, there came a variety of shorter range types. First was the French Sud-Est (now Aérospatiale) Caravelle with two rear-mounted engines and the nose and flight deck of the Comet. This was a beautiful design and 280 of various versions were built. The type entered service with Air France and SAS in the spring of 1959, and many remain in service.

The first short-range turbofan transport to enter service was the Soviet Tu-124, a scaled-down Tu-104. It went into service between Moscow and Tallinn on 2 October 1962, and more than 100 were built. It was superseded in September 1967 by the rear-engined T-tailed Tu-134, which is in wide use with Aeroflot and a number of eastern European airlines.

In 1964 two three-engined jet transports entered service. They were alike in having rear-mounted engines and T-tails but achieved very different commercial success. First to enter service was the Boeing 727, with Eastern Air Lines. It weighed up to 76,725 kg (169,000 lb) and could seat up to 94 passengers, and from it was developed the 189-seat Advanced 727-200 model with a maximum take-off weight of 94,205 kg (207,500 lb). By the late spring of 1978 more than 1,500 Boeing 727s had been ordered – the largest number of any type of jet transport – and they had carried more than 1,000 million passengers. The other three-engined jet, the Trident, was designed by de Havilland for BEA and eventually produced in several versions. A total of 117 Tridents was ordered, of which 35 were for the Chinese State airline CAAC. In the hands of BEA the Trident pioneered development of automatic landing, and the first low-visibility automatic landing with revenue passengers was made at Heathrow in June 1965. A recent addition to the three-jet field is the Tupolev Tu-154, which was designed to replace the An-10, Il-18, and Tu-104, and entered service in December 1975. It is

**Right** The Boeing 747, which entered service in January 1970, has been produced in a number of versions, including the 747SP with a short fuselage and very long range. The South African Airways 747SP shown here flies non-stop between London and Cape Town.

The first trijet wide-bodied type to enter service was this McDonnell Douglas DC-10. It was followed by the very similar Lockheed TriStar. The two airliners can most easily be distinguished by the mounting of the rear engine: in the DC-10 it is at the base of the fin, but in the TriStar's it is within the fuselage, with the air intake on top of the fuselage forward of the fin.

likely to be the main Aeroflot transport for some years. Although even larger than the Boeing 727-200, it is designed to use poor airfields and, like most Soviet transports, it has its own airstairs and ground-power supplies.

Smaller than the Trident, the BAC One-Eleven, Douglas (later McDonnell Douglas) DC-9, and Boeing 737 have two engines. The 80-seat DC-9-10 entered service, with Delta Air Lines, in December 1965. Well over 900 DC-9s have been sold in different-size versions, the largest being the 137/172-seat DC-9-80 announced in the autumn of 1977. The BAC One-Eleven, with rear-mounted engines like the DC-9, began service with Braniff and British United in April 1965. One-Eleven sales exceed 220 in several models. The Boeing 737 was a later design and went into service in the spring of 1968 with Lufthansa and United Air Lines. It has the same fuselage width as the Boeing 707 and 727 but, unlike the DC-9 and BAC One-Eleven, has under-wing engines. So far more than 530 have been sold.

Two other relatively small jet transports were the Soviet Yak-40 24/32-seat trijet, for local services linking short unpaved airstrips, and the Fokker F.28 Fellowship. Well over 800 Yak-40s have been built, the type having started carrying passengers in September 1968. Fokker's jet successor to the F.27 entered service with Braathens in March 1969; several versions are in service, with others being developed, and more than 130 have been sold.

## Wide-bodied Jets

In 1966 Boeing announced that it was going to build the very big 747, the world's largest commercial aeroplane and the first of what are now known as the wide-bodied aircraft. In overall appearance the 747 resembles the 707 but is more than twice as heavy, more than twice as powerful and carries more than double the number of passengers. The main deck, almost 6 m (20 ft) wide, can accommodate up to about 500 passengers or, in the cargo version, 100 tons of freight. The flight deck is on the upper level and aft of it is an area which can be used as an extra cabin or first-class lounge. The underfloor holds can carry as much as an all-cargo 707. The span is 59.6 m (195 ft 8 in), the length is 70.5 m (231 ft 4 in), and maximum take-off weight of the latest version is 371,945 kg (820,000 lb). The 747 entered service with Pan American on the North Atlantic in January 1970, and more than 350 have been ordered. Versions are in service with Pratt & Whitney, General Electric, and Rolls-Royce turbofans; they include the shorter 747SP which has extremely long range and is able to operate non-stop such routes as New York–Tokyo, Sydney–San Francisco, and London–Johannesburg.

Entering service respectively in August 1971 and April 1972 were the McDonnell Douglas DC-10 and Lockheed L.1011 TriStar. These are wide-bodied aircraft slightly smaller than the 747, with two wing-mounted turbofans and one in the tail. Seating ranges from about 270 to 400. The DC-10 has been produced in several versions, with 270 sold, and the TriStar has so far attracted about 170 orders.

A fourth wide-bodied type, the Airbus A300B, is a product of a company formed by several of Europe's largest aircraft builders, and it is the only twin-engined wide-bodied type. It has two General Electric or Pratt & Whitney turbofans, seats about 250, and has superb take-off and climb. It entered service with Air France in May 1974, and is now establishing itself as a major transport aeroplane. It has the most efficient wing (low-sweep supercritical) of any transport and is also quieter than almost any other. Eastern Air Lines began using the A300 between New York and Florida in the winter of 1977–8.

**Above** The fourth wide-bodied transport, and Europe's first, was the Airbus A300B, seen here in Air France colours. This is an extremely economical and quiet aeroplane, and well over 100 have been ordered, with a large number destined for Eastern Air Lines in the United States. The two large underslung engines are General Electric CF6 50C turbofans

## SSTs and the Future

The world's first supersonic transport (SST) to enter service was the Anglo-French BAC/Aérospatiale Concorde which began scheduled operations with Air France and British Airways on 21 January 1976 between Paris and Rio de Janeiro, and London and Bahrain. The Concorde is a slim, delta-wing aeroplane with four Olympus turbojets, seats for 100 passengers, and a cruising speed of Mach 2 – twice the speed of sound.

The Soviet Union produced a slightly larger aircraft, the Tu-144, and flew it on the last day of 1968, more than two months before the Concorde. But it encountered problems and went into passenger service only on 1 November 1977 between Moscow and Alma Ata, after flying that route for about two years carrying cargo only.

There have been numerous other civil transport aircraft in recent years, and several more are under development. The Russian Il-86 350-seat and Yak-42 100/120-passenger types, for example, are in the flight development stage; many projects are under way in the United States; and the European aerospace industry wants to build a new 120/160-seat twin-jet.

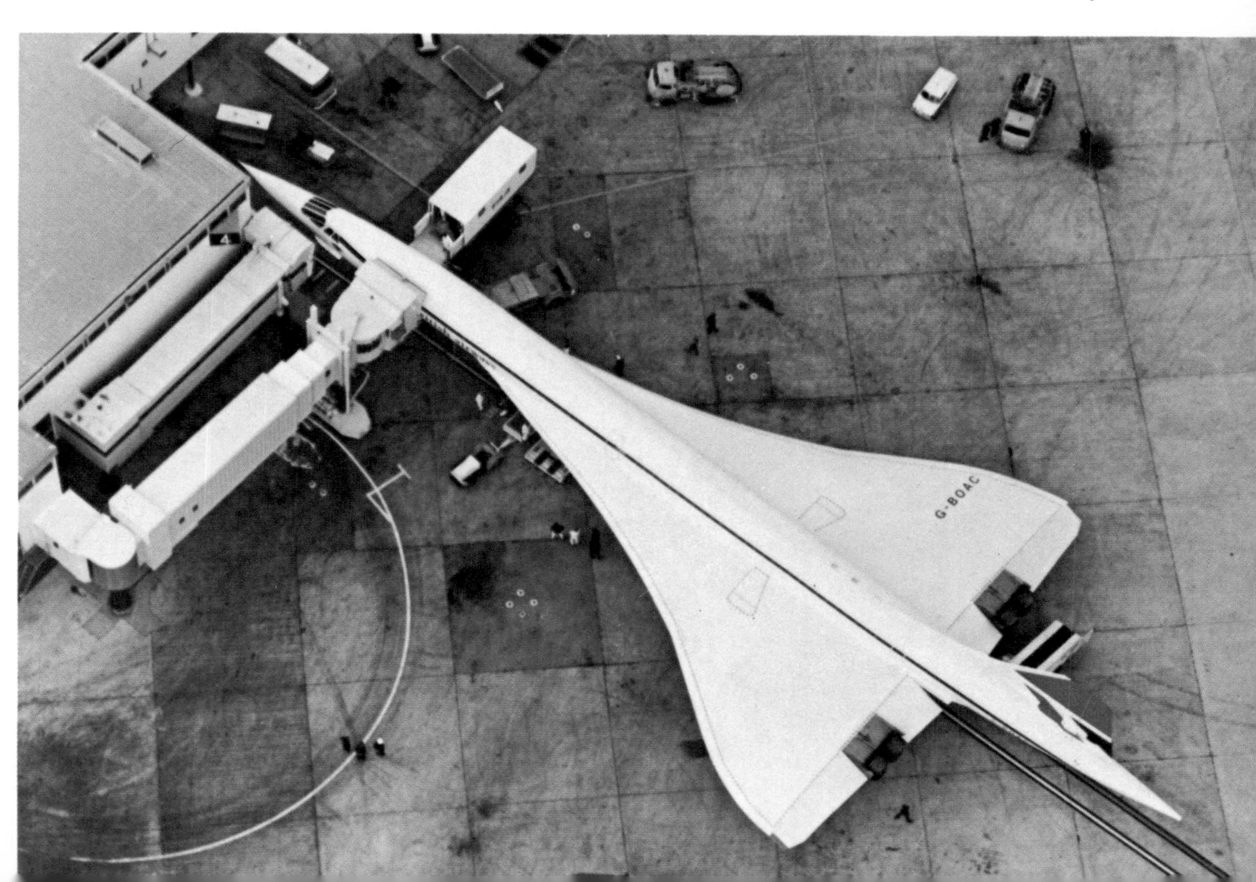


**Right** This view of the BAC/Aérospatiale Concorde supersonic transport shows to advantage its beautifully shaped slim-delta wing and the tailpipes of its four Rolls-Royce Olympus turbojets. This example is one of the British Airways fleet, seen at Melbourne International Airport during its proving trials.

# 21 General Aviation

The de Havilland Dove was the smallest of the designs selected by the Brabazon Committee to form the basis of British civil aviation after World War II. It was to be an eight-passenger airliner, but later versions were adapted for executive use and many were built as military communications aircraft. Some 550 civil and military Doves entered service, some being later modified to use turboprop engines.

Every year in the Western world 15,000 aircraft are built which are neither commercial airliners nor military machines. They are the great mass of private, flying-club, executive, agricultural, survey, and rotary-wing aircraft that are given the collective title 'general aviation'. Often thought of as being just another name for sporting flying, general aviation in fact accounts for a substantial number of working aeroplanes, some in quite unlikely roles, and they are increasing year by year.

After World War I there was an aviation doldrum in Europe. Although a few pioneers strove to create regular air services using converted wartime machines, there was little inducement to the man in the street to take up flying. He was aware of its existence, if only because of extravagant newspaper sponsorship of epic feats, but it was not for him in the drab years immediately after the war: long journeys were still taken by train or ship.

## Early Days

Aviation had to wait for its equivalent of the Austin Seven motor car, and it came with de Havilland's first Moth in 1925. Previous attempts to make an aeroplane for everyman had resulted in over-concentration on economy, with the result that the designs had been too small and basic. The Moth had a specially designed 60 hp engine and this was enough to add some worthwhile performance to the economy.

For the first time there was some encouragement to the flying clubs, and a cheap and reliable aircraft was available to the private owner. Moths began to proliferate: some acquired enclosed cabins and became Hornet Moths or (with the pilot in the open at the rear) Fox Moths. Miles and Percival joined in, all three producing beautiful aeroplanes of wood and fabric with performance which is not easy to match in the all-metal types on offer today.

Such is the present-day dominance of the United States that it is easy to forget how aviation struggled in that vast country during the

Typical of the 9,000 aircraft produced every year by Cessna at Wichita and Rheims, the Skyhawk is a four-seat design in widespread private, flying-club, and business use. Available with an engine of 160 or 195 hp, the Skyhawk can cruise at 242 km/h (150 mph) and has a range of more than 1,200 km (746 miles).

years of the Depression. In 1930, 1,937 civil aircraft were built in the United States; in 1931 the figure was down to 1,582, and the following year, when times were at their hardest, the entire US industry produced only 549 civil aircraft (valued, incidentally, at about £1 million).

Light-aircraft production was understandably slow to recover after the Depression, but even by the end of the 1920s the Americans Clyde Cessna, Walter Beech, and William Piper were beginning to point the way ahead. Piper was a shareholder in Taylor Aircraft, and when this firm collapsed in 1931 he bought it out – for $600 – simply because of its Cub design, which he was convinced was a potential winner. He was right, and Piper Aircraft went on to make 100,000 of them.

Light aircraft were almost entirely pleasure machines until the start of World War II. Although they encouraged extraordinary feats of endurance in the right hands, they were really dependent on reasonable weather. Their reliability did not inspire confidence in the unbeliever, and their timekeeping tended to be erratic.

## Post-War Revival

Britain emerged from the war with an aircraft industry that had forgotten about light aircraft. The Civil Air Guard, during the years immediately before the war, had provided a steady flow of pilots to the Royal Air Force and Fleet Air Arm and in doing so had helped keep the industry on its toes and in production. By 1945, however, the expertise that had produced the Moths, Gulls, and Hawks seemed to have dissipated.

Back into business came Auster, which had taken over manufacturing rights from British Taylorcraft of the familiar steel-tube and fabric, high-wing design based on the American Cub. Even Auster, however, relied basically on military orders, developing only slowly as different engines were tried and as the private-flying movement came, all too slowly, back to life. The Autocrat was followed by the aerobatic Aiglet and then – after amalgamation with the remains of Miles had spawned the Beagle Company – came the Airedale, a high-powered touring aircraft.

British industry was reluctant to turn away from wood or steel tube and fabric, and finally committed itself to all-metal construction only when Beagle designed its low-wing Pup trainer. The post-war Miles designs were attractive and well made, but they needed a great deal of attention if their wooden construction was to survive in, say, an equatorial climate. The Americans recognized the virtues of the light-metal, stressed structure. It was weather resistant and could be produced in very large numbers using processes very much like those of the car industry.

## Instruments and Avionics

Nevertheless, what was to turn the light aircraft into a reliable, all-weather mode of transport was less the change in construction than the transformation in instruments and radio aids. Instruments were rudimentary on the club aircraft of the 1930s: a sure way to tell a genuine Moth is by its vane-type airspeed indicator on the front left inter-plane strut; the airflow blows the vane back against the spring and the pilot reads speed off a painted scale. Wartime flying forced the pace in instruments, and by 1945 most fighting aircraft were equipped with a basic blind-flying panel. The instruments themselves were simple, reliable,

The Piper Navajo and its stretched derivative, the Navajo Chieftain (seen here), have proved to be very economical air-taxi and executive aircraft, carrying up to a maximum of 10 passengers. More than 2,000 have been produced, many with pressurized cabins. The basically similar Piper Cheyenne has two turboprop engines.

Beech began to sell its twin-turboprop King Air in 1964 and now produces this T-tailed Super 200 version at a rate of eight per month. When it is equipped as an executive aircraft it normally carries up to six passengers, but it can be used as a small airliner with less spacious accommodation for 12.

and easily adapted to immediately post-war civil aircraft.

By the early 1950s the skies of northern Europe were full of military jet aircraft, and to minimize the risk of collision an effort was made to separate them from civil aircraft. This was achieved by creating protected airspace for the civilian; but in order to enjoy it to the full he had to have the appropriate radio and navigational aids. Business aircraft could pay their way only if they could fit into the air-traffic control system, and this hastened the development of lightweight avionics. A modern lightplane radio, weighing only about 2.25 kg (5 lb) and designed to be mounted directly onto the instrument panel, has a better performance than a late-1940s unit occupying more than five cubic feet and weighing enough to affect payload. The entire avionic assembly, which will give a business aircraft the ability to fly anywhere in the world, fits easily into the shapely nose of a modern executive jet.

Once endowed with suitable instruments and comprehensive avionics, the business aircraft moved into the all-weather category. It completed the transition from being the chairman's fun aeroplane to being a day-in, day-out working tool, to be used with all the regularity of an scheduled airliner.

## Business Aircraft

The big civil jets entered service in 1958 and were followed some five years later by the first executive jets. A lot of European companies had found a use for an aircraft of their own, and had created a market for aircraft such as the twin-engined de Havilland Dove which had been undreamed of when the original light airliner version was projected in 1944. But trade is international and executive travel was coming to mean international travel. The business world is fond of equating time with money, and it demonstrated its belief by buying the business jets as fast as they became available. Some big companies set up elaborate flight-operations departments to administer their aircraft, and in at least one case a corporation was able to hive off its operating department to become an executive air-taxi business in its own right. More typically an aircraft-owning company tends to leave the whole operation to a chief pilot backed up by an engineer/co-pilot.

What is the appeal of a business jet? Why have 3,000 owners spent up to £3 million per aircraft – other than for vanity or to ease a tax problem? The answer must lie in the freedom from airline timetables which accrues to the owner. He has complete flexibility, and this includes being able to use a considerable number of airfields that are not served by airlines. At present in the United States there are about 13,000 airfields available to general aviation; of these fewer than 500 enjoy a regular airline service, and even at these there may not be a service to suit every traveller.

If the business jets are to make use of so many more airfields than the airlines they must do so with due regard for the community. During the last few years general aviation has been particularly responsive to calls for fuel economy and quietness. Neither was a strong point of the first-generation business jets, and almost all are now offered with a new turbofan engine – the Garrett AiResearch TFE731 – which has made a dramatic improvement in both respects.

The world-wide search for new sources of oil has created a demand for long-range business jets, and these are now available in the form of specially converted airliners to give great load-carrying capacity and comfort. Largest of the purpose-built aircraft is the Gulfstream 2, powered by two airline-type Rolls-Royce Spey engines. Newly developed tip tanks make the G2 (as it is universally known) into a truly transatlantic aircraft. Rivals are on the stocks, however, in the 'wide-body' shape of the Canadair Challenger (able to carry 11 passengers for over 4,000 miles) and the sleeker frame of the Dassault Falcon 50. Dassault uses three engines, arguing that the executive who wants

**Right** More than 700 Britten-Norman Islanders serve the needs of remote communities all over the world. This tough, rugged, twin-piston-engined aircraft has 10 seats when in use for carrying passengers. Its car-type doors give easy access for loading freight and are wide enough to enable stretchers to be loaded. A stretched, 17-seat version, the Trislander, is unique among piston-engined aircraft in having a third engine mounted on the fin (see page 162).

a personal aircraft to cross the Atlantic deserves the added safety factor.

Business jets cost up to £2.5 million when they leave the factory. They are then usually in the so-called 'green' state, ready for delivery to a fitting-out specialist. Furnishing, painting, and installation of radios and navigational aids may put another £500,000 on the price finally paid by the customer.

A popular step down from the business jet is the turboprop class, generally weighing less than 5.5 metric tons (12,000 lb) and costing about £500,000 when ready for service. They combine turbine reliability with reasonable running costs and offer six to eight passengers the ability to cruise above the weather at up to 485 km/h (300 mph). They are best operated over short ranges and are thus well suited to European sectors. Covering 750 miles in under three hours, and able comfortably to clear the Alps, they offer very good value. Of 3,500 business turboprops in use throughout the world, more than 1,500 are Beech King Airs, one of whose five versions leaves the Wichita (Kansas) production line every working day.

Not many executive jet fleets are available for general charter but a number of companies hire out turboprops by the day. A typical charter rate is £300 per flying hour (about £1 per mile): the charge is always for the aircraft, rather than for the number of passengers carried, so it is up to the customer to make the best use of its available seats. Regular chartering is a simple way of making sure that an aircraft is always available; it is often a stepping stone on the way to buying an aircraft.

**Below** Although special conversions of aircraft up to the size of the Douglas DC-7 are used for aerial top-dressing of crops, the bulk of agricultural flying is done by purpose-built designs such as this Rockwell Thrush Commander. Typically, aircraft in this class have engines of 300–600 hp, but the latest Thrush Commander is offered with 800 hp to ensure high performance in desert climates. It can carry 1,820 l (400 gal) of chemical, which is pumped from a spray bar below the wing trailing edge.

## Agricultural Aircraft

At the opposite end of the spectrum from the exotic business jet, the agricultural aircraft is general aviation in the raw. The Western world and the Communist bloc are thought each to have about 8,000 aircraft helping out down on the farm; to these can be added simple conversions of helicopters. The agricultural aircraft is a tough bird. It has to operate from dawn to dusk, often in extremes of heat and dust and usually in an atmosphere heavily charged with corrosive chemicals. Because of this the fuselage side panels are normally removable so that the structure can be hosed down at the end of the day's work. Ag aircraft, as they are called, have a characteristic shape, determined by the need for the pilot to be able to see the ground immediately in front and to the sides, and to be able to survive a crash. The chemical hopper is usually placed in front of the pilot for safety, but this lengthens the nose and means that the cockpit must be placed even higher to improve the pilot's view.

Spraying aircraft have to be able to take off with up to half their own empty weight of chemicals and must be easy to fly whether empty or fully loaded. They must even be easy to fly during the few seconds needed to dump their load in an emergency, when the gross weight changes almost instantaneously. Ideally suited to the vast expanses of the prairies or steppes, agricultural aircraft have to be equally at home when treating, say, the small fields of Europe or the narrow, riverside cultivations of equatorial Sudan. The manufacturers are producing about 750 new agricultural aircraft every year, as the industry expands throughout the under-developed as well as the industrialized countries.

France, Britain, and the United States all entered the executive-jet market in the early 1960s and have been in fierce competition ever since. Britain has sold more than 350 Hawker Siddeley HS.125s; this is the latest version, the -700. Quiet, economical fan engines give it a range of 3,200 km (2,000 miles).

**Right** The Learjet 35 has accommodation for a crew of two and, in this luxury version, four passengers (up to eight may be carried in less spacious comfort in other versions). It is powered by two turbofan engines and has a range of about 2,735 km (1,700 miles).

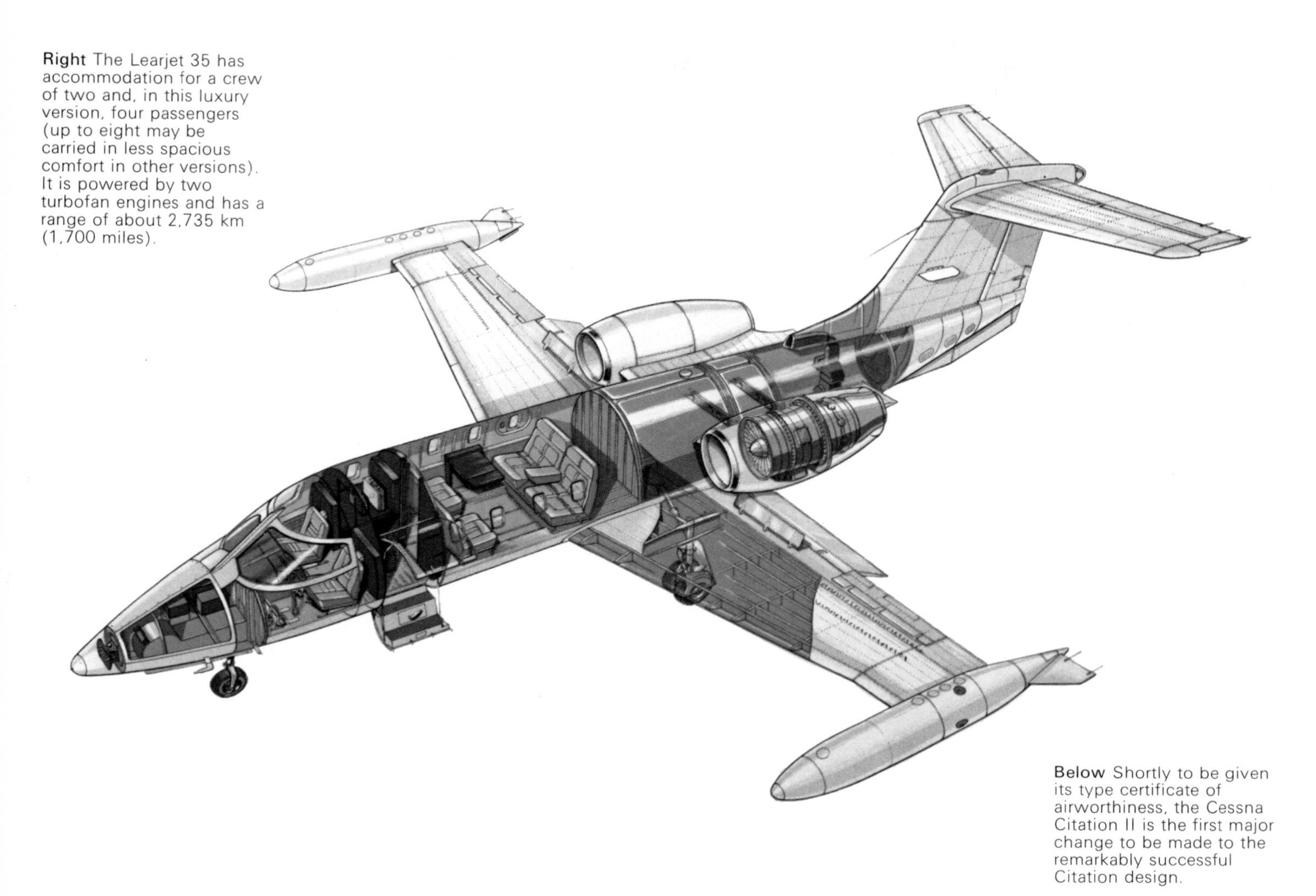

**Below** Shortly to be given its type certificate of airworthiness, the Cessna Citation II is the first major change to be made to the remarkably successful Citation design.

# 22 Sporting Aviation

## Gliding and Hang Gliding

Since a glider is always descending relative to its surrounding air, it can maintain height or climb only if the surrounding air is rising. The glider pilot calls this simply 'lift' – which is not to be confused with the basic aerodynamic force of lift – and says that the weather is 'soarable' when there is lift about.

The simplest kind of lift is caused by wind blowing against a slope or hill, and the earliest real soaring took place just after World War I in such conditions, notably at a famous German site, the Wasserkuppe, in the Rhön mountains north-east of Frankfurt. The first few annual meetings there saw flight durations progress from a few seconds in 1920 to over three hours in 1922, by which time slope soaring was being attempted at other sites in Germany and in England.

Although pilots knew of the existence of *thermals* – the vertical currents rising from heated areas on the ground – little soaring was done in them until the variometer was invented in 1928. This instrument indicates when the glider is in lift and how strong that lift is, and it effectively set glider pilots free from hills. By the early 1930s pilots in Germany and the United States had soared in thermals and even climbed in cloud – and this was in the days before blind-flying instruments. The British Gliding Association was formed in 1929, and by 1933 pilots of the London Gliding Club were regularly soaring in thermals over Dunstable Downs, in Bedfordshire.

Another, more elusive form of lift was to be discovered before World War II: the *standing wave*. This occurs when the wind is deflected up and over a range of hills. Under certain conditions the up-and-over motion can be imparted to the upper air thousands of feet above the hills and for many miles to their lee (downwind) side, producing bands of strong, smooth lift. Waves were not fully understood for years but are now known to be quite common in hilly areas. The present world altitude record of over 14,000 m (46,000 ft) was achieved in a Californian wave, and 11,300 m (37,000 ft) has been achieved over Scotland.

By the outbreak of World War II distances of hundreds of miles and heights of tens of thousands of feet had been achieved. After 1945 glider design was strongly influenced by pre-war German types. The popular British Olympia was based on the German Meise, and variants of the German Weihe were built in other countries. The Slingsby Sky, in which Britain's Philip Wills became world champion in 1952, was really a Weihe descendant with better airbrakes and a higher wing loading.

Developments in low-drag wing sections with more accurate shapes followed in the 1950s, and Germany again led the way. Soon the present age of glass-fibre, with its superbly smooth and accurate surface, was in sight. Most modern high-performance gliders, or sailplanes as they are called, are now built of glass-fibre, although several, such as the Italian Caproni A21 Calif and the Romanian IS-28, are of light alloy.

The latest type of sailplane has flaps, airbrakes to produce drag for approach and landing, and often water ballast which can be jettisoned to lower the wing loading in flight. A high wing loading is useful for high-speed transits between areas of lift, and a low one allows slow circling in weak thermals. A typical high-performance sailplane, such as the German Schempp-Hirth Nimbus II, has a span of 20.3 m (62 ft 8 in) and a maximum glide angle of 1:49, compared to 1:28 of the pre-war Weihe (in other words, if it can afford to lose a mile of height, it can travel 49 miles). Modern contests often involve races around a triangular course, the speed record for a 500 km (311 miles) triangle standing at over 143 km/hr (nearly 90 mph).

All glider pilots have a perpetual 'carrot' before them in the form of the international gliding certificates awarded by the FAI (Fédération Aéronautique Internationale). The first soaring certificate, the 'C', requires a 15-minute flight, and the system progresses through Silver C and Gold C certificates to the three 'diamonds' for a 500 km distance flight, 300 km (186.3 miles) goal flight, and 5,000 m (16,400 ft) climb. There is always something farther, faster, and higher to aim for.

Today's pupil glider pilot often has the benefit of the self-launching motor-glider, which has greatly benefited training at the bigger clubs by eliminating so much time-consuming ground-handling. A two-seat motor-glider such as the German Falke handles like an ordinary glider. After take-off its converted Volkswagen engine can be stopped, whereupon the Falke becomes a glider with a reasonable soaring performance on a good day.

The modern glass-fibre sailplane is sophisticated and very expensive; even a club glider such as the Schleicher K-18 or the glass-fibre Astir costs about three times the price of a small car. Many people form a syndicate to buy a sailplane, taking turns to fly it. Others are barred from conventional gliding because of the expense.

For the more athletic the answer is the hang-glider, which in its modern form stemmed from Francis Rogallo's delta sailwing, originally devised for military purposes and then as a space-

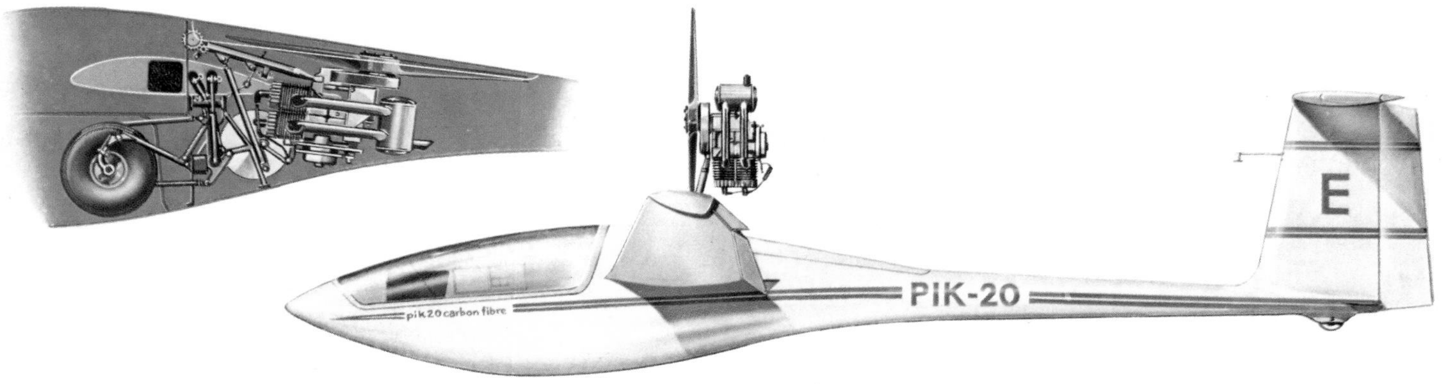


capsule recovery system. A hang-glider consists essentially of lightweight material stretched over a simple wire-braced aluminium-tube structure. The pilot sits in a harness and exerts control principally by shifting his body weight. The simple Rogallo has a poor glide angle of about 1:4, but its wing loading is so low that it can slope-soar in a light wind for as long as the pilot can 'hang on'.

Modern hang-gliding started in the United States in the 1960s, and within a few years there were over 50 American hang-glider manufacturers, building mainly derivatives of the basic Rogallo. Today this has been developed to a glide angle as high as 1:8, and it can be controlled precisely, allowing a skilled pilot to make pin-point landings, steep turns, and other manoeuvres.

Even better performance is possible with hang-gliders with a rigid aerofoil section, of which the American Icarus II, VJ-23, and Quicksilver are typical. These could almost be

**Top left** The Eon Olympia, one of the best-known British gliders of the 1950s, was based on a pre-war German Meise design. With its glide angle of 1:25 and docile handling, it could be flown by inexperienced pilots.

**Top right** A modern T-tailed glider, with a glide angle of about 1:40, capable of cross-country speeds of more than 160 km/h (100 mph).

**Above** The Pik 20 is a powered version of a high-performance Finnish glass-fibre sailplane. Its 43 hp two-stroke engine allows the pilot to take off in less than 300 m (1,000 ft). When good soaring conditions have been located, the engine can be retracted into the fuselage (see sketch at left).

**Left** Experienced hang-glider pilots usually fly with the body in the prone position, as here. It creates less drag and allows more refined control of manoeuvres generated by body movements.

The RAF Red Arrows, seen here in a typically daring manoeuvre, were formed in 1965 and are a full-time aerobatic team. Their tiny Hawker Siddeley (Folland) Gnat T.1s, which have a wingspan of only 7.3 m (24 ft), are also used by the RAF as advanced trainers. T.1s are powered by a Rolls-Royce Orpheus 101 turbojet and are capable of 1,025 km/h (635 mph).

described as foot-launched, ultra-light conventional gliders. In spite of their much better performance they have not yet appeared in large numbers because, like an ordinary glider, they need a trailer and crew, in contrast to the easily portable Rogallo.

Another development, probably with greater potential, are the collapsible semi-rigids, most of which are tailless. Several types have appeared, mostly swept monoplanes with 'tip draggers' to help directional control, such as the Fledgling Valkyrie and Easy Rider.

Hang-gliding is one of the fastest-growing aviation sports. In 1978 there are an estimated 60,000 hang-glider pilots in the world, and a similar number of hang-gliders. Distances of some 80 km (50 miles) have been flown in slope and thermal lift, and 2,450 m (8,000 ft) has been reached. Such is the performance and ambition of modern hang-gliding that the sport's international Silver C badge requires the same achievements as ordinary gliding: a 50 km (31 miles) cross-country flight, 5 hr duration, and 1,000 m (3,280 ft) height gain.

## Aerobatics

Aerobatics began in the earliest days of flying. Although hardly accepted as such today, the first aerobatic manoeuvre was probably the spin. It was originally an accidental and often fatal manoeuvre from which no one knew the correct way to recover until Lieutenant Wilfred Parke, a British naval officer, analysed his successful recovery from a spin in an Avro biplane in 1912. The first deliberate loop is reckoned to have been flown in 1913 by a Russian in a Nieuport monoplane over Kiev.

During World War I control in unusual flight attitudes became a matter of life or death, and for pilots engaged in dog fights aerobatics became a means of taking rapid evasive action to shake off a pursuer.

'Stunting' – the old name for aerobatics – developed through the 1920s and 1930s as a spectacular means of showing off aircraft and pilot capabilities at barnstorming, flying circus, and military displays. As aircraft performance increased, formation aerobatics became possible, and by 1935 formations of fighters – sometimes tied together by brightly coloured ribbons – were aerobatic attractions at all the major air displays. Their modern counterparts are, of course, the RAF Red Arrows, the USAF Thunderbirds, the USN Blue Angels, and several dozen other famous military and civil teams.

It was not until after World War II that aerobatics started to become an organized sport. Competitions were mounted at local, national, and international level, the top pilots from each nation eventually battling against each other in world championships. By 1961 the Spanish flyer Colonel José Aresti had published an aerobatic dictionary in which he identified every possible manoeuvre by means of a special shorthand symbol. This is now used all over the world by aerobatic pilots and judges, and in the latest issue of the dictionary there are no less than 8,000 manoeuvres. A competition pilot has to fly predetermined sequences of manoeuvres (which he has been allowed to practise), plus often an 'unknown' sequence (for which he is given instructions shortly before take-off), and a 'free' sequence of his own choice.

Today's complex manoeuvres call for very strong aircraft with large control surfaces and powerful engines that will run continuously when upside-down – something the normal piston engine, with gravity-fed fuel, will not do for more than a few seconds.

The biplane, with its strongly braced structure concentrating its mass into a compact space, has always been popular for aerobatics. One of the best-known types, typifying the modern sport biplane, is the American Pitts Special. Its wing span is less than 6 m (20 ft), but it has 200 hp in the two-seat version, giving a climb rate of around 600 m/min (2,000 ft/min). Still flown in aerobatic competitions all over the world, the Pitts was for several years the plane used by Britain's Rothmans aerobatic team, the only full-time civilian aerobatic team in the world.

Other prominent types today are mostly monoplanes. The best French aerobatic plane is probably the wooden CAP 20, while the Soviet

Union's finest, the Yak-50, is the latest in a long line of radial-engined Yakovlevs. Czechoslovakia has long been prominent in aerobatics with its Zlin monoplanes, the newest of which is the Zlin Z50. Britain's first aircraft to be designed purely for competition aerobatics is the Cranfield A.1, named after the Cranfield College of Technology whose staff and students created it.

## Homebuilt Aircraft

The make-it-yourself branch of aviation attracts enthusiasts throughout the world. For many it is the only possible route to owning aircraft, and flying their homebuilts (as Americans call them) is often cheaper than hiring a modern production light aircraft. There is also great satisfaction in creating an aircraft out of bare wood, metal, or glass-fibre. The average homebuilt, constructed mainly during evenings and week-ends, probably takes four years to complete and its constructor will doubtless get as much pleasure from assembling it as he does from flying it.

In most countries homebuilt aircraft can be operated under special dispensations from the normal licensing arrangements. In the United Kingdom, for instance, the Popular Flying Association has been given responsibility by the Civil Aviation Authority for the airworthiness of British homebuilt aircraft. Each new project is examined by a PFA inspector at various stages in its construction, and after it has been approved for flying it is inspected annually by the PFA before its Certificate of Airworthiness is renewed.

Many people assume that a homebuilt must be a 'minimum' aeroplane whose sole purpose is to get one person airborne as simply and cheaply as possible. In the past there have been many designs based on this principle, but only a few are still popular. Typical of this kind of design, and popular in Europe for many years, is the French Druine Turbulent, a wooden, low-wing, tailwheel monoplane with either open or closed cockpit and powered by a converted Volkswagen engine – the most widely-used engine for this sort of aeroplane throughout the world. Another popular type, and even simpler to make, is the American Evans VP-1.

Nowadays the trend in homebuilt designs is towards more sophisticated aircraft with two or more seats and often offering better performance than factory-built counterparts. The sleek, all-metal American Thorp T-18, for instance, will fly two people at 282 km/h (175 mph) on 180 hp, and in 1976 Don Taylor made the first round-the-world homebuilt flight in one of these machines. Similar designs include the Bushby Mustang and Canadian Zenair Zenith. Even the humble VW engine can provide over 240 km/h (150 mph) in the wood and plastic KR-1 and KR-2 designs of Californian Ken Rand.

American homebuilt designs vary from the bizarre to the brilliant, and two designers in particular have produced spectacular aircraft. Jim Bede's BD-5 is a tiny single-seat aerobatic aircraft driven by a pusher propeller or small jet engine. With a 70 hp engine a speed of almost 320 km/h (200 mph) is reckoned to be possible. Another remarkable aeroplane, Burt Rutan's two-seat VariEze, is a glass-fibre canard

Pilot's view of a formation loop by the Rothmans aerobatic team, which flies Pitts S-2A biplanes. The only full-time civilian aerobatic team in the world, Rothmans flew in Britain for several years in the 1970s but nowadays it performs exclusively overseas.

design, with the wing at the rear and a foreplane at the front. This arrangement gives extra-safe stalling qualities, and the VariEze will fly two people at 305 km/h (190 mph) on 100 hp over a range of more than 800 km (500 miles).

These and many more designs are being constructed in various countries from plans or kits. In the United States especially a whole industry has developed to supply plans, kits, parts, and engines. The world's biggest sport-flying association is the Experimental Aircraft Association, now 25 years old and with more than 57,000 members. Its annual fly-in at Oshkosh (Wisconsin) is the world's biggest airshow in terms of numbers. In 1977, 8,000 aircraft visited Oshkosh for the week-long event. There were more aircraft parked on the airfield, in fact, than there are on the whole United Kingdom civil-aircraft register.

The stars of the daily air display at Oshkosh are aircraft from each of EAA's four divisions: the International Aerobatic Club, the Custom-built, the Warbirds (old military aircraft), and the Antique/Classic divisions. Most old aircraft are genuine specimens rather than replicas, and they look and sound as good as new. The EAA also boasts the world's largest non-government-supported air museum in Milwaukee (Wisconsin) with over 170 sporting aircraft on display, many of them airworthy.

## Racing

Whenever someone builds a vehicle, whether it is a bicycle or a sports car, he likes to see how fast it will go. Then his competitive instinct urges him to see if it will go faster than the next man's. Not surprisingly, therefore, air racing is the oldest aviation sport.

Powered flight's first decade saw many air races. The earliest were at Rheims, in France, in 1909. The first of the American international Gordon Bennett races was flown near New York in 1910 and attracted teams from the United States, France, and the United Kingdom. Europe's first international point-to-point race was the Circuit of Europe in 1911, which lasted 19 days and was won by one of the popular Blériot monoplanes. The *Daily Mail* sponsored a 'round-Britain' air race in 1911, and the 1,625 km (1,010 miles) course was completed by the winning Frenchman in a Blériot in 22 hr 28 min.

In 1913 the first of the famous Schneider Trophy seaplane races, sponsored by a French industrialist, was held as the central attraction at a race meeting at Monaco. Over the succeeding 18 years the races created world-wide interest, attracting some of the fastest aircraft in the world and conferring great national prestige on the winners. Most of the early com-

**Right** The Schneider Trophy Race for seaplanes was instituted by the French industrialist Jacques Schneider in 1912. The last of the races, at Spithead in 1931, was won by a Supermarine S.6B (identical to the one shown here) at a speed of 547.31 km/h (340.8 mph). The S.6Bs were powered by Rolls-Royce 'R' engines, developing 2,350 hp, whose design influenced that of the Merlin.

**Left** Homebuilt aircraft have put flying within the reach of thousands of enthusiasts during the last few years. The three ultra-lightweights shown here typify the variety of homebuilts available.

**Top** The all-metal Bede BD-5, with pusher propeller driven by a 70 hp, three-cylinder, two-stroke engine, can cruise at 338 km/h (210 mph). The aircraft is also available as an unpowered sailplane or (as the BD-5J) with a Microturbo jet engine of 92 kg (202 lb) thrust.

**Middle** Probably the world's smallest twin-engined aeroplane, the Colomban CriCri has a wingspan of about 5 m (16 ft) and a maximum weight of 170 kg (375 lb). It is powered by two 125 hp two-strokes and can cruise at 195 km/h (121 mph).

**Bottom** The glass-fibre Rutan VariEze two-seater is a truly remarkable design, its rear-mounted wing (incorporating drooping 'winglets' at the tips) and canard foreplane giving it very safe stalling characteristics. With a 100 hp engine driving a pusher propeller, it is capable of 322 km/h (200 mph).

petitors were biplanes with their inevitable array of struts and wires; but by the final race in 1931 great technical advances had been made in low-drag aerodynamics, propellers, and engine cooling, and the thin-wing monoplane had become established as the best configuration for high-speed flight.

Britain won the 1914 contest and, after Italian victories, again won the 1922 trophy at Naples. Americans won the next two, the 1925 race going to Lieutenant James Doolittle in a 600 hp Curtiss R3C-2 at 374 km/h (232 mph). In 1927 Britain won with the Supermarine S.5, and thereafter Supermarine became synonymous with Schneider victories. The 1929 contest was won at 527 km/h (328 mph) by a Supermarine S.6, powered by a Rolls-Royce engine, developing almost 2,000 hp, that foreshadowed the famous Merlin engine which was to power many British aircraft in World War II. In 1931 – the race's last year, after which Britain retained the trophy – an S.6B set a world absolute speed record of 657 km/h (408 mph).

Today the largest race meeting in the world is the annual National Air Races held near Reno (Nevada). The aircraft are divided into several classes and race at low level around pylons. The biggest and fastest are in the unlimited class, and are often World War II fighters, such as Mustangs or Bearcats, capable of exceeding 645 km/h (400 mph). Other classes include the 'T-6' (for North American T-6 trainers); the Sport Biplane; and the Formula One.

Formula One has brought air racing within the reach of many, since the engine size is limited to 200 cubic inches (about 3,275 cc), which effectively restricts power output to about 100 hp, and there are various other restrictions. Nevertheless the tiny monoplanes that compete are capable of over 320 km/h (200 mph). The first Formula One racer was the American LeVier Cosmic Wind of 1948; other well-known types include the Cassutt and Britain's Rollason (Luton) Beta. Formula One racing is becoming increasingly popular in the United States and Britain and is spreading to Germany, France, and Denmark. It is an exciting spectator sport, for most of the six-sided course is visible from the base airfield. Races are often held as events in an air display.

## Epic Flights and Records

Six years after the first sustained powered flight by Orville Wright, there occurred perhaps the first epic flight to capture the imagination of people throughout the world – and, at a stroke, it challenged England's security as an 'island fortress'. Louis Blériot's crossing of the English Channel on 25 July 1909, from Cap Gris Nez to a crash-landing near Dover Castle, took 37 minutes. This was, indeed, about as long as the 25 hp Anzani engine of his Type XI monoplane would run without stopping.

In 1919 a US Navy NC-4 flying boat, flown by Commander Read, crossed the Atlantic in several stages. Later that year the first non-stop Atlantic crossing was made by Alcock and Brown, who flew a converted Vickers Vimy bomber 3,043 km (1,890 miles) from Newfoundland to Ireland in 16 hr 28 min. The same year a British airship, the R-34, crossed the Atlantic in both directions; a Vimy made the first flight from England to Australia in the hands of Ross and Keith Smith; and Van

Ryneveld and crew flew one to South Africa.

In 1924 Douglas World Cruisers of the US Army flew around the world. But even this was eclipsed in 1927 by what was probably the greatest flying feat in history: Charles Lindbergh's first solo crossing of the Atlantic, from New York to Paris. He was airborne for almost 34 hours in his Ryan monoplane *Spirit of St Louis*, flying a totel of 5,810 km (3,610 miles).

The 1930s saw many epic flights by both men and women. The American Amelia Earhart became the first woman to fly solo across the Atlantic in a Lockheed Vega in 1932; the same year James Mollison in a Moth set a speed record from London to Cape Town, and made the first east-west Atlantic crossing in a Puss Moth. Two years earlier Amy Johnson, who later married Mollison, had become the first woman to fly solo to Australia in a Gipsy Moth. In 1933 the one-eyed American Wiley Post made the first solo flight around the world – and in record time – in his Lockheed Vega *Winnie Mae*. The New Zealander Jean Batten in a Percival Gull made the first solo flight by a woman from Britain to New Zealand in 1936.

After World War II the first manned supersonic flight was made by Major Charles Yeager in a rocket-powered Bell X-1 in 1947. Two years later a Boeing B-50 Superfortress made the first *non-stop* flight around the world, covering 36,225 km (22,500 miles) with the aid of four in-flight refuellings.

Many pilots have flown around the world since the war, amongst them Sheila Scott in a single-engined Piper Comanche in 1966. The American Max Conrad has made numerous epic solo record flights, including one from Cape Town to Florida in a Twin Comanche at the age of 62.

Today, just 75 years after the first powered flight by Orville Wright, the world airspeed record stands at 3,524 km/h (2,185 mph), achieved by the Lockheed SR-71A Blackbird. In 1974 one of these reconnaissance aircraft flew from New York to London in just 1 hr 55 min – progress indeed in one human life-span.

## Restorations and Replicas

The restoration of aircraft for museum collections or private ownership is an activity that is spreading rapidly throughout the world, and reflects a growing interest in the history and traditions of aviation. 'Restoration' can mean a multitude of things, from overhauling a long-grounded aircraft in order to fly it again, to salvaging a wreck from underwater and rebuilding it – as has been done with more than 60 aircraft of World War II.

At some museums the aircraft are permanently grounded; at others some or all are kept in airworthy condition and are flown regularly. Two of the best-known British museums that feature regular flying displays are the Shuttleworth Collection at Old Warden, Bedfordshire, and the Strathallan Collection in Perthshire, Scotland. There are important aviation museums in every major European country and more than 40 in North America. The largest historical collection in the world is that of the Smithsonian Institution, at Silver Hill (Maryland) and in the magnificent National Air and Space Museum at the institution's headquarters in Washington, D.C.

In recent years the production of a number of motion pictures has involved the use of old aircraft. For *The Battle of Britain*, for instance, numerous Spitfires, Hurricanes, Messerschmitt Bf 109s, and Heinkel 111s were restored to flying or taxiing condition. In the case of several such historical films authentic examples of certain aircraft have not been available, and replicas or similar types suitably disguised have had to be used. In *Tora Tora Tora!*, the American film reconstruction of the Japanese raid on Pearl Harbor (Hawaii) in 1941, T-6 Texans and BT-13 Valiants were disguised as Japanese Zero fighters, D3A dive-bombers, and B5N torpedo-bombers; while in the British *Aces High*, set in the World War I period, Stampe biplanes were rebuilt to look like S.E.5a fighters.

A replica may be a full-sized or a scaled-down version of the original aircraft. The British firm Leisure Sport, for example, operates full-size replicas of a World War I Sopwith Camel and a Fokker Triplane, powered by radial engines, while its Supermarine S.6B is a scaled-down replica of the famous Schneider Trophy-winning seaplane.

American film makers are fortunate in having on hand in Texas the Confederate Air Force, a collection of World War II aircraft far outnumbering in size and diversity any other such collection in the world. Restorers, pilots, and other major personnel in their enterprize all hold the honorary rank of Colonel.

## Man-powered Flight

For centuries man-powered flight was a dream that inspired an astonishing and often lethal variety of machines with flapping or fixed wings. Apart from questions of aerodynamic design, the biggest problem has always been the power source – human muscles. At best, a man can develop about ½ hp, and he can sustain such an output for only a few minutes.

In 1959 a British industrialist, Henry Kremer, announced a stimulating man-powered-flight challenge by offering a series of prizes, one of which was to be awarded for a figure of eight around two pylons set half a mile apart; by

Above A half-scale Focke-Wulf Fw 190, a popular subject for makers of home-built replicas. Construction is mainly of wood and polyurethane foam. A 100 hp Continental engine gives it a cruising speed of 235 km/h (146 mph).

1973 this prize was worth £50,000. Men soon managed to get airborne under their own power in various countries, with notable success in England and Japan. At Nihon University in Japan, Professor Kimura produced 10 man-powered aircraft and his latest, the Stork, was a favourite to win the Kremer prize by early 1977, managing three-quarters of the course before a wingtip touched the ground. Most man-powered designs have a necessarily light, large, and flimsy structure, and this sort of accident usually means hours of patient re-building.

The prize was finally won on 23 August 1977 in California, when the Gossamer Condor, built by a former world gliding champion, Paul MacCready, was pedalled around the Kremer course by a cyclist and hang-glider pilot, Bryan Allen. The flight took just over six minutes, Allen averaging 17.4 km/h (10.8 mph).

Gossamer Condor is a canard design with its foreplane mounted on a thin boom ahead of the 29.3 m (96 ft) mainplane. The structure is of balsa tubing, corrugated cardboard, and foam plastic, braced by piano wire and covered in a lightweight plastic film. The pilot sits inside the transparent combined fuselage and fin. The bicycle pedal mechanism drives a pusher propeller of 3.7 m (12 ft) diameter at about 120 rpm. Limited control in roll and yaw is provided by wing-warping.

The dream of man-powered flight is now a reality, but there are still other Kremer prizes to be won. The most valuable and elusive is the £100,000 to be awarded for the first crossing of the English Channel from England to France.

## Barnstorming

The end of World War I left thousands of trained pilots and warplanes with no foreseeable prospects except unemployment. In the United States for instance, you could buy a Curtiss JN-4 'Jenny' – brand new and still in its crate – for $600. Many pilots tried to earn a living with such aircraft. They roamed from town to town, operating from any handy pasture and giving displays of 'stunt' flying – wing-walking, climbing from one aircraft to another in mid-air, and so on. They either charged admission to the field or passed round the hat, and made a bit extra by giving joy rides at about a dollar a head. They popularized the idea of flying and gave thousands of Americans and Britons their first taste of the air. Their popular name, 'Barnstormers', derived from the custom of tying their biplanes down for the night in the shelter of a barn to prevent them from being blown about by winds.

Below Paul MacCready's Gossamer Condor became the first man-powered aircraft to complete the Kremer Prize figure-of-eight course in 1977. Owing to their lightness, huge wing area, and slow speed, such aircraft are controllable only in almost windless conditions.

# Vertical Flight

16709
UNITED STATES ARMY

# 23 Rotorcraft

Although the developed helicopter came to maturity long after fixed-wing aircraft were already flying around the world, the origins of moving-wing systems were contemporary with the early fixed-wing experiments at the turn of the 20th century. The first helicopter flew only four years after the Wright brothers' epic flight in 1903; and, as already described in Chapter 1, a number of flapping-wing projects – ornithopters and the like – actually preceded the early fixed-wing experiments.

The helicopter has taken much longer to emerge as a practical flying machine because it is so much more complex in two essential design elements. Mechanically, it depends upon an elaborate transmission system designed and manufactured to a very high degree of precision. Aerodynamically, the rotor blades must produce optimum performance over a very wide speed range.

Designing an aerofoil section capable of optimum performance at both high and low speeds is difficult enough in a straightforward fixed-wing design. In a supersonic aircraft, for example, the aerofoil section of the wing must have a very low thickness ratio – the ratio of maximum thickness over chord – to minimize drag at supersonic speeds. Such a wing produces adequate lift at the plane's normal cruising speed, but at the much lower speeds involved in take-off and landing it needs to be fitted with devices such as flaps and slats to generate extra lift.

In a helicopter, however, the blades must perform well in both the high-speed and low-speed conditions *at the same time* – and all the time the machine is airborne. This is because at any normal speed of rotation the blade tips are flying at near-sonic speeds while the root ends have a much lower relative airspeed. Moreover, when a helicopter is flying forward, the blades have a relative airspeed made up of their own rotational speed plus the helicopter's forward speed on one side of the rotor, and minus the forward speed on the other side. The two sides are termed the *advancing* and the *retreating* sides of the rotor. Thus the relative airspeed of each blade not only varies greatly from tip to root but changes rapidly as the blade rotates – increasing on the advancing side and decreasing on the retreating side.

A helicopter depends on its blades moving smoothly through the airflow, with the minimum of vibration. Sub-standard engineering of the transmission system would cause mechanical vibrations, while deficiencies in blade design could result in aerodynamic vibrations such as control buffeting. This can follow from an instability of the blades and is somewhat akin to wing flutter in a fixed-wing aircraft.

The high standards required for smooth operation were quite beyond the engineering knowledge and abilities of the early helicopter pioneers, and many of their rudimentary projects almost shook themselves to pieces before ever leaving the ground – sometimes leaving the crestfallen inventor-pilot to extricate himself from the midst of a tangled heap of wreckage. Nevertheless, one or two genuine flights, albeit of short duration, were recorded before the outbreak of World War I. The Breguet-Richet No. 1 was credited with being the world's first helicopter to fly, in 1907. The flight, which took place in France and which was tethered as a safety precaution, lasted for just one minute. Even shorter was the world's first untethered helicopter flight, by the fragile Cornu tandem-rotor helicopter later the same year; this machine broke up on landing.

Experiments continued after 1918, but still with little success. It was not until mechanical engineering had advanced sufficiently with the development of mass-produced motorcars, and aerodynamics had become a much more exact science, that the first really practical helicopters began to emerge.

## The Autogyro

Meanwhile, the foundations of today's helicopters were laid with the development by a Spanish engineer, Juan de la Cierva (1885–1936), of the autogyro, or gyroplane. (Cierva chose the name 'Autogiro' for his invention, and it became the registered trade mark of the Cierva Autogiro Company. The commonly used generic term, however, is spelled *autogyro*.)

During the 15 years from the mid-1920s until the outbreak of World War II autogyros were built and flown in many parts of the world. Several hundred of the best-known type, the Cierva C-30, were built during the 1930s and they completed many thousands of flying hours. The rotor performance data so accumulated was of inestimable value in the design of contemporary and later helicopter rotors. (The differences between the operating principles of the autogyro and the helicopter are explained on pages 199–200.)

It was the simplicity of the freely rotating autogyro rotor that gave Cierva his main advantage over contemporary helicopter designers. During this early period most helicopter projects were based on the use of power-driven rotors mounted in contra-rotating pairs, so that the torque reaction due to one rotor was counteracted by the opposite reaction due to the other. Of the two helicopters that did achieve short flights before World War I, the Breguet-Richet No. 1 had originally had four rotors, but was later modified to have a

**Previous two pages** Bell OH-58A multi-role helicopters of the US Army. The remarkable development of rotorcraft during the last 30 years has been inspired largely by their importance in military roles.

twin co-axial contra-rotating (CXR) rotor system: and the Cornu helicopter had twin tandem rotors (TTR), also contra-rotating.

In contrast, the autorotating rotor of Cierva was torqueless and so needed neither duplication nor the complex mechanical transmission systems that were essential in helicopter design.

## Early Helicopters

While Cierva was making great strides in the development of his single-rotor autogyros, a group of persistent helicopter pioneers continued to experiment with a variety of contra-rotating powered-rotor designs. Their goal was sustained hovering flight, which autorotating rotorcraft could never be expected to achieve. In France the Breguet CXR configuration was followed by a project from the Marquis de Pescara, an Argentinian working in Paris. His helicopter was notable for being controlled by what was the first practical attempt at a cyclic-pitch-control system (see page 201). The machine had two pairs of contra-rotating biplane rotors driven by a 170 hp Le Rhône engine. It flew quite successfully, but it was far too complex mechanically. Nonetheless, the control system Pescara pioneered is now almost universally used on rotating-wing aircraft.

Also in France a Peugeot motorcar engineer, Étienne Oemichen, built a series of four-rotor helicopters, the best of which flew a distance of just over 500 m (1,640 ft) in 1924 – then a world record. At about the same time a Russian emigrant in America, Professor George de Bothezat, built a larger four-rotor machine. In both Oemichen's and Bothezat's helicopters the rotors were driven in two contra-rotating pairs to counteract torque, but both could make only short, low-level flights.

Another configuration tried during the 1920s had two contra-rotating rotors mounted side by side (TSR) on outriggers, one attached to each side of the fuselage. The first to use this design was Emile Berliner, in America, who fitted the two rotors to a Nieuport biplane fuselage. They were, in fact, rigidly mounted lifting propellers, and this type was never really controllable in the short flights it achieved.

The usual means of vertical-lift control was solely by use of the engine throttle to vary rotor speed: the higher the speed of rotation, the greater was the lift generated by the rotor blades. The idea of using mechanical means to vary the blade-pitch setting (see page 201) was beginning to be seen as a possibility – but how to do it was quite another matter.

In the Soviet Union the doyen experimenter Professor Boris Yuriev, who as a young man had made unsuccessful attempts to build a helicopter before World War I, was instrumental in developing several configurations. Working with design teams of the Central Aero-Hydrodynamics Institute, he made notable contributions to the design of the Russian Omega series of side-by-side twin-rotor helicopters. At the same time he was helping another team which produced a single-rotor helicopter with two small vertical anti-torque rotors, one at the nose and one at the tail. This design made a few short flights but was then abandoned. Later Yuriev was to work with Professor Mikhail Mil who, from a background of autogyro design, was destined to become the Soviet Union's leading helicopter designer.

In 1930 the Italian designer D'Ascanio followed the Breguet CXR configuration with a similar experimental machine which was reasonably controllable. Each blade in the two four-bladed rotors was fitted with a trailing-edge stabilizing surface intended to perform

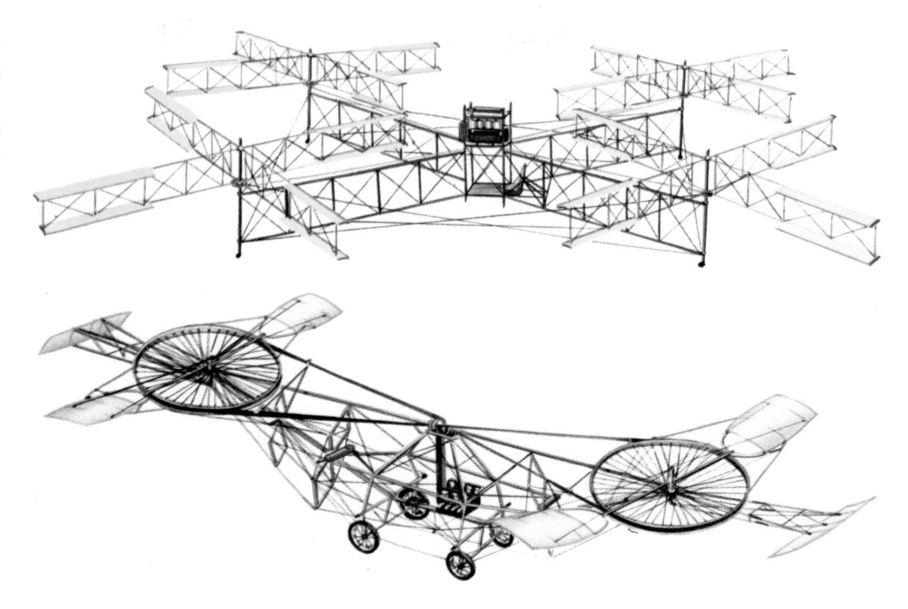

**Top** The Breguet-Richet Gyroplane No 1, with four biplane rotors driven by a 40 hp Antoinette engine, was the first helicopter to lift a man off the ground in 1907, although it had to be 'tethered' by four men to keep it stable.

**Above** Also in 1907, the Cornu tandem-rotor helicopter achieved the first untethered man-lifting flight, remaining airborne for some 20 seconds.

**Below** The Cierva C-30 autogyro. Rotor-head control was by means of the stick hanging from the pylon cowling to the pilot in the rear cockpit.

The Focke-Achgelis Fa 61, seen here piloted by Hanna Reitsch on a test flight in 1938, was the first successful German helicopter. Its two contra-rotating rotors were mounted on outriggers on either side of the fuselage.

the same stabilizing function as the tail plane of an aeroplane. These stabilizers also embodied an advance upon Pescara's early method of cyclic-pitch control. The D'Ascanio helicopter succeeded in raising the world helicopter altitude record to 18 m (59 ft), the distance record to 1.08 km (0.67 miles), and the endurance record to 8 min 45 sec. Such was the state of the helicopter art in October 1930.

A few years later Breguet came back into the picture with a more advanced version of his earlier CXR helicopter, in collaboration with another French engineer, Rene Dorand. By 1936 the Breguet-Dorand 314 helicopter, as it was designated, had taken the world helicopter altitude record to 158 m (518 ft) and was demonstrating its ability to fly at speeds of up to 105 km/h (65 mph). It was powered by a 300 bhp Hispano-Suiza engine and its twin contra-rotating two-bladed rotors, 16.4 m (53 ft 10 in) in diameter, were substantially larger than most other rotors of the period.

The Breguet-Dorand 314 was the world's first helicopter to demonstrate reasonable controllability both as a helicopter and, with the engine cut off, as an autogyro for emergency landing. It was also the first machine to make effective use of a rotor-blade cyclic-pitch control system to tilt the rotor in flight, combined with collective-pitch change for vertical control (see page 201).

The Cierva company used the tilting-hub method of rotor control in all their autogyros. In this system the rotor hub was mounted on a universal joint, with the mounting connected through a rod-and-lever mechanism to the pilot's control stick. Movement of the stick in any direction caused the rotor to tilt in that direction, and the autogyro followed. The method was mechanically simple and was well suited to the autogyro with its freely rotating rotor. No collective-pitch change was necessary because the autogyro used a fixed angle of blade pitch.

The cyclic-pitch system was more suited to the helicopter, as it would have involved considerable mechanical complexity to tilt the rotor hub while it was under constant power drive from the engine. It is interesting to note that Pescara's original cyclic pitch-change system did not use pitch-change bearings. At that time (the early 1920s) Pescara had to be content with a rudimentary system of warping the blades cyclically after the manner in which the Wright brothers had warped the wings of their *Flyer* to obtain lateral control. Success with the cyclic-pitch system had to wait for the development of more advanced engineering techniques than Pescara had at his disposal. In some helicopters today, pitch-change bearings have been superseded by torsion straps, which are smoother in operation.

After the Breguet-Dorand 314 the next big step forward came with the development in Germany of the Focke-Achgelis Fa 61 helicopter. This was still a twin contra-rotating rotor helicopter but the rotors were mounted side by side (TSR) on outriggers, as in the earlier Berliner configurations. Piloted by Hanna Reitsch, an Fa 61 demonstrated perfectly controlled hovering flight in 1938 inside the Deutschland Halle sports stadium in Berlin. Hanna Reitsch was a skilful test pilot, and she also possessed another important attribute for a helicopter pilot of that period: she weighed only about 45 kg (100 lb). The machine, a single-seater powered by a 160 bhp Bramo engine, also established a world helicopter speed record of 122 km/h (76 mph), an endurance record of 1 hr 20 min 49 sec, and an altitude record of over 3,050 m (10,000 ft). The Focke organization in Germany had had an autogyro manufacturing licence from Cierva since 1931 and the Fa 61 was based on experience gained with the Cierva autogyros it had built.

The Weir W-5, a single-seat TSR similar in many respects to the Fa 61, was built in Scotland in 1938. James G. Weir, its designer, had earlier been instrumental in introducing Cierva into England, and he became chairman of the Cierva Autogiro Company. He held somewhat different views from Cierva, however, on the relative merits of the autogyro and the helicopter. The Weir W-5 was built by G. & J. Weir in Glasgow to avoid interfering with the Cierva company's autogyro programme at its English factory. In 1939 the two-seat Weir W-6 was produced, but the outbreak of World War II brought a halt to this and most European helicopter developments. It was left to Igor Sikorsky (1889–1972) in the United States to make the breakthrough that was so greatly to advance the world helicopter industry.

Sikorsky's principal contribution was to devise a practical means of counteracting rotor torque without the complexity of twin lifting rotors. Thirty years before, in 1909, he had built two twin co-axial-rotor (CXR) helicopters, but neither had been successful. His new idea used a small vertical rotor at the end of a long tail boom. This tail rotor produced lateral thrust to counteract the torque of a single main lifting rotor – torque that would otherwise have caused the fuselage to spin around on its axis in the opposite direction to that of the main rotor. The system proved successful, and the main and tail rotor configuration (MTR) became, and remains, the most popular for all types of helicopter.

There had been two earlier attempts to use the lateral thrust of a small vertical rotor to counteract torque: one by Yuriev in the Soviet Union and another by Baumhauer in the Netherlands. Neither had been successful. Sikorsky's design was simpler and better than either, and his first prototype, the VS-300, made its maiden flight in the summer of 1940.

The VS-300 was powered by a 75 hp Lycoming engine, the drive from which was connected directly to a main transfer gearbox. Drive shafts from the gearbox transmitted the power to the main rotorhead and to the tail rotor at the appropriate speed reductions. The main rotor was controlled by a cyclic- and collective-pitch-control system, and there was also variable-pitch control of the tail-rotor blades connected to the pilot's rudder pedals for directional control.

Encouraged by its success, the Sikorsky team put in hand the design of a two-seat version, the VS-316, with an 11.58 m (38 ft) rotor powered by a 185 bhp Warner Scarab engine. This type was given the military designation R-4 and was ordered by the US Army and British Navy for training pilots. More than 100 were built and the type became well known in many parts of the world during the last year of World War II. A more powerful version, the R-5, led Sikorsky towards the design of even bigger helicopters. Acceptance of the MTR configuration was assured by its success, and military demand for these first practical helicopters led to the formation at United Aircraft Corporation of the Sikorsky Aircraft Division, with its own works and design offices at Bridgeport, Connecticut.

## Post-war Development

Between the end of the war in 1945 and 1950, helicopter development attracted the interest – and sometimes the antipathy – of every branch of the aviation industry. By 1947 there were over 70 separate helicopter development projects in the United States alone. In Europe three nationalized French aircraft constructors, SNCA du Nord, du Sud-Est, and du Sud-Ouest, engaged in a variety of new projects, while in England Cierva, Bristol, Fairey, Saunders-Roe, Westland, and others made notable efforts to re-establish the major contribution to rotorcraft made by the United Kingdom before the war. In Germany, Italy, and the Soviet Union development was initially slower but has since increased greatly.

One of the fascinating aspects of the industry's formative years was the great variety of configurations put on test. Although the single-rotor MTR designs were in the majority, there were many who regarded the tail rotor as an unnecessary source of vibration and other technical problems, and they experimented

Igor Sikorsky at the controls of his first successful prototype, VS-300, in 1940. This version had two tail rotors – one vertical for anti-torque and directional control, the other horizontal to aid fore-and-aft control. Only the vertical tail rotor was retained in the final version.

with different configurations. The American Bendix company, best known as a maker of automobile brake systems and washing machines, produced a co-axial CXR helicopter which flew quite well but was not developed to a production stage. Some Russian designs, notably by Kamov, also used the CXR layout. A variation on this theme, with two closely intermeshing rotors (TIR), was tried by Kellett Aircraft Corporation, one of the Cierva company's pre-war autogyro licencees in America. This line of development in the United States was influenced by the German designer Anton Flettner, who was attached to the US Office of Naval Research after World War II. During the war he had pioneered the TIR configuration in Germany with the Fl 282 Kolibrie, of which only a few were built because his factory was repeatedly damaged by allied air attacks. Kaman Aircraft Corporation also built a similar TIR helicopter during the 1950s, and some of this company's more recent TIR models are still in service with the US Air Forces. All later Kaman helicopters, however, have the MTR layout.

The first compound helicopter (CMP) was built by Fairey Aviation in England – the five-seat Gyrodyne. This was a compromise between Cierva's original compound rotorcraft and the pure helicopter and was designed by Professor James Bennett, who had worked closely with Cierva. Powered by a 500 bhp Alvis Leonides engine, the Gyrodyne had an offset, forward-facing propeller instead of a tail rotor to counteract the torque of the single main rotor. In this way its anti-torque thrust also contributed to forward propulsion, and the Gyrodyne demonstrated its greater speed by setting a new world helicopter record of 200 km/h (124 mph) in June 1948. A much larger variant, the Rotodyne, was built; but neither machine went beyond the prototype stage.

Another configuration originally tried in Europe, the tandem rotor (TTR), was developed after the war in America by Piasecki Helicopter Corporation. The type continues in production by Boeing Vertol, which acquired the original Piasecki interests. The first Piasecki 'Flying Banana', as it was dubbed, made its maiden flight in March 1945. At that time, it was thought – erroneously as we now know – that there must be limitations to the diameter of practical rotor systems. Piasecki worked on the basis of using two rotors of a diameter known to be practicable to provide a greater lifting capacity in helicopters larger than any others then flying. Although far larger single-rotor helicopters have since been built, the TTR configuration has proved that it possesses many advantages, particularly for troop- transport helicopters. The latest and largest of a long line of such machines is the Boeing Vertol Chinook.

The Cierva W-11 Air Horse undergoing hovering trials in 1949. This triple-rotor was the world's largest helicopter at the time, but it was not successful and only two were built.

The side-by-side rotor (TSR) helicopter, pioneered by Focke and Weir before World War II, was also further developed immediately after the war. Platt Le Page Aircraft Company and McDonnell Aircraft Corporation were two American companies which did some work on this configuration, and it was also used by one of the Russian experimental design bureaux for their Omega-series helicopters. The more recent Russian Mil V-12 is another example of the TSR layout; at present it is the world's largest helicopter.

In addition to its use in the design of pure helicopters, the side-by-side arrangement is virtually the only possible configuration for convertiplanes (CVT), in which the rotors are positioned horizontally for vertical take-off and turned forward through 90° to provide thrust for forward flight. In the latter position, lift is provided by the wing on which the two rotors are mounted. One notable example of such a type was the Bell XV-3, which first flew in 1955.

## The Turbine Revolution

It was also in the mid-1950s that the next major breakthrough gave the fledgling helicopter industry an unexpected impetus. This was the application of the turboshaft engine to helicopter use. All the early helicopters, like their fixed-wing counterparts, had been powered by conventional piston engines. But in the helicopter they had many drawbacks, not the least of which was the elaborate apparatus needed to cool the engines during extended hovering flight at high power settings. (In a fixed-wing aircraft there is no problem because the engine is cooled directly by the high-velocity propeller slipstream.) The advent of the turbine engine for helicopters changed the picture completely.

Initially turboprop engines were far too large for the helicopters then in production. In the mid-1950s, however, a French company, Turboméca, produced a range of small fixed-shaft turbines with a power output between 400 and 500 shp (shaft horse power). Their dry weight was about 135 kg (300 lb), less than half that of an equivalent piston engine – a major attraction to helicopter designers.

Turbine engines also revealed many other attributes which made them ideal for helicopters. Because the turbine is designed to operate at much higher temperatures than a piston engine, cooling problems were greatly reduced. The turbine could also deliver a much higher proportion of its rated power for long periods – essential for a helicopter engaged in lifting or rescue operations, which often involve hovering for long periods at high power settings. The lower installation weight permitted larger engines than normally required to be used; they could be derated for normal operations, while retaining a considerable power reserve which the pilot could use in emergency. The turbine's only disadvantage was that it used more fuel than an equivalent piston engine, so a greater proportion of the helicopter's payload had to be set aside for carrying fuel. This drawback was more than outweighed by the turbine's advantages, however, and also by the fact that the lower-grade fuel used by the turbine was both cheaper and safer than that used by piston engines.

Turbine-powered helicopters were an almost instant success, and it was not long before free-shaft turbines made their appearance in the smaller sizes. This had the added advantage that, because there was no mechanical connection between the compressor stages of the engine and the power-output turbine, the complex and heavy clutch mechanism between the engine and rotor drive shaft was eliminated. In some smaller helicopters, where economy was the primary consideration, piston engines were still used; but for the majority turbines became the order of the day.

The Korean War (1950–3) highlighted the ease with which helicopters could fly deep into enemy-held territory to rescue shot-down aircrew or wounded soldiers cut off from their units. Thousands of lives had been saved in this way in Korea, and military strategists soon realised that a machine which could lift out men from behind enemy positions could, equally, take assault troops into strategic positions from which they could launch attacks from the rear or flank. Consequently, military orders increased sharply and the helicopter industry grew very rapidly into one of the biggest boom industries of all time. The effect was felt mainly in the United States, where it was accentuated by the demands of the Vietnam War; but in Europe and the Soviet Union a similar sudden growth took place.

At the peak of the Vietnam War, one United States manufacturer, Bell Helicopter Company, had orders for so many helicopters that it had to sub-contract, an arrangement which fully occupied almost the entire production facilities of the Beechcraft Corporation. Bell's Iroquois helicopter has been produced in as many as 15 variants, amounting to over 10,000 helicopters of this one type alone.

The Iroquois started life as Bell Model 204, powered by an 1,100 shp Lycoming T53 turbine. It had a two-bladed see-saw rotor of 13.41 m (44 ft) diameter, a gross weight of 3,856 kg (8,500 lb), and a disposable load of 1,805 kg (3,980 lb). The original US Army designation of the military version was HU-1 – whence its affectionate nickname of 'Huey', which has stuck even though the designation was later changed to UH-1. Designed as a 10-seat utility helicopter, the Huey has been used for virtually every kind of work of which a helicopter is capable: commercial versions are equally suitable as executive transports or for lifting concrete on construction sites; military variants have been troop transports, ground-attack machines, aerial ambulances, and general maids-of-all-work.

The Kaman HH-43B rescue helicopter used the twin intermeshing rotor configuration, powered by an 825 hp free-turbine engine. Its clamshell doors gave easy access for loading stretchers into the rear cabin. The large ball slung underneath the helicopter contained a fire-fighting chemical.

The latest civil version, the Model 214, is powered by a 2,930 shp Lycoming T55 turbine, derated to 2,050 shp for take-off; the reserve power of 880 shp is thus almost as much as the total power of the Model 204. Rotor diameter is 15.24 m (50 ft) and the maximum gross weight of 7,258 kg (16,000 lb) is almost twice that of the original. The enlarged cabin has 16 seats, including the pilot's, and the maximum disposable load has risen to 3,629 kg (8,000 lb).

There is also a twin-engined version, the Model 212, powered by a pair of coupled Canadian Pratt & Whitney PT6T turbines rated at 900 shp each. This version has a rotor diameter of 14.69 m (48 ft 2 in), a gross weight of 5,080 kg (11,200 lb), and seats for 15 including the pilot. In the 212, as in all the Iroquois series, the passenger seating is quickly removable, leaving the cabin clear for loading cargo through the wide doors; bulky cargoes can be slung on a cargo hook below the fuselage.

Most recent versions of the type are capable of 200–250 km/h (125–150 mph). A high-speed gunship version, the AH-1G HueyCobra (*see* Chapter 15), designed with a two-seat, slim-line fuselage to meet a special US military requirement in Vietnam, can exceed 350 km/h (217 mph) in level flight. Although in outward appearance the HueyCobra seems completely different from the rest of Iroquois series, all types have substantially the same dynamic systems. A twin-turbine version, the AH-1J SeaCobra, has a coupled power unit similar to that in the Model 212.

Originally the Iroquois series was fitted with the gyroscopic stabilizer bar developed by Bell for all their early helicopters, but later variants dispensed with this system. The short stabilizer bar was fitted to the rotor head at right angles to the two blades. If the rotor's flight attitude was affected by turbulence or other disturbance, the stabilizer tended to restore the attitude without the pilot having to make adjustments to the rotor pitch-change controls. The stabilizer was not an automatic pilot, but it did ease the handling problems in rough weather. A somewhat similar system was used on early Hiller helicopters, but in this case it was more of a servo-control system than a stabilizer.

Such control-stabilizer systems are still used on some helicopters, but in the more sophisticated types they have largely been superseded by electronic equipment – the so-called 'black boxes'.

## Naval Helicopters

In the United States it is not only the army and air force that have used vast numbers of helicopters: the navy has also been well to the fore. Rotorcraft have been used mainly in anti-submarine operations, in which a helicopter with a crew of four can often do the work of a surface escort vessel with a crew of 100 – and sometimes do it more effectively.

The Sikorsky Sea King, designed specifically for this anti-submarine role, has become familiar throughout the world during the past decade and is in service with many of the world's principal navies; it is built under licence in England by Westland Aircraft, in Italy by Giovanni Agusta, and in Japan by Mitsubishi. The Sikorsky company developed rapidly after its initial successes and its progress has always been towards construction of the larger types of helicopter. The Sea King, following the classic Sikorsky MTR configuration, has a five-bladed main rotor of 18.9 m (62 ft) diameter and is powered by twin General Electric T58 turbines rated at 1,500–1,800 shp each; its gross weight is 8,617 kg (19,000 lb) and the disposable load is 2,966 kg (6,541 lb). Normal cruising speed is 224 km/h (139 mph).

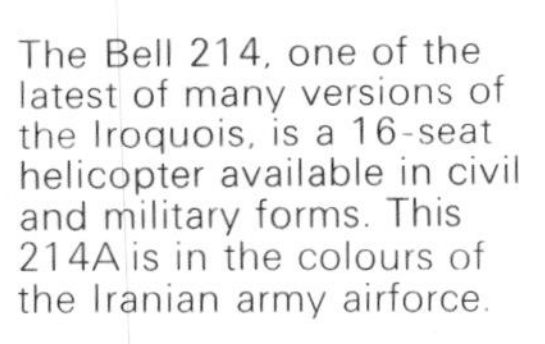

The Bell 214, one of the latest of many versions of the Iroquois, is a 16-seat helicopter available in civil and military forms. This 214A is in the colours of the Iranian army airforce.

Derived from the Sikorsky S-61 series, the first of which made its maiden flight in March 1959, the Sea King has many features to equip it for maritime operations. The most obvious is its watertight hull, which is designed on the lines of a flying-boat. With this and its retractable main landing wheels, it is equally at home in waterborne and overland duties. The Sea King has become familiar to millions of television viewers as the helicopter used to retrieve American astronauts after splash-down into the Pacific Ocean. Two civil versions, the S-61L and S-61N, are used mainly for public-transport airline work and for supplying offshore oil rigs. Their passenger accommodation is 24 to 28 seats, according to cabin arrangement. The versions built under licence in England, Italy, and Japan may have minor design variations.

The Sea King is not the only type of naval helicopter. There are many naval duties, other than anti-submarine work, that call for the use of medium-sized helicopters. Some of these duties require helicopters to operate from the decks of small naval craft at sea.

## Jet-driven Rotors

The first recorded suggestion that jet reaction at the blade tips could be used to power a rotor system seems to have been made in 1842 by an Englishman, W.H. Phillips; but it was not until just 100 years later that practical experiments with this method began. The prospect of a jet-driven rotor was of great appeal to many of the early helicopter pioneers, who believed it would make possible an extremely simple machine. In principle they were right: the jet rotor, being driven from within itself, has no torque to counteract, so it obviates the need for either contra-rotating rotors or a tail rotor. It also avoids the complexities of gearboxes, clutches, transmission shafts, and their accessories, and so reduces design weight to a minimum.

The drawbacks are only few – but they are formidable. The greatest difficulty is to design a leak-proof rotating seal at the rotor head to convey compressed gas, or air and fuel, from the fuselage to the rotor-blade tips. The other main problem is that fuel consumption by the jets is extremely high, so that the weight saved by the absence of a transmission system is offset by the greater weight of fuel needed to provide an equivalent endurance. Noise is also a serious problem.

The first successful single-jet-rotor (SJR) helicopters were built in Austria by Baron von Doblhoff during World War II. His first experimental machine, a single-seater, was powered by a 90 bhp piston engine driving a supercharger unit from a larger aero-engine as the compressed-air source. The compressed air and vaporized fuel was ducted through the rotor head to nozzles at the blade tips, where it was ignited by a conventional sparking plug. Later, a slightly larger version was built and flown successfully before being captured by American forces in 1945. It was taken to the United States, where Doblhoff himself was attached to McDonnell Aircraft Corporation and helped further development of pressure-jet helicopters.

At the same time, some members of Doblhoff's design team were attached to the French nationalized company SNCA du Sud-Ouest, while others were attached to Fairey Aviation in England. With their guidance, both companies embarked upon jet-rotor development

**Above** The Westland Sea King is the Royal Navy's principal submarine-hunter. A development of the Sikorsky S-61, this British version is powered by two 1,500 shp Rolls-Royce Gnome turboshaft engines, giving it a maximum speed of 250 km/h (155 mph) and a range on maximum fuel of 1,500 km (930 miles).

**Below** The two-seat SNCA SO-1220 Djinn, which flew in 1953, was the first helicopter powered by cold-jet (compressed air) to reach quantity production. It was built in both civil and military versions, and was relatively easy to fly.

**Right** The Bell 206 JetRanger, a five-seater introduced in 1967, has proved one of the most popular light turbine-powered helicopters. Suitable alike for construction work (as here) or as an executive transport, it has also sold widely in various military versions. Its use of a 'nodalized beam' to connect the rotor hub and the fuselage greatly reduces vibration within the cabin.

programmes. In England this led to the construction of the Fairey Rotodyne and the Fairey Ultra-light helicopter. Both had a pressure-jet-driven rotor, but neither went beyond the prototype stage. In France, the SNCA SO-1220 Djinn helicopter was developed after a few false starts, and this type was put into quantity production. It was the only SJR helicopter in the world to reach full-scale production; even so, only about 100 were built before production was discontinued.

The Djinn (Goblin) was a two-seat helicopter with an empty weight of only 360 kg (794 lb). It was powered by a Turboméca Palouste turbine modified to deliver compressed air only (it had no output drive shaft), and this was ducted through the rotor head to be ejected through nozzles at the tips of the two-bladed rotor. The system was termed a warm-air jet because no combustible fuel was added; systems in which fuel is burnt at the blade-tip nozzles are termed hot jets. The Djinn was reasonably successful for pilot training and for light crop-spraying work, but it was never popular enough to justify further development.

## The New Generation

A constant process of refinement over several decades has turned the helicopter into a sophisticated means of transport, fully comparable in its own sphere to fixed-wing aircraft.

**Below** One of the new generation of helicopters, the Messerschmitt-Bölkow-Blohm Bo-105 has an advanced hingeless rotor conferring great manoeuvrability. The version shown here is in Netherlands airforce colours; other military versions are equipped with anti-tank missiles.

The overwhelming majority of all types produced has been used for military purposes, but the enormous variety of uses that have been developed for helicopters in war has begun to disclose important new applications in the civil field.

An obvious example is the petroleum industry and the delivery of urgent spares to off-shore drilling rigs. The cost of a specially equipped helicopter that can guarantee delivery in all weather conditions is much less than the cost of having several rigs out of action for days on end.

In this and many other ways helicopter manufacturers are diverting technological progress enforced by the needs of war into commercial channels. And while military orders continue to form the bulk of the market – and represent the major area of research and development – the use of civil helicopters is expanding steadily.

Typical of the new generation of commercial helicopters are the latest types under development by Bell Helicopter Company in America, Messerschmitt-Bölkow-Blohm in West Germany, and by Giovanni Agusta in Italy. The Bell Model 222 is a medium sized, 8–10 seat, twin-turbine MTR helicopter designed for high-speed communications in all weathers. Its two-bladed rotor, following Bell's standard design of a see-saw flapping hinge, is 11.89 m (39 ft) in diameter. The 222 rotor is similar to that on the latest Iroquois, having a special vibration-absorbent suspension to ensure smoother flight, and pitch-change bearings that need no lubrication. Power is derived from coupled twin Lycoming LTS101-650C turbines of 600 shp each; the gross weight is 3,039 kg (6,700 lb), and normal cruising speed is 259 km/h (161 mph).

The MBB Bo-105 from West Germany is rather smaller, with 5–6 seats and a gross weight of 2,097 kg (4,630 lb), but is also twin engined to provide safe all-weather operation. Power units are Allison 250-C20 turbines of 400 shp each, which give a cruising speed of about 235 km(h (146 mph). The MTR configuration of the Bo-105 is conventional, but the advanced four-bladed main rotor is the result of many years' research and development by the company. The four blades, of 9.83 m (32 ft 3 in) diameter, are fabricated in glass-reinforced plastics and have no flapping or drag hinges. The hingeless system is designed for smoother flight and improved handling. Pitch-change bearings are embodied in the rotor head to accommodate a conventional cyclic-pitch-control system. The cyclic-pitch system itself is designed to compensate for unequal lift generated on the advancing and retreating sides of the rotor, and so replaces flapping hinges.

**Above** The Aérospatiale/ Westland Lynx is another type with four hingeless rotor blades. Powered by two 900 hp turbines, the Lynx is capable of 322 km/h (200 mph). It is seen here landing on HMS *Sheffield*, a guided-missile destroyer.

Slightly larger, at 2,450 kg (5,402 lb), the Agusta A-109 is a 6–8-seat MTR helicopter powered by twin Allison turbines similar to those in the Bo-105. The A-109 also has a four-bladed rotor, but it is fully articulated, not hingeless, and has a diameter of 11 m (36 ft 1 in). Normal cruising speed is 233–272 km/h (145–169 mph). All three of these advanced designs can fly on one engine in the event of partial power failure, and all are equipped for all-weather and night flying under 'instrument-flight rules'.

The Westland Lynx is another advanced design, but at present it is being produced mainly to military specifications. Of conventional MTR configuration, the Lynx has a hingeless four-bladed rotor similar to that of the Bo-105. The rotor head is a single forging of titanium, and the blades are fabricated in a combination of stainless steel and glass-reinforced plastics; rotor diameter is 12.80 m (42 ft). Power is supplied by twin Rolls-Royce

**Below** The Aérospatiale SA.318C Alouette (Lark) II, a turbine-powered French helicopter that first flew in 1961, is widely used by police forces and other civil as well as military customers.

Gem turbines of 900 shp each, and the gross weight is 4,309 kg (9,500 lb). Normal cruising speed is 270 km/h (167 mph), but the type has achieved 322 km/h (200 mph). The Lynx is being developed jointly as a high-performance helicopter by Westland and the French company Aérospatiale.

In the heavy-lift category, Sikorsky and Boeing Vertol are the present leaders, apart from the giant Russian Mil helicopters. The latest version of Sikorsky's largest current model, the S-65, has a maximum gross weight of 31,631 kg (31 tons), a maximum load capacity of some 16,270 kg (16 tons), and a total power of 13,140 shp in a triple General Electric T64 turbine installation. Its six-bladed main rotor is about 24 m (almost 80 ft) in diameter. (The number of blades, incidentally, does not necessarily reflect any degree of advancement in rotor design). The Boeing Vertol Chinook tandem-rotor helicopter has a gross weight of 20,865 kg (20.5 tons) and a disposable load of about 12,200 kg (12 tons).

The Russian Mil V-12, with a lifting capacity of about 40,700 kg (40 tons), is substantially larger than any of these. In 1956 the Mil design bureau produced what was then the world's largest helicopter, the Mi-6, with a single main rotor of 35 m (114 ft 10 in) diameter powered by twin TV-2BM turbines of 5,500 shp each. A helicopter with a larger diameter rotor of 39.62 m (130 ft) had been built in 1948–50 in the United States by the Hughes company but this was soon abandoned. To produce the V-12, which was flying in the late 1960s, the Russians mounted two Mi-6 rotors, each with its twin-turbine power pack, on outriggers from a large new fuselage with an aeroplane-like tail. The result was a four-engined TSR helicopter with a total installed power of some 22,000 shp. It is believed that the Mil V-12 is used for carrying heavy cargoes in the remote developing regions of the Soviet Union, but it is not known how many have been consutructed.

In all, there are more than 100 different types of helicopter in current civil and military service, and of these by far the most numerous are the 5--10 seaters. Helicopters such as the Bell JetRanger, Hughes 300 and 500, and the little Enstrom three-seater are a common sight going about their daily affairs. One promising newcomer in this category is the five-seat Aérospatiale Ecureuil (Squirrel). Efforts have been made to revive the autogyro at various times over the past 30 years but never with more than spasmodic success, except for a few simple sporting single-seaters. Although the helicopter is considerably more expensive to build and to operate than the autogyro, its ability to hover gives it a decisive advantage.

**Above** The Boeing Vertol CH-47 Chinook, a tandem-rotor design, is powered in its latest versions by two 3,750 shp turboshafts. Its lifting capacity is demonstrated by this large cargo container, suspended from two hooks to prevent it twisting in flight.

**Above** The Russian Mil V-12, which first flew in the late 1960s, is the largest helicopter in the world. Its two massive twin-turbine rotors are mounted on outriggers. Its freight compartment, loaded via front clamshell doors and rear ramp, is 28 m (92 ft) long and 4.4 m (14.5 ft) square.

**Left** A Sikorsky S-64 Skycrane demonstrates its capacity by lifting a damaged seven-ton Sikorsky CH-3E rescue helicopter out of the battle zone during the Vietnam war. The Skycrane's landing gear can be lengthened or shortened hydraulically, making it easier to secure loads of different heights.

# Autogyro and Helicopter Rotors

Although the autogyro and helicopter are superficially similar, their rotors operate in quite different ways. In the autogyro the pilot's control stick is in the central position during level forward flight and the rotor is tilted slightly backwards. There is no tendency for the autogyro to fly backwards because the forward thrust of the propeller counteracts any backward drag of the rotor. Since the helicopter has no propeller, the rotor itself must provide forward thrust as well as lift. For this reason it is tilted forwards in level forward flight, the tilt being induced by the pilot moving the control stick forward.

It follows, then, that in forward flight air flows upwards through the autogyro rotor and downwards through the helicopter rotor. As the diagrams show, the *blade-pitch angle* (or angle of incidence) is the angle between the chord line of the blade section and the line representing the plane of rotation. If the leading edge of a rotor blade is above the plane of rotation, the angle is said to be positive; if it is below the plane, it is said to be negative. (The angles in the diagram on page 200 have been exaggerated for clarity; the actual values would be about 2° positive in the autogyro and 10° positive in the helicopter.)

Now, in terms of the aerodynamic forces that affect a rotorblade, the important factor is not its angle of pitch but its *angle of attack* – that is, the angle between the blade's chord line and the direction of the relative airflow. In both the autogyro and the helicopter the relative airflow is a combination of (a) airflow due to the blades' own rotation, and (b) airflow through the rotor due to movement of the whole machine through the air.

As we have seen, in the case of the autogyro that flow is upwards through the rotor. In the autogyro blade-section diagram the airflow due to rotation is shown (to the left of the blade section) by the line in the plane of rotation that is marked with a single arrow, and the upward-flow component is shown by the shorter line, also single-arrowed, at right angles

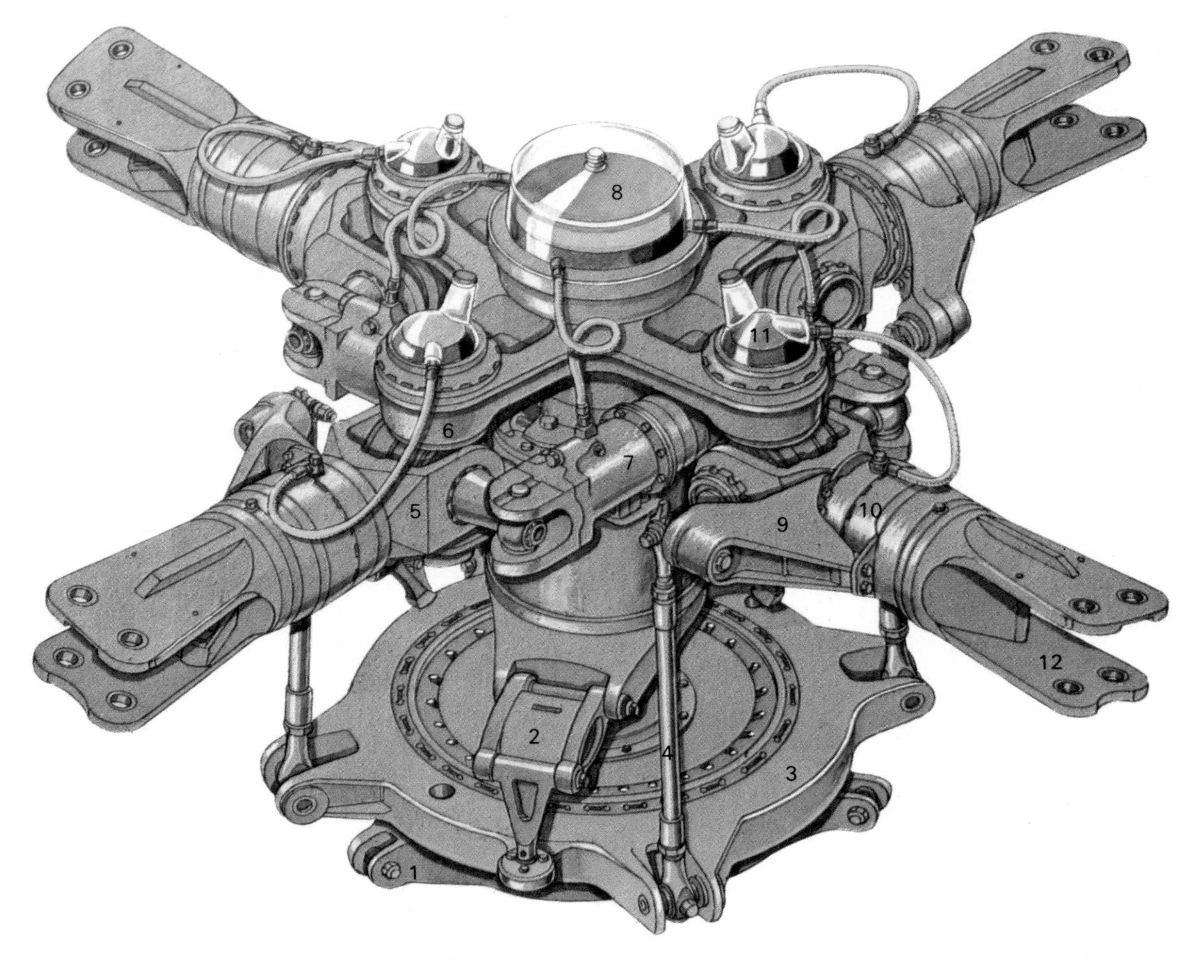


Rotor head of an Aérospatiale/Westland SA.330 Puma, a combat-support transport helicopter powered by twin turbine engines. The numbered components are: 1 Lower swashplate, 2 Scissors link, 3 Upper swashplate, 4 Connecting rod, 5 Flapping hinge, 6 Drag hinge, 7 Drag hinge hydraulic damper, 8 Drag hinge hydraulic damper reservoir, 9 Pitch-change arm, 10 Pitch-change bearing, 11 Pitch-change bearing lubrication reservoir, 12 Blade attachment point.

to it. The combination of the two, double-arrowed, completes the triangle of forces and defines the angle of attack. In the triangle drawn above the blade section, lift generated by the blade is shown as a single-arrowed line at right angles to the angle-of-attack line, and drag is shown by the short line at right angles to lift. The total resultant force, a combination of lift and drag, is shown by the double-arrowed line completing the triangle of forces. This triangle lies *in front* of the axis of rotation, which means that the blade would tend continuously to increase its rotational speed.

Although it is the angle of attack which directly affects the aerodynamic forces, it is clear from the diagrams that this, in turn, is governed by the blade-pitch angle. If the pitch angle shown in the diagram were slightly increased, so that both lift and drag were greater, a point would be reached at which the double-arrowed line showing the combined force would lie exactly in line with the axis of rotation. This point represents a state of equilibrium in which the blade's rotational speed will tend neither to increase nor decrease. This is the state of *autorotation* discovered by Cierva, and it occurs when the blade-pitch setting is fixed at approximately 2° positive. Above this setting the blades will tend to slow down; below it they will tend to speed up progressively until they approach the 'windmill' state, in which negligible lift is generated and only rotational energy is produced.

In contrast, the equivalent triangle of forces in the helicopter diagram lies *behind* the axis of rotation, indicating that the helicopter blades would tend to slow down if they were not driven by engine power. The combination of the helicopter blade's much higher pitch angle and the downward flow of air through the rotor has the effect of shifting the direction of its resultant lift force behind the axis of rotation. Its angle of attack, however, is almost the same as that of the autogyro.

If a helicopter's engine fails the blades would quickly slow down and stall if left at a high pitch angle, so that the rotor would fold upwards like an umbrella turning inside out. The pilot prevents this happening by reducing the blade-pitch angle. The helicopter's variable-pitch mechanism allows blade pitch to be changed from a maximum of about 15° to a minimum of 2° – the same as that of the fixed-pitch autogyro blade. In effect, the helicopter is converted into an autogyro and descends gently with the rotor in autorotation.

## Rotor-blade Hinges

During forward flight of both the autogyro and the helicopter, blades on the advancing side of the rotor generate more lift than those on the retreating side, owing to the greater relative airflow. This could cause lateral instability if it were not compensated for, and one solution is to insert a hinge bearing between each blade root and the hub. This allows the blade to hinge upward in response to greater lift on the advancing side and then to hinge downward on the retreating side. The effect of the blade rising is to reduce its effective angle of

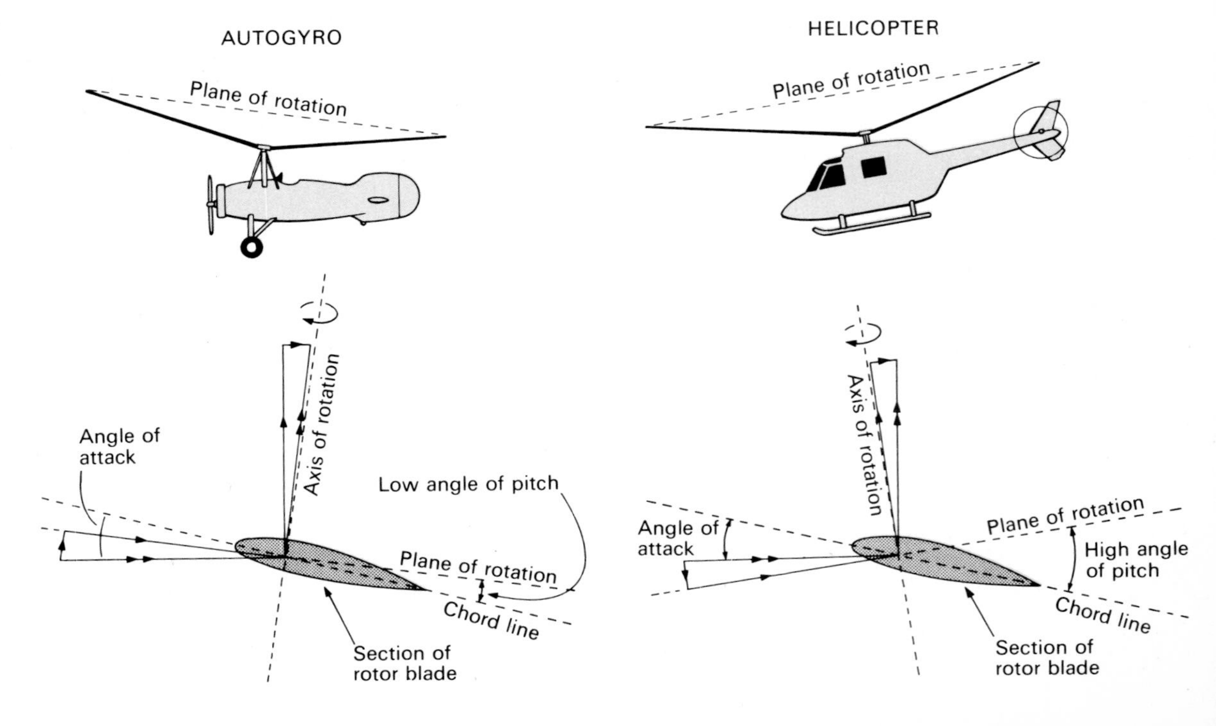


The operation of the rotors in an autogyro and a helicopter. An explanation of the forces represented by the arrowed lines is given in the text.

Pitch sectors on a rotating helicopter blade. Each blade goes through a cycle of increasing and decreasing angle of pitch during the course of a single rotation.

Rotor-blade hinges. The flapping and drag angles have been greatly exaggerated for clarity.

attack, and thus to equalize the lift on the two sides of the rotor. The bearing is termed the *flapping hinge*.

The operation of this hinge, however, causes the blades to bend slightly in the plane of rotation, successively widening and narrowing their normal angle of separation from each other. The effect is only slight, but because this bending could cause fatigue it has to be relieved by incorporating a second hinge, termed the *drag hinge*, at right angles to the flapping hinge. These two hinges in combination provide the equivalent of a universal joint between each blade and the rotor hub; rotors so equipped are termed fully articulated. On two-bladed rotors some helicopter manufacturers (notably Bell and Hiller) use a 'see-saw' hinge, which is the equivalent of two flapping hinges, and dispense with the drag hinges; such rotors are called semi-articulated.

## Pitch-change Systems

The means to enable the helicopter pilot to vary blade pitch for control in flight is provided in the rotor head mechanism. The key component of this assembly is the *swashplate*, which is fitted below the head on a universal mounting, so that it can be tilted in any direction. It can also be moved up or down.

The swashplate is made in two halves on a common bearing. The upper half turns with the rotor head and is linked to it by a *scissors link*, which ensures that its rotation is constant with the head, irrespective of any angle of tilt or up-and-down movement. The lower half is connected to the pilot's control levers.

Connecting rods between the rotating upper half of the swashplate and the blade-pitch arms translate swashplate movements induced by pilot control into the required blade-pitch changes. For instance, when the pilot pulls up the *collective-pitch lever* the whole swashplate is moved upwards and the connecting rods actuate the pitch arms in unison. These twist all the blades simultaneously into higher pitch angles about their pitch-change bearings. The collective-pitch lever is also connected to the engine throttle, so an increase of collective pitch produces an increase of power.

Tilting the swashplate in any direction is done by movement of the control stick. This is sometimes called the *cyclic-pitch stick* because it varies the blade pitch cyclically. When the helicopter pilot holds his stick central, the swashplate is level and the rotor remains horizontal for stationary hovering flight. Movement of the stick away from the central position tilts the swashplate in the direction of stick movement, and the connecting rods are drawn downward by the tilt on one side of the plate and pushed upward on the opposite side. This twists the blades about their pitch-change bearings so that, in the course of a single revolution of the rotor, each individual blade is made to fly low as its pitch is reduced and high as its pitch is increased again. For example, if the blades have a 10° pitch angle in hovering flight, a 5° swashplate tilt will cause them to reduce pitch to 5° on one side of the rotor and increase it to 15° on the opposite side.

The overall effect is to tilt the rotor's plane of rotation to the same angle as that of the swashplate, and the helicopter moves into the direction of the tilt. Tilting the swashplate and moving it up and down are quite independent functions of the same mechanism. For some manoeuvres, a combination of the two is required at the same time.

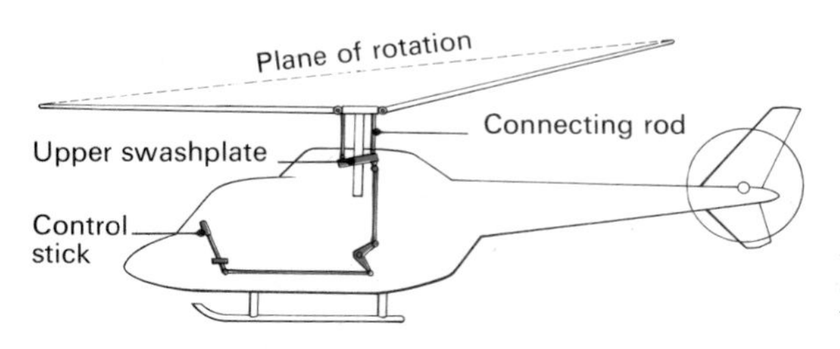


Linkage, in simplified form, of the control stick and rotor blades via the upper swashplate and connecting rods. Movement of the control stick forward (as here) has tilted the upper swashplate forward; this, in turn, alters the cyclic pitch of the blades to a forward-flight setting.

# 24 V/STOL

Even before the helicopter had matured as a practical flying machine, the beginning of a new, *convertiplane* philosophy was becoming evident with the development by several design teams of various novel forms of vertical take-off (VTO) aircraft.

Cost effectiveness in aviation is closely related to cruising speed, but because such a large proportion of the helicopter's available power is absorbed in sustaining lift, the proportion of power available for propulsion was, and still is, limited. In contrast, the principle of the convertiplane is that after vertical take-off, by whatever means, all the power can be converted to horizontal thrust as soon as the machine is wing-borne in its cruising-flight altitude.

In quite a different category, there has also been a nucleus of aircraft designers who, from the outset, have regarded it as a dangerous trend that aeroplane take-off and landing speeds have increased progressively with the advance of higher-performance aircraft.

Ignoring the extremely high landing speeds of supersonic fighters and the like, which are a special case, the effect of simple streamlining improvements on typical civil aircraft has been to increase landing speeds considerably over the years. In 1930, for example, the average landing speed for such an aircraft would have been in the region of 80 km/h (50 mph). Today an equivalent modern aircraft may well land at around 160 km/h (100 mph) or even faster.

From these two distinct motivations there have emerged two distinct types of aeroplane: the VTOL (vertical take-off and landing) aircraft and the STOL (short take-off and landing) aircraft. Both, in their respective spheres, can be eminently successful, but the two categories have little in common. STOL aircraft, which depend on maximum wing lift for take-off, have never been intended as a kind of half-way house towards the VTOL type, which uses wing lift only in cruising flight. Nor have VTOL designers ever seriously regarded the STOL capability (which most VTOL aircraft possess) to be in direct competition with the quite different STOL designs.

## VTOL Tail-sitters

During and immediately after World War II there were several attempts at building VTO aircraft. Most were 'tail-sitters', designed to lift off vertically by means of thrust from large-diameter propellers. Their landing wheels, on which they sat poised vertically before take-off, were behind the tail unit. Once airborne, the tail-sitters could be manoeuvred into the horizontal, wing-borne attitude and continue as high-speed, fixed-wing aircraft. In the United States Lockheed and Convair were engaged in such projects, while in Germany Focke-Wulf also had an advanced tail-sitter design intended as a high-speed fighter. None could be described as practicable; the American projects did fly but the maximum propeller thrust was hardly sufficient to maintain adequate control when landing. Development of the true VTOL aircraft had to await development of the turbojet engine but, in the meantime, STOL aircraft were beginning to emerge.

**Above** A Fieseler Storch demonstrating STOL capability in the centre of Berlin in 1939.

## Origins of STOL

In the period between the two world wars, when the science of aerodynamics was progressing rapidly from its try-it-and-see beginnings to an advanced level of sophistication, various lift-improving devices had made their appearance. Flaps, slats, and other ingenious systems were developed and incorporated in aircraft designs of the day (*see* Chapter 7). All were designed to improve wing efficiency and handling characteristics at low airspeeds.

The need of the military authorities for light aircraft capable of operating with an army from forward field positions further stimulated development. The operating conditions of these army aircraft were similar in some

respects to those experienced in bush flying. Their duties were mainly for reconnaissance and artillery spotting, and the pioneer in this category was the German Fieseler Storch (Stork), which was in service before World War II. Aptly named, with its exceptionally long-legged undercarriage, the Storch was designed to take off and land in an accentuated nose-up attitude, with the wing at a high angle of attack. The wing was fitted with leading-edge slats and wide-span flaps which gave it a landing speed in the region of 48 km/h (30 mph). When operating in a moderate breeze, the Storch could demonstrate a VTO capability almost equivalent to that of a helicopter.

Since the war one of the leading companies engaged in the design of more conventional types of STOL aircraft has been de Havilland of Canada. With a bush-flying market on its doorstep this company soon had an intimate knowledge of the requirements. The DHC-2 Beaver, with slotted flaps and slotted ailerons, appeared in 1947. Many are still giving valiant service in various parts of the world. Powered by a 450 hp Pratt & Whitney Wasp engine, it has accommodation for eight, including the pilot, or for equivalent cargo. It can take off at full load with a run of only 170 m (560 ft). From the DHC-2 the company has developed a whole range of STOL aircraft with progressively improved performance and payload. The types include the DHC-3 Otter, DHC-4 Caribou, DHC-5 Buffalo, DHC-6 Twin Otter, and the four-engined DHC-7. This last, known as the Dash-7, has accommodation for 50 passengers or almost 5,100 kg (5 tons) of cargo, but it can take off in about 300 m (1,000 ft) at full load in still air. With reversible-pitch propellers, its landing run may be only half this distance.

## Augmentor-wing Concept

The development of STOL aircraft has reached the stage where it is a fully fledged branch of aeronautics in its own right. One of the latest design techniques involves the use of what is known as the augmentor-wing concept. A DHC-5 Buffalo has been rebuilt with this system to an American order from NASA (National Aeronautics and Space Administration), the prototype being designated XC-8A. It is fitted with full-span slats and full-span double-slotted flaps. The inboard flaps are capable of downward deflection to 75°; the outboard sections, which also serve as ailerons, can be drooped to 30° during take-off and landing, when the main flaps are also down. For lateral control they can have a differential movement – when one turns up the other turns down – of plus or minus 17° on either side of the drooped datum position.

To increase wing, flap, and aileron efficiency still further during a slow approach or climb-away, air can be bled from the Rolls-Royce Spey turbofan engines and expelled over and through the trailing-edge surfaces. Chokes in the ducting linked to the pilot's controls regulate this system to permit adequate control when the aircraft is in exaggerated nose-up attitudes near the stalling position. In addition, some of the air is used to activate a boundary-layer control (BLC) system to delay flap airflow separation.

To see an aircraft of this size – over 10,170 kg (10 tons) at full load – approach to land down an extremely steep glide path, and then come to rest almost immediately after touchdown, is a truly remarkable sight.

## Jet Lift

In a completely different category are the VTOL aircraft, which rely on pure-jet thrust to achieve vertical take-off. Their wings – like wings on most conventional aeroplanes – contribute nothing to lift until the machine has moved into forward flight. The development of jet-lift aircraft followed logically from the development of jet engines themselves, when it became apparent that even the early jets were capable of producing sufficient thrust to lift many times their own weight.

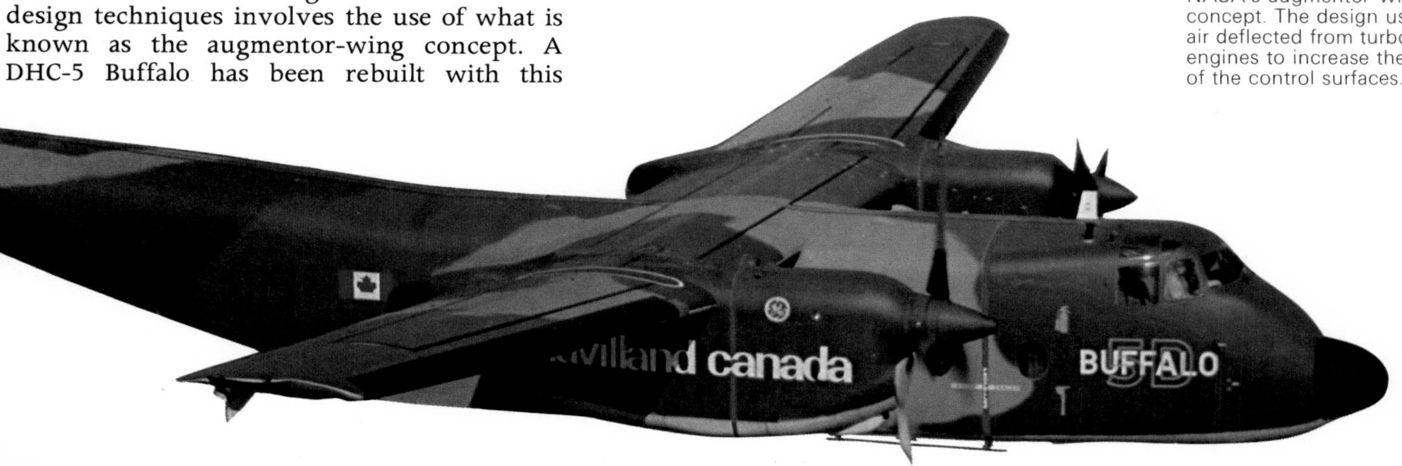


The De Havilland Canada DHC-5 Buffalo, an STOL tactical transport that first flew in 1964, has more recently been the subject of experiments with NASA's augmentor-wing concept. The design uses air deflected from turbofan engines to increase the lift of the control surfaces.

**Left** Rolls-Royce's four-legged Flying Bedstead, powered by two Nene turbojets, achieved vertical jet-lift by deflecting the jet efflux downward. Once airborne, the craft's attitude could be altered by swivelling four downward-pointing nozzles fed with air ducted from the engines' compressors.

Two significant advantages followed. On the one hand, there was a far greater power reserve than in the old tail-sitters to control the rate of descent in the landing approach. On the other hand, there was no longer a need for the aircraft to sit on its tail. Apart from one project (see below) the new jet-lift aircraft were all designed for horizontal take-off and landing, which is much easier for pilot control.

The first public demonstration of jet-lift was made in 1953 with the 'Flying Bedstead', a tubular-steel framework built by Rolls-Royce to mount two Nene turbojets horizontally, with their jet pipes facing inward. Deflector plates turned the jet efflux downward, to combine in one central downward blast which could lift the whole framework into the air. A ducting system fed a proportion of the compressed air bled from the engines to four downward-pointing nozzles at the ends of arms extending to left, right, front, and rear. These 'puffer pipes' could be swivelled by the pilot to control attitude in hovering flight.

A later jet-lift VTOL was the Short SC.1 single-seater of 1957, powered by five Rolls-Royce RB.108 turbojets designed specifically for this purpose with a high thrust to weight ratio. The engines were each rated at 1,002 kg (2,210 lb) take-off thrust for a dry weight of 122 kg (269 lb). A battery of four, ejecting downward, provided lift; the fifth, mounted horizontally, was used for propulsion. Since the aircraft's gross weight was 3,538 kg (7,800 lb), the SC.1 had a good reserve of thrust at take-off.

The SC.1 could not use all its installed power for propulsion: four-fifths of the power had to be shut down as soon as the machine moved into forward flight and became wing-borne. This is not the case with the vectored-thrust system, which was developed by Bristol Siddeley (now Rolls-Royce) for the Hawker P.1127 prototype. This system, which began flight trials in 1960, was based on the BS.53 Pegasus turbofan. The entire efflux is ejected through two pairs of swivelling nozzles. The forward pair eject air from the outer chamber of the engine, compressed by the fan, while the rear pair of nozzles eject the hot exhaust gas from the combustion chamber.

For vertical take-off the four nozzles are rotated to direct the jets vertically downward, while for forward flight they are turned rearward through 90°, so that the full thrust is used for propulsion. The difference in speed is pronounced. Whereas the SC.1, with only one-fifth of its installed power usable for horizontal thrust, had a maximum speed of 396 km/h (246 mph), the P.1127 could fly well over twice as fast. A developed version, named the Kestrel, could attain supersonic speed in a shallow dive, and the latest versions of the Hawker Siddeley Harrier – the culmination of this design principle – can be classed as truly supersonic aircraft.

**Right** The Short SC.1 used lightweight Rolls-Royce RB.108 turbojets directly inspired by the success of the Flying Bedstead experiments. The SC.1 was the first true aeroplane to make the transition from vertical take-off, via hover, to forward flight. Its engine arrangement – four turbojets for VTOL and one for propulsion – meant that it could use only one fifth of its installed power for forward flight.

The exciting growth of jet lift sparked off the development of a whole crop of VTOL aircraft. The French Dassault company embarked upon an ambitious programme to produce a VTOL version of their Mirage fighter. This started with the conversion of an existing machine which, in its modified VTO form, was known as the Balzac. Then a much-enlarged re-design of the Mirage III resulted in the Mirage III-V, with eight Rolls-Royce RB.162 lift jets for take-off and a SNECMA TF-306 afterburning turbofan for propulsion. It was intended that the development would lead to a Mach 2 VTOL fighter-bomber, but the programme was abandoned owing to cost and other problems.

A German company formed by Bölkow, Heinkel, and Messerschmitt in 1960 and known as Entwicklungsring (EWR) Sud, used a novel engine layout to avoid the complexities of a puffer-pipe control system. Their VJ-101C was a small high-wing monoplane with two wing-tip pods, each housing a pair of Rolls-Royce RB.145 lift jets. The pods could be swivelled through an arc of 94°, from horizontal to just beyond the vertical. For take-off, they would be vertical, with the jet efflux expelled directly downward. Also expelling directly downward was a third pair of RB.145 engines mounted vertically in the forward fuselage. The three pairs of lift jets formed a triangle, and their throttles were linked to the pilot's flight controls so that normal control movements varied the lift-jet thrusts to control the hovering attitude. For control in yaw the tip pods were swivelled differentially by the rudder pedals, a differential tilt of a few degrees being enough to turn the machine.

To move into forward flight, the tip pods were tilted slightly forward together. When forward speed had increased sufficiently for the aircraft to become wing-borne, the fuselage lift-jets were shut down and the tip pods tilted fully forward to the horizontal position. Although the VJ-101C was thus able to use two thirds of its installed power for propulsion, and although the prototype achieved Mach 1 in a dive, the project was later abandoned.

Another German project, by Vereinigte

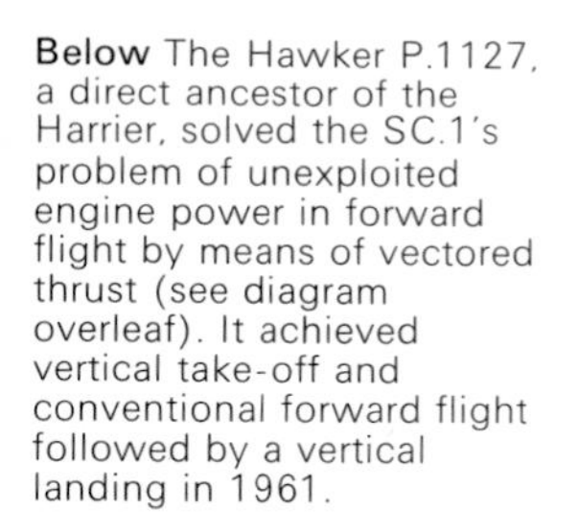

**Below** The Hawker P.1127, a direct ancestor of the Harrier, solved the SC.1's problem of unexploited engine power in forward flight by means of vectored thrust (see diagram overleaf). It achieved vertical take-off and conventional forward flight followed by a vertical landing in 1961.

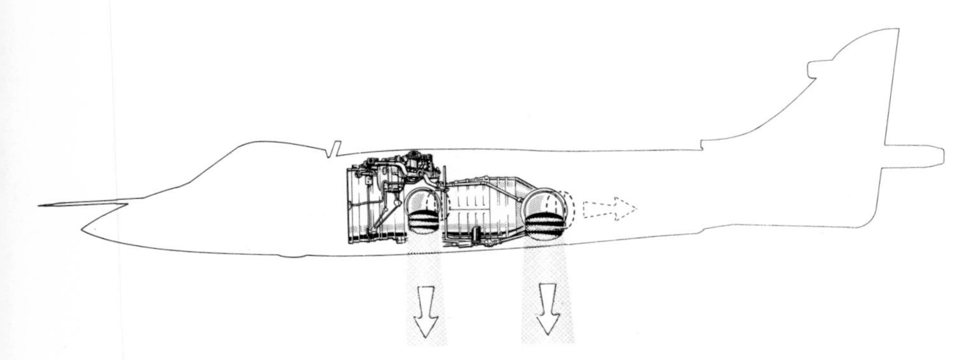

**Above** The vectored-thrust system (seen here against the outline of a Hawker Siddeley Harrier) involves the use of jet-efflux nozzles that are pointed vertically downward for take-off and then swivelled through about 90° to point backward for conventional forward flight.

**Below** The Hawker Siddeley Harrier, which entered squadron service in 1969, has four vectored-thrust nozzles. It is powered by a Rolls-Royce Pegasus 103 turbofan developing 9,750 kg (21,500 lb) thrust, and can operate as either a VTOL or STOL aircraft. An American version, designated AV-8B, with a larger wing and various lift-improving devices, is being developed under licence by McDonnell Douglas.

Flugtechnische Werke (VFW), was for a subsonic tactical-reconnaissance aircraft using a combination of the lift-jet and deflected-thrust systems. It was designated VAK-191 and intended to be the German replacement for the Fiat G91. This, too, was abandoned, but a similar configuration is used today in the Yak-36 of the Russian naval air arm aboard the carrier *Kiev*. The Yak-36 and the British Harrier are at present the only jet VTOL aircraft in operational service in the world.

The world's largest jet VTOL project was a 25,000 kg (24.6 tons) German military transport developed by the Dornier company between 1965 and 1970. Designated Do 31 it was designed to carry a payload in the region of 6,000 kg (13,250 lb), including wheeled vehicles which could drive on over a rear loading ramp. Its small, high wing carried two Bristol Siddeley Pegasus 5 engines, supplemented by eight Rolls-Royce RB.162 lift jets installed vertically in tip pods. The prototype flew satisfactorily until the early 1970s, when development was discontinued – one factor influencing the decision being the thunderous noise at take-off.

Most of the pioneer lift jets were built in North America. Ryan Aeronautical Company, who, like Rolls-Royce in England, had been experimenting since the early 1950s, was one of the leaders. The Ryan X-13, an improved, turbojet version of earlier tail-sitters built by Lockheed and Convair, first flew in 1955. The X-13 was designed to take off from and land on a mobile gantry which could be raised to a vertical position. Although it had a greater margin of power with a Rolls-Royce Avon engine, the jet tail-sitter proved hardly more practicable than its propeller-lifted predecessors and it was, in fact, the last of its type to be tried – but it was the first successful jet VTOL aircraft to fly.

Ryan then introduced a new design which used ducted fans to give more efficient vertical take-off thrust. The XV-5A was a fan-in-wing aircraft. Prototypes flew well (apart from one serious crash) but could not be developed into the hoped-for versatile operational machine.

Lockheed's XV-4A Hummingbird was powered by twin Pratt & Whitney JT12A-3 turbojets, mounted horizontally in nacelles alongside the centre fuselage. For take-off, the efflux from both jets was diverted through ducting into rows of downward-facing nozzles in a centre-fuselage compartment (termed the nozzle chamber) between the two engines. Above and below this chamber, long doors in the upper and lower fuselage skin could be opened to allow the free passage of air past the nozzles.

In this way, the take-off thrust from the nozzles was greatly increased by an induced flow of air downward through the chamber. The angle of the nozzles was designed to give a slight forward impetus to the aircraft after take-

off. When forward speed reached about 145 km/h (90 mph) the jet efflux from one of the turbojets could be diverted back through the normal tailpipe; the resulting acceleration soon established wing-borne flight, when the second turbojet could be diverted aft and the nozzle chamber closed. Designed maximum speed was 837 km/h (520 mph).

## Convertiplanes

In addition to the jet-lift projects, the VTO fever in the United States extended to other types of aircraft. At the instigation of military research authorities, numerous design studies were made to give VTO properties to military support aircraft, and several of these reached the 'hardware' stage. The largest prototype built was the XC-142A, produced by Ling-Temco-Vought (LTV) as a transport with a payload of more than 7,000 kg (15,400 lb). This was a four-turboprop aircraft in which the wing was positioned vertically for take-off, then gradually tilted forward by hydraulic power to the horizontal position as the machine moved into forward flight. A similar, although somewhat smaller tilt-wing aircraft was Canadair's CL-84.

Bell Aerosystems Company – a separate organization from Bell Helicopter Company – built a prototype designated X-22A to determine the practicability of using four tilting, ducted propellers for lift and propulsion. The company had earlier experimented with the first American deflected-thrust jet-lift aircraft, the X-14, which was powered by twin Bristol Siddeley Viper turbojet engines. Similar in some respects to the X-22A was the Curtiss-Wright X-19A, in which tiltable propellers without ducts were used for vertical lift and forward propulsion. All these projects proved to be capable of flight, but none was developed beyond the prototype stage.

## The STOL Mode

At present, apart from rotorcraft, the VTOL field is held almost exclusively by the vectored-thrust system of jet lift, represented predominantly by the Hawker Siddeley Harrier.

One of the advantages claimed for this system is that the aircraft can be used either for VTOL operations or, if very heavy loads are carried, as an STOL aircraft. To give the aircraft longer range, for example, it can make a running take-off with extra fuel aboard and with a total weight greater than its normal VTO lifting capability. By the time it reaches the combat area, the extra fuel will have been consumed and the aircraft will be able to land and take off vertically again.

The vectored-thrust system also has a role in normal wing-borne flight. Known as VIFFing (vectoring in forward flight), a sudden change of thrust-direction at high forward speeds, from horizontal towards vertical, can reduce speed dramatically within seconds, or tighten a turn to aid combat manoeuvres. For shipboard take-off the best technique is to use full forward thrust for high acceleration along a limited length of deck, then to switch to vertical thrust for what is called a 'ski-jump'. This catapults the aircraft upward, whereupon the pilot sets the nozzles to the 60° down position. The resulting thrust helps to lift and accelerate the aircraft into a continuing climb-away.

Whether VTOL systems will ever be used to supplement lift at take-off for civil airliners is a question for the future. In the present stage of knowledge, there seems to be little hope of lowering the extremely high noise levels that would be involved without lessening lift efficiency to an unacceptable degree. Although jet-lift VTOL aircraft are also capable of STOL operations, the distinctive differences between them and the true STOL aircraft are still as pronounced as ever.

**Above** The Lockheed XV-4A Hummingbird, which first flew in 1963. The angle of its downward-pointing lift nozzles gave it a characteristic nose-up attitude at take-off. Clearly visible are the large open doors above and below the midships nozzle chamber. These induced a downward flow of air through the nozzle chamber, increasing lift at take-off.

**Above** Two Canadair CL-84 tilt-wing aircraft. The one in the foreground is in the STO mode, with wing and tailplane partially tilted. The CL-84 on the right has fully tilted wings and engines for VTO. For hovering flight in this mode there is a small horizontal tail rotor to provide fore-and-aft control.

60
61
Ausgang
Exit

# Airports and Airspace

# 25 Airports

Since the earliest days of aviation the basic requirement for aeroplane take-offs and landings has been a reasonably flat area free from obstructions beneath the approach and climb-out paths. Most such areas can be grass fields, sandy beaches, and even stony deserts, ice, or water. These sites have been called by many names, such as flying grounds, landing grounds, and 'fields', but the most common terms are 'aerodrome' and 'airfield', the latter being usual when referring to a military aerodrome. The term 'airport' was at first reserved for a Customs aerodrome serving international flights, but it is now in general use to describe an aerodrome used by passenger, cargo, and mail services.

## Early Aerodromes

Military aerodromes in the pioneering days were mostly grass areas with a few hangars, a windsock (an open cone indicating wind direction), and often men's living quarters. Many of the earliest airports had been military aerodromes during World War I, and when civil flying began a few austere passenger amenities were provided – generally a wooden hut and an office for Customs. When London–Paris air services began in the summer and autumn of 1919 three terminal airports were used.

London's first airport was Hounslow Aerodrome on Hounslow Heath, to the west of London and close to the south-east corner of the present Heathrow Airport. Hounslow was a small wartime grass aerodrome with several hangars and a collection of huts. One of the hangars was taken over as the passenger terminal and Customs shed. A weather map was provided on a hangar door, and the name Hounslow was laid out in white capitals on the landing area to help incoming pilots to identify the 'airport'.

The cross-Channel Handley Page Transport services operated from the company's factory aerodrome at Cricklewood in north-west London. This aerodrome was small, provided a mainly uphill take-off into the prevailing wind, and was bordered on its west by the London and North Western Railway.

In Paris the military aerodrome of Le Bourget served as the terminal aerodrome. It still exists and ceased to be one of the main Paris airports only during 1977.

Hounslow was not particularly suitable as a terminal airport and the decision was taken to develop a 'permanent' terminal at Croydon, south of London. During World War I two adjoining aerodromes had been established, their dividing line being Plough Lane, which ran north-south. The western aerodrome was known as Wallington or Beddington and the eastern as Waddon. On the west side of Plough Lane was a number of large hangars; a hutted village on Plough Lane's east side formed the civil terminal, which included a wooden control tower on stilts, and the aircraft taxied between the terminal and hangars over a 'level crossing' across Plough Lane. Air traffic was transferred from Hounslow to Croydon Aerodrome – The London Terminal Aerodrome, to give it its official name – on 29 March 1920; later Handley Page transferred its operations from Cricklewood to Croydon.

The landing area was undulating, and always remained so, and the traffic area primitive, so the decision was taken to build a single, larger airport with a terminal area on the eastern boundary. A large terminal building, a control tower, hotel, and new hangars were built, the two landing areas were joined (removing a stretch of Plough Lane), and the old buildings demolished. With the title Airport of London, Croydon, the new terminal came into operation at the end of January 1928, and the improved aerodrome provided a maximum take-off and landing run of about 1,500 m (5,000 ft). Apart from the erection of a few additional hangars and the installation of lighting, the airport remained virtually unchanged until it was closed at the end of September 1959.

Croydon in its early period was representative of most of the world's major city airports, although many remained more primitive. Large numbers of aerodromes and airports were established in Europe, while difficult terrain forced the Scandinavian countries and some others to rely mainly on seaplanes for their air services.

In the early days of air transport almost all flying was confined to daylight and there were few navigational aids. In view of the unreliability of aero-engines and the dangers of bad weather, numerous emergency-landing grounds were laid out along the routes – there being several between London and Paris.

When night flying began, paraffin flares were laid out in the form of a letter L, the long arm being parallel to the landing run. Gradually improved lighting was installed: the landing area was defined by boundary lights, obstructions were marked by red lights, an airport beacon flashed the airport's recognition signal, and eventually floodlights illuminated the landing area. Illuminated wind-direction indicators were also installed. At Croydon a line of lights was sunk into the aerodrome surface to act as a landing guide at night or in poor visibility. On some air routes light beacons were installed at regular intervals. Radio was used for communication and a simple form of radio direction finding introduced (*see* Chapter 26).

**Previous two pages**
Düsseldorf/Lohausen airport at night.

## Paved Runways

From the early days of air-transport operation until the mid-1930s most aircraft were relatively small and light. They took off with runs of only a few hundred feet and landed at only about 80–95 km/h (50–60 mph), so modest grass fields were adequate. As aircraft weights and speeds increased, longer runs were required. Increased traffic and other factors forced the introduction of paved runways, and these allowed the use of heavier aircraft. They also reduced regularity to some extent, because in bad weather pilots had to aim for a narrow runway not necessarily aligned with the wind.

Paved runways were used during World War I, when German heavy bombers were operated from two concrete runways at an aerodrome in Belgium. Newark Metropolitan Airport (now Newark International and one of the New York group of major airports) was opened in October 1928 and it was claimed that its hard-surfaced, 490 m (1,600 ft) runway was the first at a US commercial airport – although there is evidence that the Ford Airport at Dearborn had two concrete runways in 1925. It seems likely that Stockholm's Bromma Airport, opened in May 1936, was the first European airport with paved runways, and that Schiphol, the Amsterdam airport, was the second; at about the same time Fornebu Airport (Oslo) and Malmi Airport (Helsinki) were opened and these too had paved runways. The three northern airports were blasted out of rock and were major feats of engineering. In the United Kingdom there were no paved runways before World War II; the first RAF station to have them was the wartime Dalcross, now Inverness Airport.

One major development before World War II was the German Lorenz instrument-landing system which enabled aircraft to be landed in poor visibility by flying down a radio beam; but this was not in widescale use when the war started.

A start had also been made on much more advanced terminal buildings. At Le Bourget, a very large terminal was opened in 1937, and in Germany the new terminal planned for Berlin's Tempelhof was on a massive scale.

Along the trunk routes to Africa, India, and Australia, aerodromes were laid out at frequent intervals. Most of these were simple, with only essential equipment, but Britain built modern combined land and marine airports at Basra and Singapore. A major airport was also built at Johannesburg.

In the United States a string of landing grounds was constructed from coast to coast for the use of the transcontinental air mail service. They had identification beacons, boundary and obstruction lights, and floodlights, but did not have radio. From the mid-1920s numerous United States cities built their own airports, which were progressively improved.

Contradicting the requirement that a landing area should be flat, an aerodrome was built on a steep hillside at Wau in New Guinea. In the late 1920s this was prepared for goldfield operations; it is nearly 1,065 m (3,500 ft) above sea level, and measures 850 m (2,790 ft) in length. Aircraft have to be parked across the slope to prevent their running away. Many other mountainside aerodromes have been built since, including that at Omkali in the New Guinea highlands, which is perched on a mountain ridge at an elevation of 1,675 m (5,500 ft), with a runway gradient of 13 per cent; along one side of the runway there is a sheer drop of 365 m (1,200 ft) into the valley.

When World War II ended a vast number of aerodromes had been built throughout the world for military purposes. Many were

Imperial Airways' Handley Page 42 *Hannibal* flying over the Airport of London, Croydon, in 1931. Although the aerodrome was used as the London airport from March 1920, the terminal area seen here was brought into operation only in January 1928. From left to right alongside Purley Way are the administration building and control tower and the main hangars and workshops.

destined to serve as civil airports, but in some places new airports were planned to cope with the rapidly increasing air traffic and the much larger and heavier aircraft coming into service. All the major airports had paved all-weather runways of asphalt or concrete, and many have become small cities with working populations of over 50,000.

It is impossible to say how many aerodromes and airports exist. The International Civil Aviation Organization (ICAO) reported that in 1976 in the ICAO-member states there were 27,751 civil aerodromes of which at least 15,371 were open to public use. Most countries are members of ICAO but the figures are unlikely to include Russian and Chinese aerodromes, although the Russian airline Aeroflot is known to serve some 3,500 communities in the Soviet Union. ICAO's figures also showed that at the end of 1976 there were 988 regular and alternate aerodromes serving international operations and another 24 under construction or projected; of these 642 were designated primarily for international scheduled air transport. The US Federal Aviation Administration listed 13,062 'airports' in that country in 1974, stating that 2,747 had paved runways, 2,878 had runway lights, and 61 were designated ports of entry.

## Modern International Airports

New York International will serve as an example of a major postwar airport. New York Municipal Airport – LaGuardia Field – had been opened as the city's airport in December 1939, but it was realized that a new and much larger airport would be required. A 445 hectare (1,100 acres) site was chosen on the shores of Jamaica Bay and work began on filling the marshy tidelands of the Idlewild golf course. The name 'Idlewild' continued in general use after the airport was opened in July 1948, although it was officially New York International. Today its title is John F. Kennedy International; it is widely known simply as JFK. A series of parallel runways was constructed around a large terminal area; some were never used because of opposition by local residents, but others have been built and development has been continuous over the past 30 years.

The airport now covers an area of 1,995 hectares (4,930 acres) and has five active runways: 04L/22R of 3,460 m (11,352 ft) and 04R/22L of 2,560 m (8,400 ft), the southwest-northeast parallel pair; 13L/31R of 3,048 m (10,000 ft) and 13R/31L of 4,441 m (14,572 ft), the southeast-northwest parallel pair; and 14/32 of 842 m (2,762 ft). Except for 14/32 the runways are of concrete and are 45 m (150 ft) wide. The 04/22 runways and 13L/31R have transverse grooving to prevent aquaplaning. The short 14/32 runway is a section of a taxiway marked out for general aviation use.

John F. Kennedy International Airport is the main airport serving New York. This view shows the large central terminal area with the 3,048 m (10,000 ft) 13L/31R runway running across the background. On the right of the terminal area is the Port Authority's International Arrivals Building and Airline Wing Buildings with, immediately adjacent, the circular-roofed Pan American terminal. Then, reading clockwise, are the terminals of Northwest/Delta airlines, Eastern Air Lines, United Airlines (beyond the main access road), American Airlines, British Airways/Air Canada, National Airlines, and TWA. In the centre of the area can be seen the lagoon and three chapels.

Originally the airport had a temporary terminal, a control tower, a cargo terminal, and some maintenance hangars. Now it has a complete terminal city, with the large Port Authority of New York and New Jersey International Arrivals Building, the adjoining Airline Wing Buildings, and the following airline-owned terminals: American Airlines, British Airways (shared with Air Canada), Braniff (shared with Northwest), Eastern, National, Pan American, TWA, and United. Most are orthodox rectangular buildings, some with traffic piers; but Pan American's has a large oval cantilever roof, and TWA's, designed by Eero Saarinen, has four barrel vaults and suggests a large bird. Several of the terminals have had to be considerably enlarged since they were first built. Extensive areas of the airport are given up to the maintenance bases of major airlines, and there is a large cargo-terminal area with its own loading aprons and taxiways. Some of the passenger terminals have aerobridge jetties connecting them with the aircraft and, as at Washington/Dulles, there are some mobile lounges which drive between the buildings and aircraft on remote stands; there are about 145 passenger boarding gates. In 1977 the airport was served by 72 scheduled airlines; in 1976 the airport handled more than 300,000 aircraft movements, more than 21 million passengers, and over a million tons of air freight; it provided parking space for 13,000 cars.

Although John F. Kennedy is fairly typical of a major airport, it is not the busiest. In 1976 Chicago's O'Hare International, with seven runways and three main passenger terminals, handled 41.75 million passengers and 718,000 aircraft movements, including 641,541 transport aircraft movements. The second busiest was Atlanta with more than 27 million passengers.

London's Heathrow, a little over half the area of JFK and with three active runways, three passenger terminals, and a cargo terminal complex, handled 23,654,444 passengers in 1976, with 414,606 metric tons of cargo and 278,108 aircraft movements; but although it handles less traffic than some US airports it deals with more international traffic than any other airport.

Large as airports such as London, Chicago, and New York may seem, they are dwarfed by the new Dallas/Fort Worth Regional Airport in Texas, which has an area of 7,082 hectares (more than 27 sq miles), and which in its final form will have a terminal spine 6.4 km (4 miles) in length with 13 semi-circular terminal build-

**Above** One of the world's biggest airports in total area is the new Dallas/Fort Worth Regional Airport in Texas. At present it has three main concrete runways in use and a 1,219 m (4,000 ft) STOL runway. Four of its projected 13 semi-circular passenger terminals are working and have 71 passenger boarding gates and one general aviation gate serving 77 aircraft parking stands. Automatic trains link the terminals, hotel, and car parks. This view shows the spine terminal area and, beyond, one of the main runways.

**Left** Amsterdam's Schiphol Airport was originally prepared as a small military aerodrome in 1917 and became the city's airport in the early 1920s. The airport has been continuously enlarged and improved and was the second European airport to have paved runways. This view shows the terminal area and beyond it the 3,250 m (10,663 ft) 06/24 runway. Schiphol now handles more than 8 million passengers, 250,000 metric tons of cargo, and 176,000 aircraft movements a year.

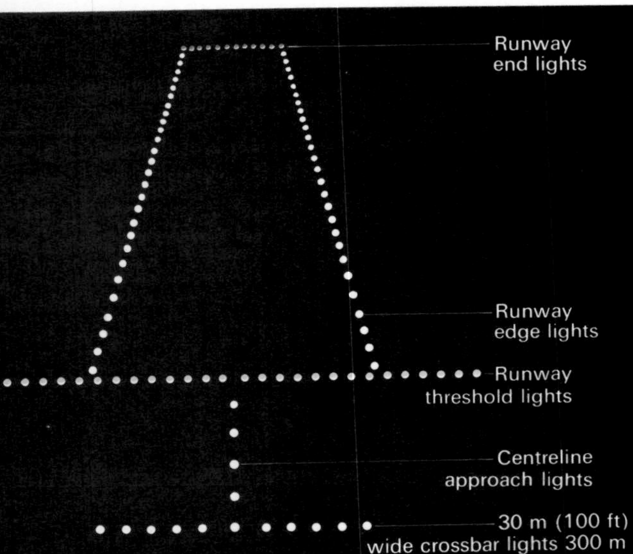


ings. This may well be the pattern for the next few decades. The reconstructed Galeão, now Rio de Janeiro International, is being built on the same lines but to a smaller scale.

Although these major airports handle the bulk of the world's air traffic and are the best known, there is a whole range of lesser airports, some consisting of a single rough grass or gravel runway and handling one or two services a week. Another class of airport coming into use is the STOLport, designed to bring air transport close to the cities. It is served by short-take-off-and-landing (STOL) aircraft, is of simple layout, and it is possible that some will be built in elevated positions over rivers, roads, or rail-

**Above** Airport lighting for use under reasonable weather conditions, with runway, threshold, and end lights, and simple approach lights with one crossbar. The threshold is the beginning of the touch-down zone.

**Right** A precision-approach runway with lighting approved for Category II and III approaches. This includes high-intensity runway-edge and centre-line lights, threshold, end, and touchdown-zone lights, and high-intensity approach lighting with supplementary lights over the inner 300 m (1,000 ft). The approach lights shown are the Calvert system, with distance-coded centreline and five crossbars. VASI approach-slope indicators are also seen. In the United States and some other countries the centreline consists of barrettes and there is only one crossbar. Barrettes are closely spaced groups of lights that from the air appear as a bar of light.

ways. In the French Alps a variation of the STOLport is the altiport. These are single runways high on the slopes of mountains (mainly to serve winter sports traffic); they are often snow covered, and landings have to be made uphill and take-offs downhill.

Most of the major airports share the same kind of problems. They have to have adequate runways to handle the enormous volume of traffic, adequate lighting, radio and radar aids to ensure safe operation, passenger and cargo terminals with associated roads and car parks, and large areas for aircraft maintenance and overhaul and fuel storage. They must cater for all these activities without annoying the neighbours, and must make every effort to keep noise and pollution to a minimum. In some cases noise has become so serious that restrictions or even complete bans have been put on night operations, while daytime operations are strictly regulated and involve special routings and operating techniques. Added to all these requirements is the need for comprehensive firefighting, rescue, and emergency services, and – the latest problem – security against hijacking and terrorist attacks.

## Airport Lighting

Lighting standards are laid down internationally for different categories of airport and for operations in all weather. To allow landings in the worst conditions an airport must have a high-intensity centreline and five-crossbar approach-light system, with supplementary lights over the inner 300 m (1,000 ft) of the approach; VASIS (Visual Approach Slope Indicator System); high-intensity runway centreline lights, runway touchdown zone lights, high-intensity runway threshold and end lights; taxiway centreline lights; a light beacon; obstruction lights; and a secondary power supply. This standard of lighting meets the requirements for ICAO Category II and III precision approaches. Cat II requires 30 m (100 ft) cloud base and 400 m ($\frac{1}{4}$ mile) runway visual range (RVR), while Cat IIIa requires no cloud base limit and only 200 m (700 ft) RVR. The RVR will be reduced to 45 m (150 ft) in Cat IIIb, and in Cat IIIc to no external visibility – in other words, dense fog.

For the private pilot in a light aeroplane, provided he or she flies only under visual-flight rules, a small grass field with boundary markings and a windsock still serves.

## Military Airfields

Military aerodromes have developed in much the same way as civil, except that instead of passenger and cargo terminals they have various operational buildings, living quarters, and military stores such as bomb dumps. At many such airfields there are also arrestor devices on the main active runway to stop aircraft overrunning in case of brake failure or some other malfunction. A heavily loaded bomber requires as much runway as a heavy civil airliner, but developments in fighter and strike aircraft have made them much less dependent on airfields. The Harrier VTOL aircraft can be dispersed in woods and can take off from small clearings. In Sweden, it is common for fighters to be kept in underground hangars or dispersed in woods and to take off from stretches of motorway or specially improved farm roads. Unlike civil airports, military aerodromes have to be defended against air attack and it is common to find missile sites in their vicinity.

## Heliports

The heliport can be a small piece of ground, an elevated platform, or even the roof of a tall building such as the Pan Am Building in central Manhattan. The main requirements are sufficient clearance around the helicopter's rotor and safe approach and climb-out paths. A wind direction indicator is required, and for night operations boundary, obstruction, and other lighting. If the helicopters are being used for public service, heliports, like airports, must have adequate fire-fighting and rescue services. In marine oilfields large-scale helicopter operation takes place from small platforms attached to the decks of oil and gas exploration rigs and production platforms.

This view of the Pan Am Building in the heart of Manhattan shows the rooftop heliport, which was opened in 1965. The landing area measures 37 m (121 ft) by 34.4 m (113 ft) and is 244 m (800 ft) above street level. In the background is the Chrysler Building and the East River. The helicopter is a Vertol 107, formerly used by New York Airways.

# 26 Air Navigation

At some time in his career every pilot has had his ego bruised by the awful realization, 'I'm lost!' Even that experience was accepted as part of the fun of flying by daring aviators in the early days, but for aeroplanes to become practical machines, either in peace or war, it was essential that pilots were given the means to identify their position and heading. The techniques of air navigation are now one of the most important aspects of aircraft operation.

Knowing one's position at any given moment is central to the art of navigation, and at one time pilots determined their whereabouts with the aid of merely a map, a compass, and a watch. Knowing their speed, they would calculate the time it would take to cover certain distances, fly that amount of time in the required direction, and watch for landmarks shown on their maps. This form of navigating is called *dead-reckoning*. It is not easy because winds can alter the aircraft's ground speed (speed over the ground as distinct from through the air), or make it drift off-track.

A nautical technique adopted by some early long-distance fliers was *astro-navigation*, which requires a chronometer (a very accurate watch), a set of tables giving details of star or planet positions and their apparent movements, and a sextant. By day the Sun was used as a navigational reference, and by night the stars. A good astro-navigator could steer an aircraft across thousands of miles, with few glimpses of the ground, and arrive within sighting distance of his goal. The only time that astro-navigation let aviators down was when the sky above was clouded – and early aircraft flew much lower than modern ones. Some early pilots were forced to descend into cloud, not knowing whether the ground would be seen in time if the cloudbase should happen to reach down to the surface. Many radio navigation aids (navaids) available today reduce this type of risk.

Navigation can be assisted by communications. Pilots first had routine two-way radio contact with aerodromes in the 1920s. The pilot could at least be informed in flight of weather conditions at his destination. If it was cloudy he could be told when he was heard passing overhead. 'Motors overhead' was a common cry from early control towers, and on hearing that call the pilot could fly in a direction where he knew there were few obstructions. He would descend until he broke through the clouds, and then fly back to the aerodrome and land.

The procedure was simplified by the advent of ground-based *direction finders*. Initially the D/F radio in the control tower used signals sent out by the aircraft's radio. The crew would transmit a message lasting for several seconds, and the ground operator would turn a directional aerial to detect the radio bearing. This information could then be transmitted to the aircraft. It was much easier to fly overhead this way – discounting wind, the pilot merely flew the reciprocal, 180° different from the radio bearing – but descending was usually by the same method as before.

Pilots were given their first chance to experience true radio-guidance when the *radio-range* was introduced in the United States about 1930. Radio-range stations were erected at strategic points along major routes, or adjacent to aerodromes. They consisted of large aerials arranged in such a way that Morse A (·–) and N (–·) signals could be transmitted across large arcs. Where they overlapped these signals formed narrow beams. Along the centre of each beam the A and N merged to produce a continuous tone heard in the pilot's headphones. Now, at any time and in any weather, a pilot could tune his radio receiver to the appropriate wave length and 'fly the beam'. If he deviated left or right of track a Morse A or N was heard. To give positional information away from the

Sequence of operations of the Autoland instrument-landing system with the BAC Trident.

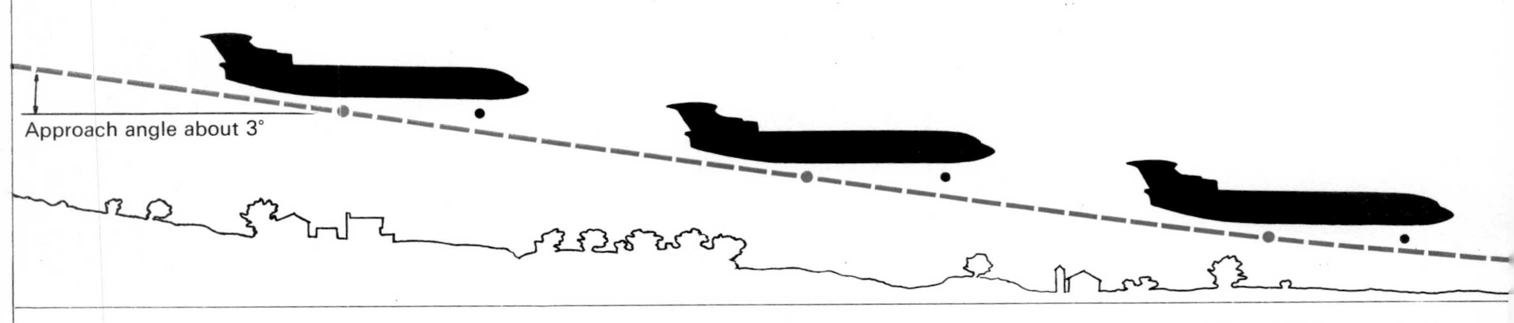

aerials it was common to install marker beacons. These pointed directly upwards and caused a light to flash as the aircraft passed overhead.

A development of the radio-range was the *non-directional beacon* (NDB). Although radio-ranges have disappeared many hundreds of NDBs are still in service. They use medium-frequency (MF) transmissions and are effective up to almost 113 km (70 miles) in the most favourable circumstances. The NDB transmits a continuous tone in all directions, broken at about 30-second intervals to inject identifying Morse letters. It is a so-called omni-range. An *automatic direction finder* (ADF), linked to the aircraft's receiver, shows the pilot which way he has to fly to reach the NDB. It can be difficult to fly a straight course without experience at counteracting the effects of winds, but even so the NDB was a big step forward.

During World War II engineers perfected the stable oscillators needed to generate and detect signals that beat about 100 times faster than those in NDBs, and a new navaid, the VHF (very high frequency) omni-range (VOR) was born. Large numbers of VORs are in use today, and they form the backbone of the world's airline navigation systems. VHF provides line-of-sight range – more than 160 km (100 miles) at high altitude – relative freedom from signal interference, and the ability to *modulate* the transmitted signal with a very accurate reference frequency. A VOR beacon operates at a frequency between 112 and 118 MHz (1 MHz = 1 million cycles per second) and onto this is superimposed a 9,960 Hz reference frequency. Phase differences occur at a lower frequency and give the aircraft's bearing from the beacon. Using VOR aircraft can fly accurate tracks at almost any altitude and in any weather. VOR was adopted as a standard navigation aid in 1957.

Throughout the world there has been an increasing acceptance of *distance-measuring equipment* (DME). Although a complicated device, it does a simple task. Co-located with a VOR (i.e., the VOR and DME beacons are at the same place) it gives an accurate indication of distance to the beacon. Combined with the VOR bearing information this tells a pilot his actual position.

One of the most useful aids, related to VOR, but derived from earlier 'beam-approach systerms' is the *instrument-landing system* (ILS), which guides aircraft down to a particular runway. ILS has two transmitters adjacent to the runway it serves. The *localizer* beams are transmitted from a broad aerial at the far (upwind) end of the runway. There are two beams, both at the same VHF frequency but one modulated at 90 Hz and the other at 150 Hz. The aircraft localizer receiver measures the difference in depth of modulation between the two beams. If there is no difference the aircraft is directly in line with the runway, but deviating either side of the centreline causes a difference signal to appear, the dominant modulation indicating which side of the centreline the aircraft has strayed.

The second part of ILS is the *glideslope* system, again with two beams modulated at

A VOR beacon at Ibsley, near Southampton. The VHF navigation signal is emitted from the ring of dome-shaped aerials; the large horizontal frame (called the counterpoise) makes the radio waves travel into space as if they were being produced by a beacon situated on level ground. The vertical aerial behind the beacon is on distance-measuring equipment.

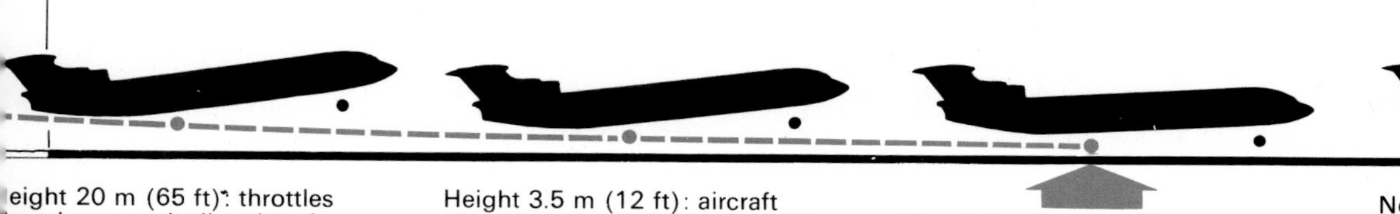

90 Hz and 150 Hz, but this time at a radio wavelength in the UHF (ultra-high-frequency, about 300–400 MHz) waveband. The beams are transmitted from an aerial set just to one side of the runway touchdown point, about 300 m (1,000 ft) from the 'threshold' at the near (downwind) end. The aircraft's receiver is able to tell whether the aircraft is above, below or on the glide-path set at a particular angle. Most systems are inclined at 3°, but STOL approaches can be much steeper. Guidance information is usually presented to the pilot by an instrument with vertical and horizontal pointers which swing up and down or left and right. ILS is the standard landing aid at every major airport, runways so equipped being called 'instrument runways'.

*Autopilots* can be made to respond to VOR and ILS guidance signals, thereby allowing a pilot to let his aircraft steer itself automatically, even down the landing approach. Today almost every airliner is equipped with a system that can bring it to within about 800 m (½ mile) of the landing runway without any pilot intervention and in any weather.

It was soon after this system had been proved that *automatic-landing* became a reality. At about 45 m (150 ft) above the ground, special electronics take over from the glideslope-coupled autopilot. They close the engine throttles, and 'flare' the aircraft by smoothly pulling up the nose to arrest the rate of descent, exactly as a pilot does. The manoeuvre is controlled by height and rate-of-descent signals from *radio altimeters*. Normally the ILS localizer beams provide steering guidance throughout an automatic landing, even after touchdowns. Aeroplanes equipped with automatic-landing systems can operate in thick fog.

No matter what the weather, pilots like to be able to see as much of the runway as possible. Well-equipped runways not only have lights along their edges and centreline but also special patterns which assist orientation extending perhaps 400 m (1,300 ft) back under the approach path (glidepath). Early lighting was just a set of petrol-soaked rag wicks in cans, and the naked-flame gooseneck flares developed during the 1930s are still common on small fields. Modern airports have high-intensity lights, flush with the ground that remain undamaged by jet blast or aircraft tyres.

Although ILS might appear to be the perfect landing aid, it has several shortcomings which could be alleviated if it operated at higher radio frequencies. Currently there are plans to replace ILS with a *microwave landing system* (MLS) which operates at frequencies of about 5,000 MHz. Guidance information will be more accurate and available over a wider area, allowing the execution of curved approaches that will avoid obstacles and built-up areas and greatly increase traffic-handling capacity. MLS should also be cheaper than the existing system.

Alongside point-source radio aids, many area-coverage (R-nav) systems were developed. The first was Gee, used by bomber crews to find targets while flying above cloud at night, but nowadays area-coverage aids are used extensively for navigation everywhere, especially away from the VOR/DME route network.

The most common are Loran (Long-Range Navigation) and Decca, both of which are called *hyperbolic aids* as will be explained. Decca and Loran can work down to ground level and thus

**Above** A BAC Trident autolanding in fog at London Airport. Under such conditions of low visibility a pilot cannot rely on his own skill to land a plane safely, but he is now able to use the ILS system shown on pages 214–5.

**Left** This experimental microwave landing system (MLS) is at the end of the runway at the Royal Aircraft Establishment, Farnborough, from which a Vickers-Armstrongs Super VC-10 has just taken off. Microwave landing systems are likely to replace the instrument landing system (ILS) now in use at hundreds of airports throughout the world. The lattice-like framework behind the MLS is the ILS localizer aerial which serves the same runway.

**Right** Navigation equipment used on modern airliners. Like other transport aircraft, the Airbus A300B seen here would have most but not all the equipment shown. Key to numbers: 1 VHF communications aerial, 2 HF communications aerial, 3 Satellite communications aerial, 4 ADF aerial, 5 VOR receiver, 6 DME aerials, 7 ILS localizer receiver, 8 ILS glidescope receiver, 9 Marker-beacon receiver, 10 Radio altimeters, 11 Transponder aerials, 12 Loran aerial, 13 Omega aerial, 14 Weather radar, 15 Doppler radar, 16 Avionics bay.

are used by helicopters, ships, and even submarines, and differ mainly in the wavelengths at which each operates. Loran has been through several development stages, and almost all the equipment in use today is called Loran-C.

Each radio-chain comprises a master transmitter and three slave transmitters located many miles around it. The slaves are synchronized with the master in such a way that the interaction of radio-waves from the stations produces *interference patterns* in the sky which, if drawn on a map, are shaped like hyperbolae. A special receiver in the aircraft or ship can determine its position on the lattice-like interference pattern. Decca can either plot position continuously on a roller-blind map or indicate exact latitude and longitude. The advantage of these long-established R-nav systems is that pilots can fly where they like: they do not have to stay on the crowded routes but can go direct to their destination.

Recent developments which are challenging the traditional navigation techniques use very-low frequency (VLF) signals. The most accurate radio aid now available on a worldwide basis is called *Omega*. This system covers the world with only eight stations, all radiating synchronized transmissions and producing a complex wave pattern that a receiver can correlate with its own position once it has been given the co-ordinates of its starting point. The heart of an Omega receiver is a micro-processor, a tiny computer about 10 millimetres (0.4 in) square, which is programmed to do in a matter of seconds calculations that would have taken a human navigator several hours. This integration of modern radio technology and electronics is now being taken a stage further with navigation satellites which orbit the Earth and provide exact navaid-carrying stations.

Two widely-used navaids are self-contained in the aircraft: the *inertial navigation system* (INS) and *Doppler*. INS measures the exact

**Top** An inertial navigation system (INS) detects aircraft movements by using accelerometers mounted on a platform which is maintained perfectly level throughout flight by a set of gyroscopes. The photograph shows four different types of INS platform, ranging in weight from 39 kg (86 lb) to a recent model weighing less than 2.3 kg (5 lb). Up to three INS platforms are installed in most modern long-range airliners.

**Above** This Omega navigation system receiver, no bigger than a shoe-box, is one of the most ingenious pieces of electronics found on a modern airliner. The heart of the equipment is a tiny microcircuit, about 10 mm square, which is virtually a complete computer. Data electronically stored in the equipment is compared with radio information to calculate an aircraft's exact position anywhere in the world.

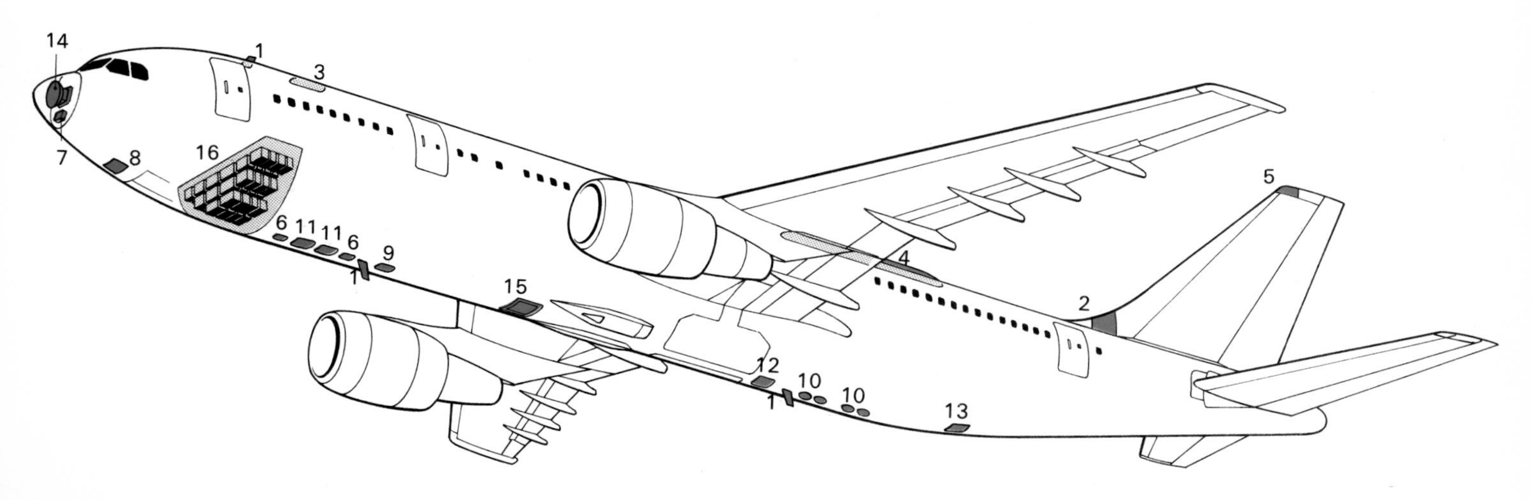

accelerations imparted to it, so that a computer can convert these to velocities and distances relative to a given starting point. It is much more expensive than other systems, but it is very reliable and accurate. Doppler is a radar which emits several (usually three or four) beams pointing obliquely down to Earth. By measuring the changed frequencies of the reflected beams it gives an aircraft's along-track and across-track velocity, which a computer converts to position measurements.

Radars that sweep the sky from the ground are much more than just navaids. One important use is to help maintain safe separation between aircraft in the same portion of sky. Other kinds of radar are used by air traffic controllers, and in military operations.

Primary radars emit concentrated pulses of radio energy from a special transmitter. Each pulse radiates outward at the speed of light and is reflected by objects in its path. A sensitive receiver detects echoes from these objects, and, based on the time it takes the pulse to make the return journey, it can calculate the object's range. By emitting several hundred pulses per second, and scanning (rotating the aerial) several times per minute, it is possible also to tell the exact direction of objects. On some systems the radar operator watches a glowing tube on which a revolving line of light shows the aerial's rotation. When an object is detected it causes a 'blip' (spot of light) to appear on the screen, at the appropriate range and bearing.

Typical examples of the above are the *surveillance* radars used by air-traffic controllers; but other types are also used in the control tower. *Approach* radars scan, sometimes with two criss-crossing beams, along runway approaches, providing their operators with sufficient information for them to 'talk-down' aircraft which want to land in bad weather – a technique called GCA (ground-controlled approach).

In order to avoid turbulent clouds, many airliners carry a weather radar in the nose. This is usually a simple primary radar and, although it has relatively poor bearing and range resolution, it is able to detect storm cells. Such radars can be aimed downwards to operate in a *ground-mapping* mode to assist navigation. Military aircraft radars are much more powerful, because they have to detect very small objects with pin-point accuracy. They often feed target-position information directly to automatic weapon-release systems.

An important but very different ground-based system is *secondary surveillance* radar (SSR). This transmits low-power pulses which trigger devices called *transponders* on board many aircraft, each of which sends back a coded reply at a slightly different wavelength. Only transponder-equipped aircraft can be seen by SSR, but the coded message contains aircraft identity and height, which is valuable to air-traffic controllers.

Surveillance radar at Nairobi Airport, Kenya. It scans a radius of 435 km (270 miles) at altitudes up to 30,500 m (100,000 ft).

There are about 500,000 powered aircraft in the world. At any one time perhaps one-tenth of them are airborne, and most of these are over land – about 168 km$^2$ (65 million sq miles) of the Earth's surface. That is equivalent to one aircraft to every 1,300 square miles. This may seem to be a scattered population, but 1,300 square miles is only 36 miles square, and some aircraft can cover 36 miles in less than a minute. As nearly all the aircraft fly over much smaller areas of the Earth's landmass, giving typical densities of one per 50 square miles, the true problem is acute.

Altitude differences help to provide separation, but in airspace around cities, where all types of aircraft congregate, collision risks can rise to unacceptable levels. Given that radio beacons strongly tend to funnel all traffic towards specific points, and that pilots cannot see in all directions (and, of course, are often flying in cloud) the need for an aircraft policing service becomes obvious. This is the function of *air-traffic control* (ATC).

Air-traffic controllers work either in airport control towers or in radar-filled rooms called air-traffic-control centres. Each controller is responsible for a block of airspace, called a *sector*. Within his sector there will be defined flight paths along which he will separate, horizontally and vertically, all the aircraft under his control. If he cannot fit in all the aircraft that want to use his airspace, without reducing the separation below safe limits, he must delay or re-route traffic. Often this is made unnecessary by intently monitoring on radar any *short-lived risks* – for instance, aircraft climbing and descending close to one another. The controller's job is complicated by the fact that aeroplanes, once airborne, cannot be stopped, and they have a limited endurance. He often routes aircraft to their destination airport via an intermediate radio beacon, usually a VOR. If a landing queue develops he tells pilots to join a *stack*, a racetrack pattern over the beacon, allocating a different level (altitude) for each aircraft. Near airports the controllers have large volumes of airspace, usually called a *control zone*, in which they sort out arriving and departing traffic. The routes between airports are called *airways*. Within each airway aircraft flying in opposite directions are allocated separate levels; these are usually 300 m (1,000 ft) apart, but at high altitudes, where barometric-altimeter errors are greater, the levels are often 600 m (2,000 ft) apart. The airways and control zones are called controlled airspace. The rest of the sky is uncontrolled airspace, within which are *restricted areas* that are either reserved for military-pilot training or surround places such as nuclear power stations, wildlife sanctuaries, and gunnery ranges.

At present the controller is responsible for preventing mid-air collisions, but collision avoidance radars may soon be fitted to commercial aircraft. They were previously too expensive to be worth adopting. In addition, many ATC organizations are developing *conflict-detection* programmes to be included in future radar-system computers. They will warn controllers if any aircraft are on a collision course. As a first stage, many radars in the US already have computer programmes which warn if an aircraft is flying too low or seems to be in danger of hitting high ground.

The ground is the aeroplane's worst enemy. Every pilot aims to land at the end of his flight, but landings made inadvertently are recorded in the statistic books as crashes. Inadvertent descents into the ground may be reduced by the *ground-proximity warning system* (GPWS). In this system a radio altimeter provides height data, relative to terrain, which is mixed with rate-of-descent information and analysed by a computer. Any dangerous combination of circumstances triggers a synthetic-voice generator, which cries 'Pull up!' on the flight deck.

The flight-deck of a modern airliner presents an array of dials and switches that is utterly baffling to the average passenger. But, although numerous, such instruments are carefully designed to present information in the simplest and clearest manner possible. Modern computer-based long-range-navigation systems, for instance, usually have a set of windows in which numerals light up to show electronically derived information; many also have a keyboard on which the pilot enters data for the system. In the future there will almost certainly be more cathode-ray-tube (CRT) displays on flight decks. A CRT display already in use shows weather-radar information, and even on simple systems this can be used to show the aircraft track through weather or to list take-off or emergency checks.

**Above** Air-traffic controllers watch aircraft movements on their circular radar displays, and write details about each aircraft on strips placed in racks below the displays. Above the radar, maps and radio-frequency lists are illuminated in back-lit panels. Most controllers wear radio and telephone headsets through which they can talk to aircraft and to other controllers.

**Below** This air-traffic controller has two displays. The circular one is a radar display showing aircraft, in symbolic form, with identifying labels; unwanted information has been electronically removed from the screen. The small television-like display shows computer-stored details about the aircraft and their flight plans.

Beyond Planet
Earth

# 27 Rocketry

No one knows who invented the rocket. Its history begins with primitive tribes who used materials such as sulphur, saltpetre (potassium nitrate), pitch, and animal fats in fire-making – and later as weapons. Long before the Christian era burning materials were being catapulted into besieged towns, and tubes of incendiary mixtures tied to the shafts of arrows had set fire to many a wooden sailing ship. By AD 850 some form of 'gunpowder' must have existed in China, where fireworks were used to celebrate religious festivals.

From China rocketry spread throughout Asia, and in 1242 the English philosopher Roger Bacon produced a 'secret' formula which established the basis of modern gunpowder (sulphur, charcoal, and saltpetre).

An early mention of the rocket in Europe appears in the *Chronicle of Cologne* (1258); and an Italian historian credits a rocket with setting a defending tower alight during the battle for the Isle of Chiozza in 1379. In 1645, during the Thirty Years' War, black-powder rockets were fired in large numbers against the German city of Philippsburg.

In the Battle of Paniput, India, in 1761, the Rohillas are said to have fired volleys of two thousand rockets at a time against their enemy, the Mahrattas, which not only terrified their horses but 'did so much execution, also, that they could not advance to the charge'. Spike-nosed rockets used in India in 1792 against British cavalry by special rocketeer troops were iron-cased, 20 cm (8 in) long by 4 cm (1.5 in) diameter, with a 2.5 m (8 ft) long balancing stick of bamboo.

## Military Rockets in Europe

Examples of these rockets brought to England inspired the artillery expert William (later Sir William) Congreve to begin a private study aimed at improving rocket performance, mainly by refinement of the gunpowder charge and redesign of the motor case and nozzle. Later he was allowed to use the Royal Laboratory at Woolwich with the objective of doubling the range of the Indian rockets. His iron-cased rockets were eventually to achieve ranges of almost 4,600 m (about 5,000 yards). Another Englishman, William Hale, developed the spin-stabilized rocket, which eliminated the balancing stick.

The Congreve and Hale rockets were used extensively by the British in many campaigns. In 1806, during the Napoleonic Wars, incendiary rockets designed by Congreve played a major role in the fall of Boulogne. They were volleyed into the harbour from specially constructed projector boats which parent sailing ships had conveyed to the scene.

Members of the Royal Artillery at rocket-firing practice in 1845. Although they could not be aimed with any great accuracy, such 19th-century rockets could cause enormous damage by fire when used against wooden ships and buildings.

Some of the Congreve rockets, specially made for marine warfare, contained a liquid incendiary in a pointed nose which enabled them to stick into whatever they hit. Upon impact, the flaming liquid squeezed out through holes drilled in the head. The destruction which such projectiles inflicted on wooden sailing ships and buildings must have been considerable. In 1807 thousands of rockets fired on Copenhagen set fire to large parts of the town.

By 1818 the British had an official rocket brigade, and other countries followed their example. For a long time, however, the rocket remained a relatively inefficient device compared with conventional artillery. The big advances were to start modestly in the minds of men who lacked the facilities to bring them about.

## A Russian Visionary

One of these, a Russian village schoolteacher, Konstantin E. Tsiolkovsky (1857–1935), laid the foundations for the modern liquid-fuel rocket. As early as 1883 he had grasped the principle of rocket motion – that a rocket is propelled by the recoil effect of its exhaust gases and not because these gases push against the air (it works better in space). In 1903 he sketched a design for a spaceship to be powered by liquid oxygen and liquid hydrogen, the propellants that 65 years later were to take men to the Moon. Tsiolkovsky suggested that a rocket chamber could be cooled by passing one of the liquids through a double-wall or 'jacket', and went on to propose using valves to adjust the flow of propellants and thereby to control

**Previous two pages**
Destination Moon: Apollo 17 lifts off from Cape Canaveral just after midnight on 7 December 1972 – the sixth and final lunar-landing mission in the Apollo programme.

the thrust. He advocated placing movable vanes in the exhaust for steering purposes, and proposed that rockets could be stabilized by the gyroscopic effect of a rotating flywheel.

To reach the high speeds necessary for space flight, Tsiolkovsky expounded the virtue of step rockets or (as he called them) 'rocket trains'. He reasoned that step rockets opened the way to artificial satellites. 'When the velocity reaches 8 km/sec [about 17,850 mph],' he wrote, 'the centrifugal force cancels gravity and, after a flight which lasts as long as oxygen and food suffice, the rocket spirals back to Earth, braking itself against the air and gliding without explosions.'

## Rocketry Comes of Age

Tsiolkovsky's conclusions were soon to be verified and expanded by others, notably Robert H. Goddard (1882–1945) in the United States and Hermann Oberth (*b.* 1894) in Germany. Goddard, a physics professor, began in 1913 experimenting with solid fuels, and quickly concluded that liquid propellants offered better prospects. On 16 March 1926 he launched the world's first liquid-fuelled rocket at Auburn, Massachusetts. He chose gasolene (petrol) as the fuel and liquid oxygen as the oxidizer, using pressure from the latter to force both liquids simultaneously into the combustion chamber in which they mixed and burned. When the rocket was fired it made a brief flight and fell to the ground some 56 m (185 ft) away. It was estimated to have ascended 12.5 m (41 ft) and to have reached a speed of about 96 km/h (60 mph).

In 1923 Hermann Oberth set out the mathematics of space travel in his book *The Rocket into Interplanetary Space*. This slim volume produced a burst of enthusiasm for rocketry, which led on 5 July 1927 to the founding of the Verein für Raumschiffert e.V. (VfR), the (German) Society for Space Travel.

In 1930 Max Valier, who had experimented with gunpowder rockets, teamed up with Dr Paul Heylandt, whose company prepared industrial gases including liquid oxygen, to produce a small $LO_2$ gasolene motor. This had an annular combustion chamber through which fuel was circulated for cooling before being injected into the chamber. Test firings lasted several minutes at a time, producing thrusts of 18–22 kg (40–50 lb).

So far only Goddard had launched a rocket in

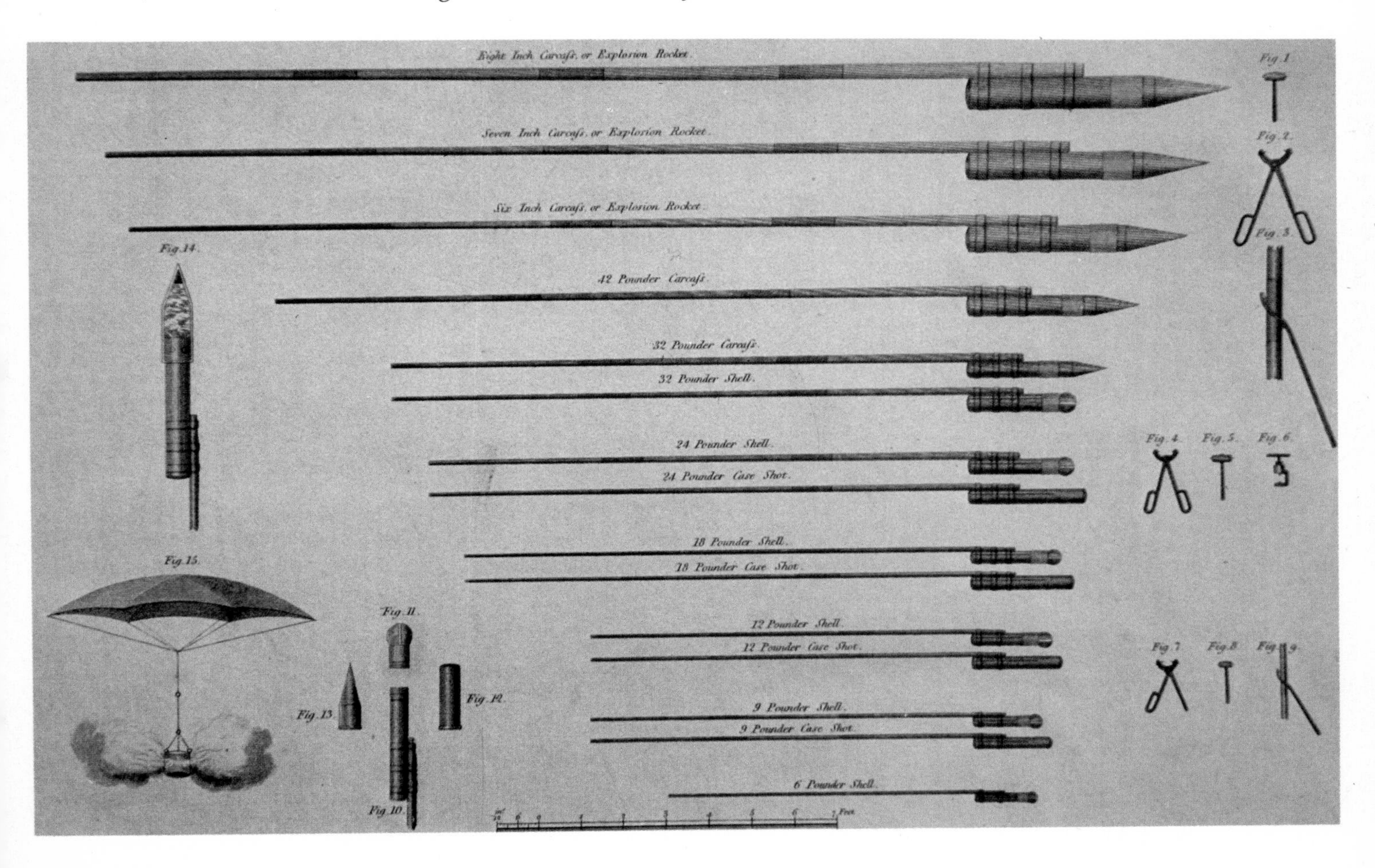


An engraving from William Congreve's *Details of the Rocket System* (1814), a pamphlet on the types and uses of rocket-powered weapons. The rockets shown vary in size from the 2.7 kg (6 lb) shell, at the bottom, to the 20 cm (8 in) calibre 'carcass' with a charge of more than 23 kg (50 lb) of powder and a range of 2.3 km (2,500 yards). The umbrella-like device (bottom left) was intended to be released by a rocket, and would carry a flare or a delayed-action bomb.

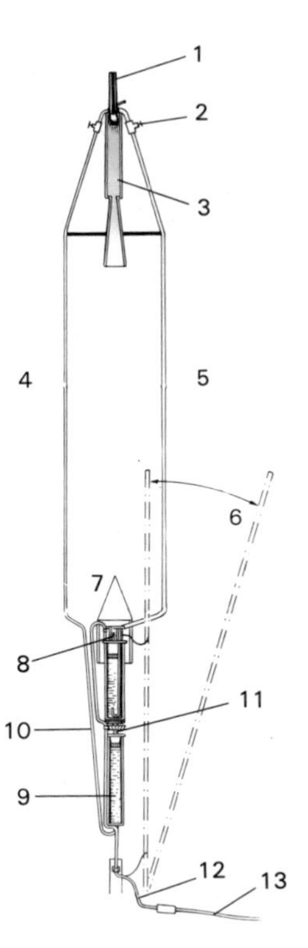


**Above** Sketch of the world's first liquid-fuel rocket, flown by Robert H. Goddard on 16 March 1926. Key to numbers: 1 Igniter, 2 Fuel-line valves, 3 Rocket motor, 4 Gasoline line, 5 Liquid-oxygen line, 6 Hinged rod with pull cord, 7 Exhaust shield, 8 Liquid-oxygen tank, 9 Gasoline tank, 10 Oxygen pressure line, 11 Alcohol burner, 12 Detachable starting hose, 13 Oxygen supply line.

**Left** Goddard with another of his early rockets.

free flight. The next person to do so was the German Johannes Winkler, a founding member of the VfR. Although his first attempt on 21 February 1931 was less successful than Goddard's – his rocket made a brief hop of some 3 m (10 ft) – three weeks later the same vehicle climbed more than 300 m (985 ft). It was fuelled with liquid methane and liquid oxygen.

The next country to make progress was the Soviet Union. On 17 August 1933 on the Nakhabino Polygon, near Moscow, the GIRD 09 rocket rose to a height of around 400 m (1,300 ft). This time the fuel was a solidified gasolene (gasolene and colophony, a resin obtained from turpentine) and liquid oxygen. Built under the direction of M. K. Tikhonravov, the team included the young Sergei P. Korolev, who 24 years later was destined to launch the world's first artificial satellite. On 25 November 1933 another Russian rocket, the GIRD 10 powered by an $LO_2$ gasolene engine designed by F.A. Tsander, ascended more than 4.8 km (3 miles).

In the meantime, events were moving swiftly in Germany. Members of the VfR decided to embark upon their own experiments. They took a *Kegeldüse* (cone motor) design of Oberth's, built it in steel, and copper-plated the interior. The uncooled motor ran successfully on 23 July 1930 for $1\frac{1}{2}$ minutes, producing a constant thrust of about 7 kg (15.4 lb), burning 6 kg (13.2 lb) of liquid oxygen and 1 kg (2.2 lb) of gasolene. As a result Rudolf Nebel, who had worked with Oberth on a previous rocket, suggested that a small liquid-fuel rocket should be built as a cheaper method of gaining experience. Rockets of this type would be known as *Minimumraketen*, later abbreviated to Mirak.

Mirak I, built by Klaus Riedel, Nebel, and Kurt Heinisch, had a cylindrical tank with a bullet-shaped nose containing liquid oxygen. At its base was the cone-shaped motor made of copper surrounded by the supercold (−183°C) oxygen on the theory that combustion heat would maintain feed pressure by increasing the evaporation rate.

It was at about this time that members of the VfR acquired a disused arms dump at Reinickendorf, near Berlin, which they developed as their *Raketenflugplatz* (rocket-flying field). The group was now able to progress to Mirak II – basically a larger version of the original except that it had a cylindrical combustion chamber and a pressure-relief valve. Thrusts of about 32 kg (70 lb) were achieved during the spring of 1931. However, Mirak II suffered the same fate as the original when its $LO_2$ tank burst.

Mirak III was started, but it was never completed. Instead, Riedel took two lengths of magnesium tubing of the same section as the Mirak tail tubes to serve as $LO_2$ and fuel tanks, and placed an aluminium rocket motor in the nose, which had a cylindrical water jacket. It was a heavy construction and not really meant to fly. However, when Riedel fired the rocket – named the Repulsor – on 10 May 1931 it took off and slowly ascended to about 18.5 m (60 ft); it fell back and merely broke a feed pipe.

This partial success led to a second Repulsor of much reduced weight, with four aluminium tail fins for stability. Launched on 23 May, it roared up almost vertically to about 61 m (200 ft) before curving over and ending its flight in a tree about 600 m (2,000 ft) away. However, in August 1931 a rocket ascended more than 1,000 m (3,300 ft). This was the famous 'One-Stick' Repulsor, with tanks arranged in tandem instead of side by side, the rocket motor still being at the front in a streamlined water jacket, and the parachute container between the tail fins.

## Wernher von Braun

In the meantime the VfR had fallen on hard times. The Depression was beginning to bite, the society's membership began to decline, and funds ran low. If the pioneer work was to continue it was obvious that more substantial funds would be needed than the VfR could muster in its best days. In a last despairing effort in the summer of 1932 Nebel, Riedel, and Wernher von Braun (1912–77) put on a demonstration of a large 'One-Stick' Repulsor at the army proving ground in Kummersdorf, near Berlin.

Although no one realized it at the time, this was the turning point. Von Braun, then just 20 years old, was given the chance to do experimental work for his doctor's thesis on rocket combustion under sponsorship of the army. He and a mechanic began to develop a series of small test rockets burning liquid oxygen and alcohol. The first of these, given the name Aggregate One, or A-1, was just 1.4 m (55 in) long and 30 cm (1 ft) in diameter; it had a thrust of 300 kg (660 lb). Propellants were fed to the combustion chamber by compressed nitrogen, and the firing time was 16 seconds.

Its successor, the A-2, included engine improvements and a gyro located between the tanks which operated aerodynamic controls to keep the vehicle on a straight path. Two successful firings carried out from the North Sea island of Borkum, near Emden, in December 1934 reached heights of 1,980 m (6,500 ft).

It is interesting to compare this work with American achievements of the period. Goddard, working at Roswell, New Mexico, had begun to test rockets employing pressure-feed systems and stabilized by gyro-controlled vanes working in the exhaust. On 28 March 1935 a gyro-

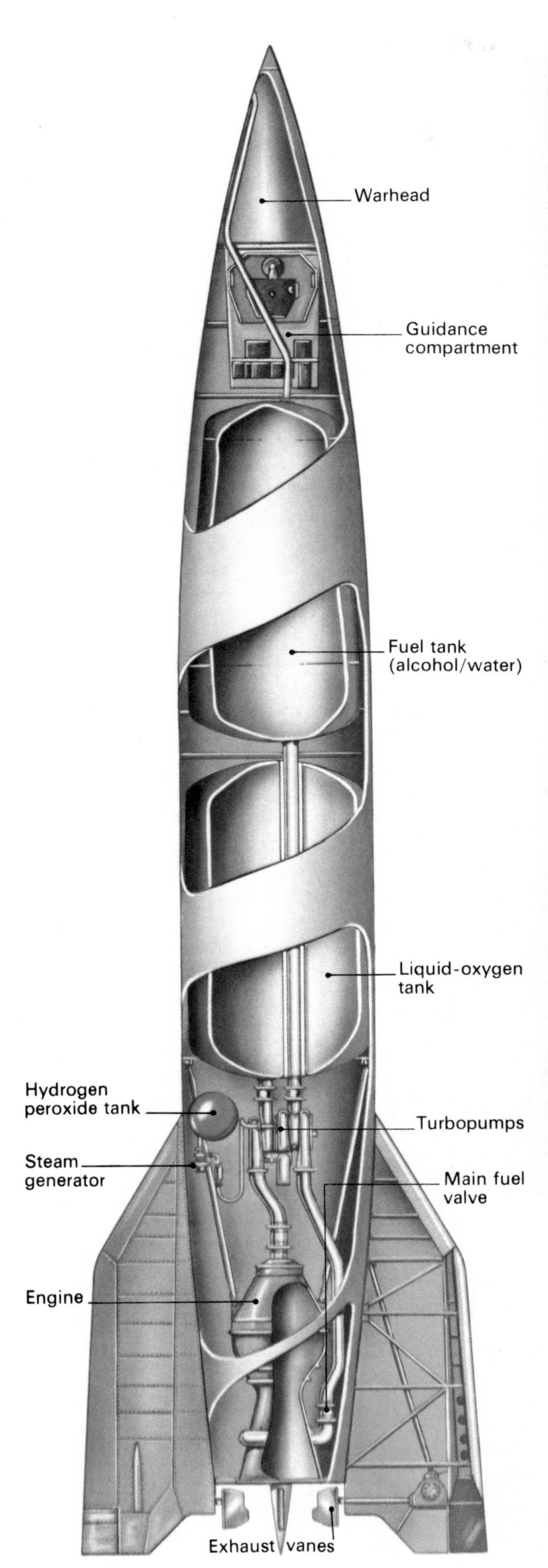


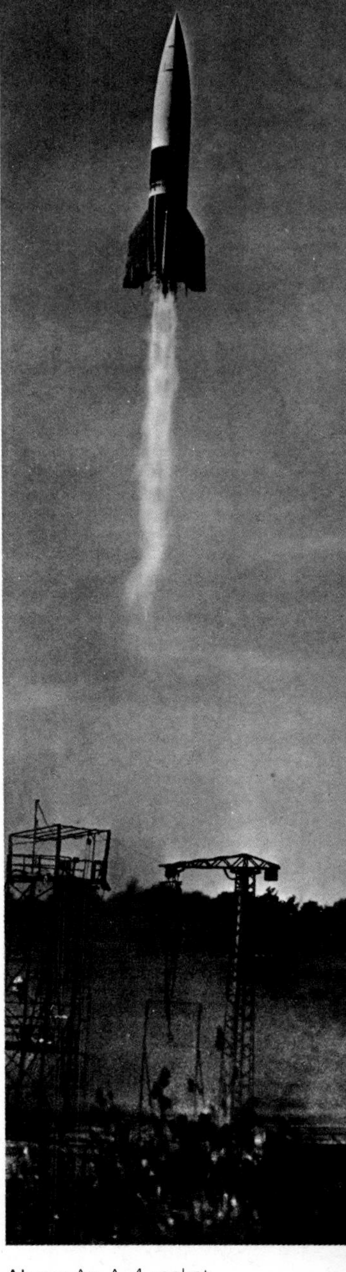

**Above** An A-4 rocket blasts off from launch pad No. 7 at Peenemünde in the summer of 1943.

**Left** The A-4, originally developed as a field weapon for the German army, was launched against London and Antwerp in World War II. The 12,500 kg (12.2 ton) rocket was 14 m (46 ft) high and had a diameter of 165 cm (65 in). It was highly mobile in operation, being erected vertically on a small launch table by its own transporter.

A Vanguard rocket on the launch pad at Cape Canaveral in June 1959.

An Intelsat 4A communications satellite in the nose of an Atlas-Centaur rocket blasts off from Cape Canaveral in September 1975.

stabilized rocket 4.5 m (14.7 ft) long travelled a distance of 3,950 m (2.5 miles) and reached a zenith of 1,463 m (4,800 ft). The rocket was seen to correct its flight automatically several times during the first 20 seconds of its journey.

It was at about this time that von Braun had the A-3 on the drawing board. This rocket is noteworthy in being the first German example to employ gyro-stabilized steerable vanes in the exhaust. Such vanes allowed a rocket to obtain control from the moment of lift-off, whereas aerodynamic rudders worked only when a vehicle had picked up speed.

The A-3 stood 7.6 m (25 ft) tall, had a body diameter of 76 cm (2.5 ft), and a take-off weight of 748 kg (1,650 lb). Its engine gave a thrust of 1,500 kg (3,307 lb) for 45 seconds.

Huge funds now began to flow to the rocket team. The Luftwaffe allocated five million marks for the construction of test facilities, to which the army added six million to support work on long-range rockets. 'Henceforth,' wrote von Braun later, 'million after million flowed in as we needed it.'

The first priority was a new research centre which would allow rockets to be launched over distances of 520 km (200 miles) or more, and von Braun chose a coastal site near the village of Peenemünde on Germany's Baltic coast because his grandfather had once gone duck-shooting there. It took two years to build the huge rocket station, to which the team transferred in April 1937 with von Braun appointed as its technical director.

Hardly had the rocketeers moved into their splendid new quarters than test firings of the A-3 began from the small Baltic island of Greifswalder Oie. The rocket had a new gyro-control system and was launched from a vertical position 'sitting on its fins'; but early versions tumbled and crashed. The fault was traced to the new gyro-control system. This entailed fundamental re-design, which led to a much-improved rocket known as the A-5. However, it was not until the autumn of 1939 that the new rocket was tested successfully.

This was the beginning of the V-2 story. Over the next two years some 25 A-5s were launched, some of them several times. The ones which were launched vertically attained heights of some 13 km (8 miles); others were made to follow inclined paths to achieve maximum range, and gave confidence in the ability to achieve directional control.

In the design offices at Peenemünde the A-5's 'big brother' was already taking shape in response to an Ordnance Department requirement for a field weapon which exceeded the range of conventional artillery. The range was determined by the largest rocket that would pass through railway tunnels. Thus was born the A-4 with a design range of 275 km (170 miles). It would burn liquid oxygen and alcohol and have a rocket engine of 25,000 kg (55,125 lb) thrust fed by turbopumps driven by steam generated from hydrogen peroxide and permanganate.

The rocket stood 14 m (46 ft) high on the launch pad at Peenemünde in the spring of 1942 – but the first launch attempt ended in disaster. Although the engine burst into life with a tremendous roar, the A-4 rose precariously from its platform for about one second; then the thrust subsided, the fins crumpled beneath it, and the rocket toppled over and was destroyed in a tremendous explosion.

On 3 October 1942 the third rocket made a perfect flight, with an engine burn of 63 seconds. The rocket reached its zenith at 85 km (52.8 miles) and landed some 190 km (118 miles) downrange.

The German High Command now ordered the still imperfect A-4 into mass-production and military personnel began training to launch missiles operationally. Allied air raids on Germany had raised Hitler's enthusiasm for rocketry, and in 1944 the A-4 was brought to bear on London and the Home Counties by mobile batteries; another key target was Antwerp, the important Allied supply port in Belgium. The rocket's name, V-2, was an abbreviation of *Vergeltungswaffe zwei* (Vengeance weapon no. 2).

One last development remained in the big league. This was the A-4b, a winged version of the V-2 designed to extend its range by gliding in the upper atmosphere. On 24 January 1945 the second A-4b flew faster than sound, reaching a cut-off speed of 1,207 m/sec (3,960 ft/sec) and a maximum height of over 77 km (48 miles). However, as the vehicle made its return to denser levels of the atmosphere, it spun out of control, falling far short of its planned goal of 434 km (270 miles).

The end of Peenemünde was now in sight. Within the months the Allies were exerting a pincer grip on the German heartland, and in the face of the Russian advance from the east more than 4,000 people were evacuated from Peenemünde after they had blown up the rocket installations.

Wernher von Braun, who had fled westward in a convoy of vehicles carrying essential files, surrendered to the US Army on 2 May. The Americans also captured the enormous Mittelwerke underground rocket factory in the Harz mountains. They removed 300 railway wagons of V-2 components, which were later used to make 68 working rockets that von Braun and his team helped to launch in the United States.

## Post-war Rocketry

The Soviet Union missed out badly on the V-2 booty. The Russians found two V-2s at a test range in Poland, but Wernher von Braun and a large number of his best people eluded them.

Unknown to the West, however, the Russians had already made much progress on their own account, largely through the efforts of Sergei Korolev and the engine designers Valentin P. Glushko, L. S. Dushkin, and A. M. Isayev.

Korolev was basically an aircraft engineer. He started work in the aircraft industry in 1927 and three years later graduated from the Aeromechanics Department of the Bauman Higher Technical School; he also completed a course of study at the Moscow School of Aviation. In 1937–9 he led development of the Project 212 'flying bomb', a small pilotless aircraft which took off from a ramp under rocket assistance. The integral rocket engine was the ORM-65 designed by Glushko, and bench-tested by him in November 1936, which burnt nitric acid and kerosene and had a thrust which could be varied between 50 and 175 kg (110 and 386 lb). Korolev also designed a 17 m (55.7 ft) span rocket glider, the PR-318-1, which he powered by a modification of the ORM-65.

Next to appear was a liquid-fuelled rocket fighter designed by a team under V. F. Bolkovitinov, the BI-1. This low-wing monoplane had a 7.5 m (24.6 ft) span and carried 500 kg (1,102 lb) of propellants for a D-1-A-1100 engine designed by A. M. Isayev and L. S. Dushkin. The prototype took off under its own power on 15 May 1942 with Captain Grigori Y. Bakhchivandzhi at the controls, and completed a successful flight.

In 1941 Glushko was given his own experimental design bureau (OKB) for liquid-fuelled rocket engines. The first of these was the RD-1, flight-tested in 1943. It was the forerunner of a famous series which, in post-war years, was to find application in Russian-designed ballistic rockets and space boosters.

Rocket groups under Korolev started by building improved versions of the V-2 (NATO code names Scunner and Sibling), and on 30 October 1947 they began test launchings from a site at Kapustin Yar, south-east of Stalingrad (now Volgograd). In 1949 came launches of the Pobeda (Victory-class) ballistic missile (Shyster), which had a range of several hundred miles. It was followed by a much-improved missile, known by the NATO code name Sandal, which burned nitric acid and kerosene and was intended to carry a nuclear or conventional warhead up to 1,770 km (1,100 miles). It was the unexpected discovery of Sandal in Cuba that led to the 'missile crisis' of 1962.

## Missiles and Research Rockets

Meanwhile, in the United States, the ex-Peenemünde group had taken part in V-2 launchings at White Sands, New Mexico, where opportunity had been taken to send research instruments into the upper atmosphere. Another key step was the Bumper Project which involved mounting small American WAC-Corporal rockets on the nose of V-2s. One of these two-stage rockets launched on 24 February 1949 reached a record altitude of 393 km (244 miles) and a speed of 8,286 km/h (5,150 mph).

In 1950 von Braun's group, together with several hundred Americans, moved into Redstone Arsenal in Huntsville, Alabama, to blaze a new trail in rocketry as the Army Ballistic Missile Agency (ABMA). First priority was given to developing the US Army's Redstone medium-range ballistic missile which, like the Russian Pobeda, owed something to V-2 technology although it embodied new structural techniques and engine technology.

While other parts of US industry contended with the giant problem of designing intercontinental ballistic missiles (ICBMs), ABMA was busy with the army's Jupiter intermediate-range ballistic missile (IRBM) for use in allied countries within range of Moscow. A US Air Force IRBM of similar range – 2,400 km (1,500 miles) – and capability was the Thor which, between 1958 and 1963, was based in eastern England. It was on the basis of Redstone, Jupiter, Thor, and the big Atlas and Titan ICBMs that the United States was to enter space.

Already the Russian-built V-2 and Pobeda had been adapted as geophysical rockets to carry research instruments into space. In May 1949 the first had lifted a 120 kg (264 lb) payload to a height of 110 km (68 miles).

In the summer of 1955 the Russians began building a major rocket base in a remote area of Kazakhstan north of the town of Tyuratam. It was here that the Soviet Union's first ICBM, known as the R-7 (NATO code name Sapwood), was brought for testing in great secrecy. Fuelled with kerosene and liquid oxygen, it comprised a central core and four strap-on boosters. Firing at lift-off were no fewer than 20 main-thrust chambers and 12 small, swivel-mounted motors for thrust-vector control and fine (vernier) adjustment of velocity. The engine of the central core was an RD-108, and each of the four boosters contained an RD-107. Total lift-off thrust exceeded 450 metric tons. On 3 August 1957 a near-perfect launching of the 30.5 m (100 ft) rocket sent a dummy warhead some 6,437 km (4,000 miles) to land in a target area off the Kamchatka peninsula.

The day of the Sputnik was at hand.

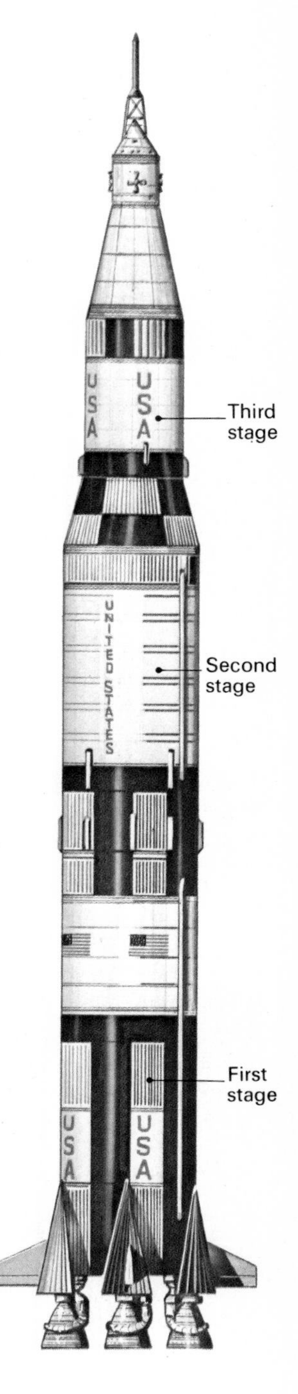


**Above** The huge three-stage Saturn 5 rocket that took men to the Moon (see pages 238–9).

# 28 Sputnik and After

The success of the big ICBM gave the Kremlin a golden opportunity. President Eisenhower had already announced that the United States would launch a small artificial satellite as a contribution to the International Geophysical Year (1957–8). The project would be based on a rocket developed for peaceful scientific purposes, and would orbit a 9.75 kg (21.5 lb) satellite of 50.8 cm (20 in) diameter; it would be called Vanguard.

Imagine the surprise, therefore, when the world awoke on 4 October 1957 to the bleeps of a Russian satellite, Sputnik 1, which swung round the Earth every 96 minutes! In a triumphant communiqué, the Russian news agency Tass gave its weight as 83.6 kg (184 lb), more than eight times the Vanguard payload. While the United States was left wondering how such a major advance could possibly have been made by a 'technically backward' country, steps were immediately taken to get Vanguard to the launch pad at Cape Canaveral (Florida). On 6 December 1957 the slim, three-stage rocket stood ready for lift-off with a tiny 1.47 kg (3.25 lb) test satellite. But when the engine ignited it failed to develop sufficient thrust and, instead of heading into space, the vehicle toppled over and exploded in flames. At the edge of the inferno was the tiny satellite – still bleeping.

Fortunately, the ex-German rocket team at Redstone Arsenal was ready to step into the breach. Wernher von Braun had already offered to launch a satellite using the Jupiter-C rocket which had been built to test sub-scale nose-cones for the Jupiter IRBM, but he had been turned down – perhaps a little late in the day – on political grounds.

The basis for this launcher was a modified Redstone rocket to which was attached a spinning drum of solid-fuel Baby Sergeant rockets as second and third stages. The addition of one more rocket in the nose, plus a scientific payload, offered the opportunity, with minimum modification to the guidance system, to achieve the 28,157 km/h (17,500 mph) velocity required for orbital flight. The task was completed in 84 days by ABMA and the Jet Propulsion Laboratory (JPL), which supplied the solid rockets.

Launched on 31 January 1958, the final stage with its attached instruments achieved a high elliptical orbit which ranged between 356 and 2,548 km (221 and 1,583 miles). The satellite itself, designed and built by JPL under the direction of Dr William H. Pickering, consisted of a Baby Sergeant rocket motor (the launch vehicle's top stage) and a canister of instruments supplied by Dr James Van Allen of the State University of Iowa. It weighed just 13.97 kg (30.8 lb), including its scientific payload.

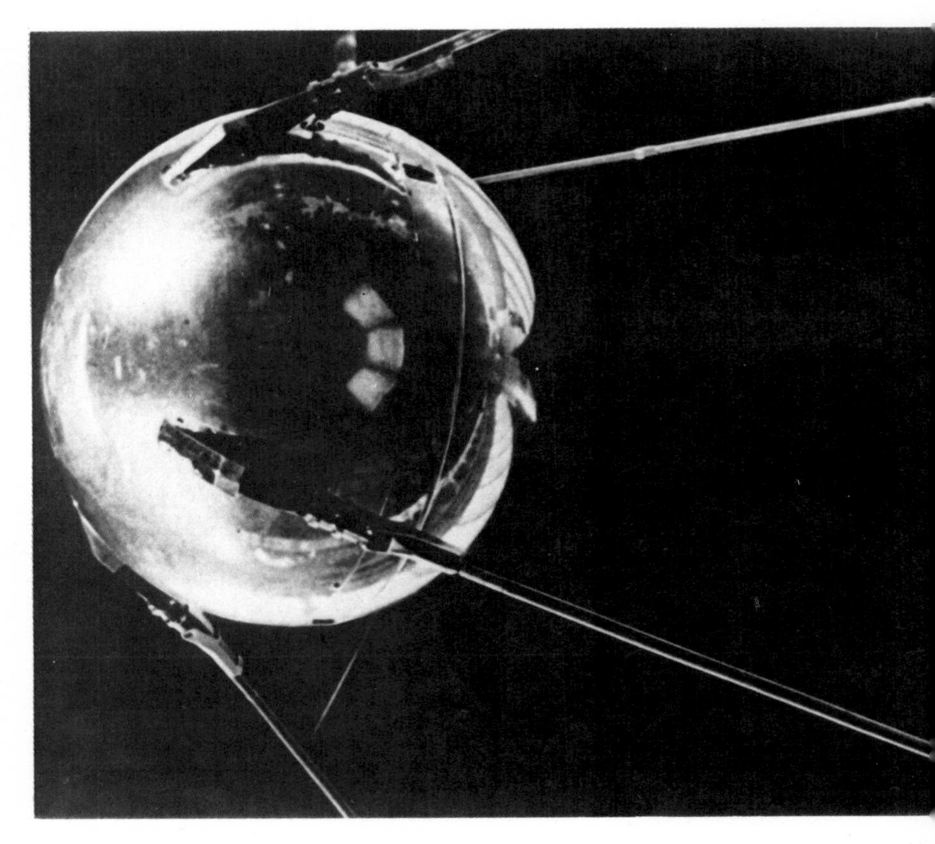

**Above** On 4 October 1957 the Soviet Union opened the 'Space Age' with Sputnik 1. It was basically a radio transmitter with chemical batteries encased in a polished metal sphere of 58 cm (22.8 in) diameter with four 'whip' aerials. It weighed just over 83 kg (184 lb).

**Left** Vanguard satellites prepared for the International Geophysical Year (1957–8) had a diameter of 50.8 cm (20 in). This one is attached to the final stage of its launch vehicle. When it was launched it was protected by a jettisonable nose cone.

In spite of its small size compared with Sputnik, America's first satellite – Explorer 1 – was responsible for the major discovery that the Earth is girdled by a belt of charged particles trapped in space by the planet's magnetic field. Today this region is known as the Van Allen radiation belt after the scientist who instrumented the satellite.

It was the Soviet Union, however, that continued to dominate the space scene. On 3 November 1957 the big ICBM scored again by orbiting Sputnik 2 with the dog Laika. The animal was fed and watered from an automatic dispenser inside a pressurized compartment so that biomedical data could be received on Earth for as long as possible. There was no provision for recovery, and Laika lapsed into unconsciousness and died as oxygen ran out. An even bigger payload appeared with Sputnik 3 on 15 May 1958, in the shape of a 1,327 kg (2,926 lb) geophysical laboratory, which confirmed the existence of the Van Allen belt.

Although the United States responded with successful shots of small satellites of the Explorer and Vanguard types, it was evident that the much larger Soviet rocket could outstrip America's capability by a handsome margin.

## The Space Race

In response to the Soviet challenge the United States began a major reorganization of its aerospace establishments, and on 1 October 1958 the National Aeronautics and Space Administration (NASA) came into being, taking over the research centres of the former National Advisory Committee for Aeronautics (NACA). Within NASA were also to be grouped such establishments as JPL and ABMA, the latter reconstituted as the Marshall Space Flight Center.

What emerged from the quickening space race between the Soviet Union and United States has passed into history. Artificial satellites opened our eyes to a universe to which we had been practically blind because earthbound observation of much of the incoming radiation from space is blocked by the atmosphere. With the help of orbiting observatories such as Uhuru, Ariel 5, and the High Energy Astronomical Observatory (HEAO), we are beginning to study some of the most intriguing objects in the cosmos, including supernovae (exploding stars), pulsars (pulsating stars), the super-dense neutron stars, and 'black holes'.

The Sun – our home star – has received special attention. Several observatory satellites have focused instruments upon it to measure its influence upon the Earth's environment, especially during periods of sunspots and solar flares. Two remarkable West German space probes, Helios 1 and Helios 2, have actually come within 45 million km (28 million miles) of its flaring surface.

No less remarkable has been the pace at which space technology has been applied on Earth. Thanks to communications satellites poised in orbit some 35,880 km (22,300 miles) above the Atlantic, Pacific, and Indian oceans, nation has been linked to nation by telephone, telegraph, and television. Just one Intelsat 4A satellite can relay 20 colour television programmes or more than 11,000 telephone calls.

In 1978 a thousand million people – one in every four on Earth – can see an international event 'live by satellite' as it happens. The space age has even reduced communications costs. Today a telephone call between New York and London costs only half as much as it did 10 years ago – quite apart from the effect of inflation, which makes it less than a quarter the price in real terms.

People in developing countries have been among the first to reap the benefits. A powerful American satellite, ATS-6, in stationary orbit above East Africa was able to re-broadcast educational TV programmes transmitted from Ahmedabad directly to thousands of scattered towns and villages throughout India. The Russian geostationary satellite Ekran makes a similar link with remote communities in Siberia. All that is needed to pick up the broadcasts is a television set, a simple converter, and a cheap aerial pointed at the satellite. Such satellites are called comsats (communications satellites).

Another big advance has come from satellites, such as the US Navy's Transit, which serve as 'radio stars', allowing ships to fix their position with great accuracy in all weathers. This technique is now being extended to allow commercial aircraft to obtain their navigational 'fix' from space using aerosats (aeronautical satellites).

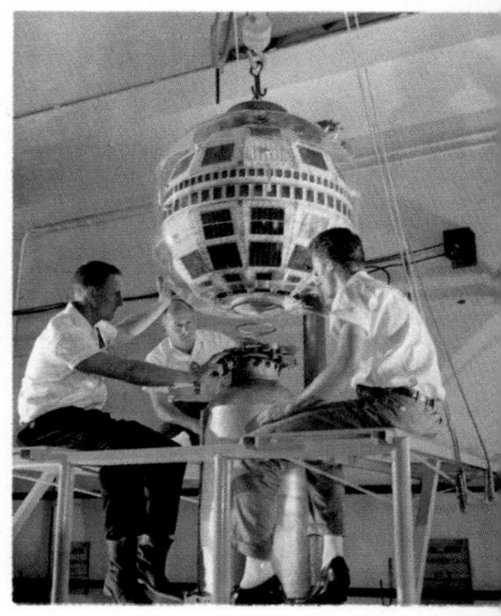

**Above left** Helios 1, built in West Germany, was launched by a Titan-Centaur rocket from Cape Canaveral in December 1974. The 370 kg (816 lb) probe flew within 45 million km (28 million miles) of the Sun on 15 March 1975, enduring temperatures hot enough to melt lead. One of its discoveries was that micrometeoroids relatively close to the Sun were more intense than those nearer Earth; they also appeared to move in different directions at different times.

**Above right** The 77 kg (170 lb) Telstar 1 satellite, which made possible the first television exchange between the United States and Europe in the summer of 1962. In its relatively low orbit of 952 by 5,632 km (591 by 3,500 miles), existing ground stations were able to track Telstar across the sky for periods of only about 20 minutes, after which it dipped below the horizon and communication was lost.

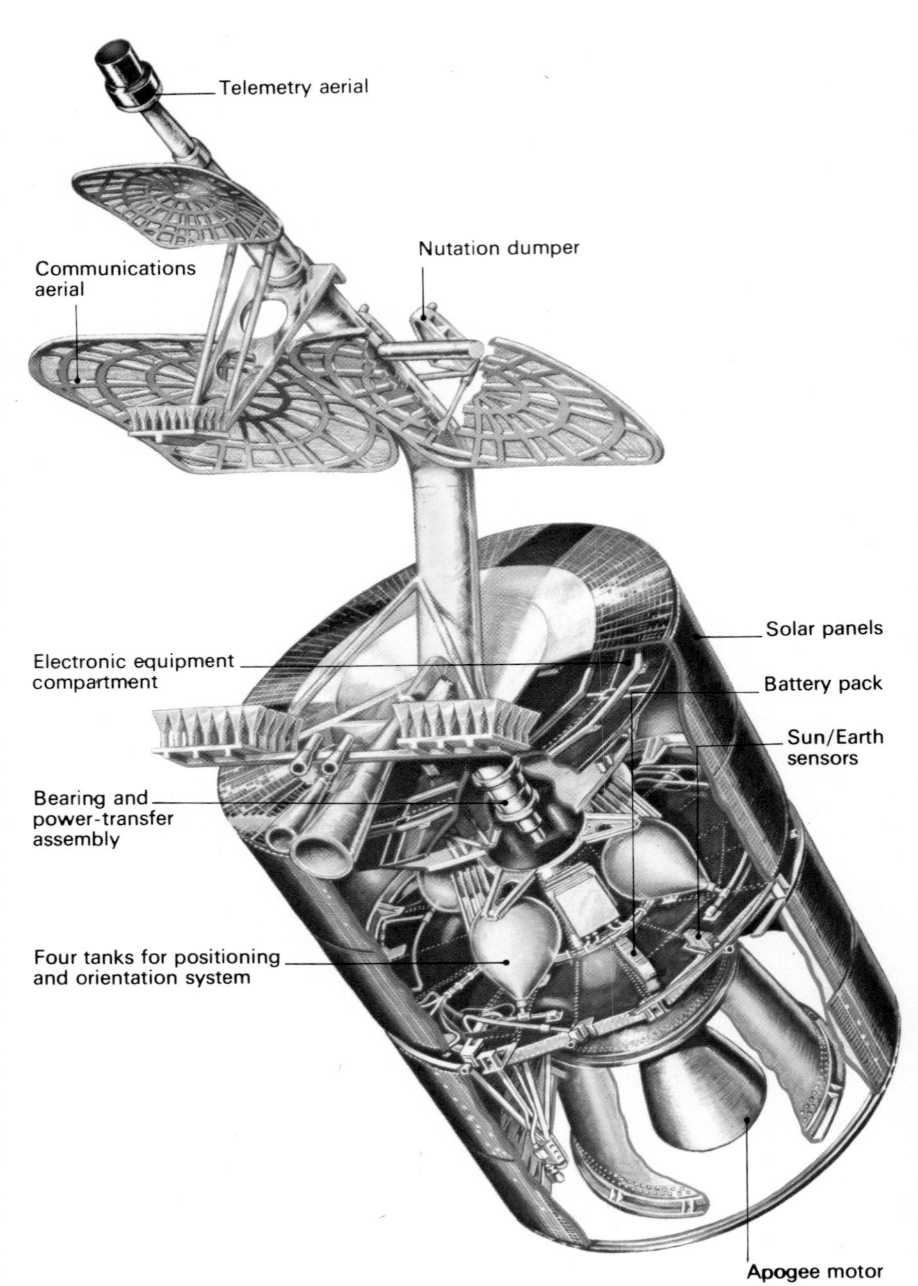


Observation satellites, circling our planet in north/south polar orbits, keep watch on the world's weather, giving advance warning of destructive storms such as hurricanes and typhoons. They also play a vital role monitoring the breakup and formation of sea ice, and map the snow cover on mountains. Typical examples are the United States' Nimbus, the Soviet Union's Meteor, Japan's Geostationary Meteorological Satellite (GMS), and Europe's Meteosat.

Another breed of observation satellite keeps watch on the world's natural resources of minerals and fossil fuels. The US satellites, called Landsat, take 'electronic' pictures of the ground in which false colours emphasize subtle differences unnoticed by the human eye. In agricultural areas, for example, blighted crops show up grey-black and healthy crops bright pink or red. Even differences in soil constituents and moisture content can be identified in the photos. Food production can reap immense benefits from this kind of routine survey, and the developing countries of Africa and South America have been among the first to set up ground stations so that satellite data can be applied on the spot without delay.

In this way farmers are being helped to find the best places to cultivate their crops, and the best times to plant and harvest them for maximum yield. It is even possible to obtain an early assessment of total agricultural production, so that food supplies can be properly set against future need. The space survey can also provide timely warning of drought, floods, and erosion. Close study of space photos of the terrain, often in remote and little explored parts of the world, also give geologists vital clues which can lead to the discovery of new oilfields and mineral deposits. The same photos can reveal the spread of industrial and other man-made pollution affecting the air we breathe and the waters of our oceans, lakes, and rivers. The Soviet Union is making similar observations from the manned Soyuz spacecraft and Salyut space stations.

## Beyond Earth's Orbit

In a flurry of space activity in 1959 the Soviet Union passed, hit, and photographed the Moon. Luna 1, launched on 2 January 1959, carried instruments for measuring magnetic fields, cosmic and solar radiation, gas composition, and micro-meteoroids. It missed the Moon by 5,955 km (3,700 miles) and passed into orbit around the Sun. The following September a similar capsule struck the Moon between the craters Archimedes, Aristillus, and Autocycus. In October came the spectacular circumlunar flight of Luna 3, which obtained the first

**Above** Ten years after Telstar large, drum-like satellites such as this Intelsat 4A, about 6.7 m (22 ft) high, were established in geostationary orbit 35,880 km (22,300 miles) above the Atlantic, Pacific, and Indian oceans.

**Left** Soviet Meteor satellites – the first was launched in March 1969 – photograph Earth's changing cloud cover by day and night, keep watch on the formation and break-up of sea ice, and give advance warning of dangerous storm conditions.

pictures of the Moon's hidden side. The 278.5 kg (614 lb) robot, which passed within 7,886 km (4,900 miles) of the surface, took pictures on film, developed them internally, and televised the results to Earth.

The United States replied by sending Ranger spacecraft on a collision course aimed at investigating surface features. The first series failed in its task of landing instrument capsules on the Moon, and in 1963 the project was revised to obtain television pictures of surface features down to the point of impact. Ranger 7 obtained no fewer than 4,316 frames during the last 13 minutes of its descent onto the Sea of Clouds.

Once again, however, it was the Soviet Union that took the top prize with Luna 9, the first vehicle to land instruments in working order. On 3 February 1966 the 99.8 kg (220 lb), egg-shaped landing capsule, ejected from a mothercraft a short distance above the surface, bounced and rolled on the Ocean of Storms, west of the craters Reiner and Marius. Four petal-like covers opened, setting the capsule upright and revealing a small television camera which sent a panoramic picture of the surrounding landscape showing rocks 10–20 cm (3.9–7.8 in) across. In 1966 Luna 10, 11, and 12 followed into orbit around the Moon.

Then the Americans took a major step towards demonstrating soft-landing techniques in support of the challenging Apollo manned lunar programme. On 1 June 1966 the three-legged Surveyor 1 landed in the Ocean of Storms, using radar reflections to gauge its height above the lunar surface and a retro-rocket to brake its speed with vernier jets acting for final touchdown. Its television camera sent no fewer than 11,150 high-resolution photos.

In December the Russians landed another of their capsules, Luna 13, on the Ocean of Storms. Apart from a small television camera, the capsule had two hinged arms which flopped open after landing. On the end of one was a soil-density meter. This took the form of a gun-powder device which punched a rod into the lunar soil to investigate its density. On the other arm, also in contact with the surface, was a radiation-density meter which probed the soil with gamma rays. The soil at the landing site was found to have a density of under 1 $g/cm^3$ – much less than the density of terrestrial ground.

Surveyor 3, which reached the Ocean of Storms in August 1967, contributed more vital data. Scientists in California operated a mechanical scoop by radio control, trenching the lunar soil and photographing the results with an on-board camera. This craft returned nearly 6,320 photos. The United States rounded off the first phase of exploration, 1966–7, with five Lunar Orbiters, built to photograph the Moon and allow suitable landing sites to be selected for Apollo astronauts.

**Above** The Soviet Union achieved a major triumph in October 1959 when Luna 3 looped behind the Moon to photograph the side permanently hidden from Earth. Pictures processed automatically inside the probe were scanned electronically and the images transmitted to the Russian space centre. The launch weight of Luna 3 was 278.5 kg (614 lb).

**Right** The Russian Luna 9 became the first spacecraft to land instruments on the Moon on 3 February 1966. A capsule ejected from the mother craft bounced and rolled on the Ocean of Storms. When it came to rest, four petal-like panels opened to reveal a small television camera, which sent pictures to Earth. Total weight of Luna 9 was 1,583 kg (3,490 lb) including the 99.8 kg (220 lb) capsule.

The Russians turned to Zond circumlunar spacecraft – which they are thought to have been preparing for later manned flights – but, lacking the ability to challenge Apollo to an actual landing, their final lunar challenge was to depend on a new generation of ingeniously designed robots.

In the Soviet Union, another ambitious programme was taking shape alongside the Luna Moon programme and the effort to get a man into Earth orbit. Sergei Korolev had convinced the Soviet authorities that it was possible to launch space probes to Mars and Venus.

Premier Khrushchev evidently regarded this as an opportunity to score another propaganda victory over the United States. Rockets, again based on the original Soviet ICBM, were prepared to take advantage of the 1960 Mars

**Below left** The American Surveyor Moon robots, operated by remote radio control, were pathfinders for Apollo astronauts. They made bearing tests of the lunar soil and carried out trenching operations with a mechanical scoop, giving scientists their first look beneath the lunar surface. The Surveyors also carried seismometers to detect underground disturbances (Moon-quakes) and meteoroid impacts. The picture shows a Surveyor test vehicle. The white turret to the left of the mast houses a television camera; the rectangular flaps on the mast are the solar-cell array and high-gain aerial.

**Below right** The Venera 7 capsule which soft-landed on Venus by parachute in December 1970. Faint signals from the surface, lasting about 23 minutes, indicated a temperature of 475°C ± 20° and a crushing pressure of 90 atmospheres ± 15.

launch 'window' – the period when Earth and Mars are in the correct relationship in their respective orbits.

Khrushchev apparently timed his arrival in New York aboard the ship *Baltika* in October 1960 to coincide with launch preparations at Tyuratam, expecting to announce the flights. In fact, both attempts failed and one seems to have ended tragically. (It was nearly two years before the US authorities, on 5 September 1962, officially stated that the USSR had attempted to launch space probes to Mars on 10 and 14 October 1960 but that they had failed to reach orbit.) It seems that when the start button was pressed at the end of the countdown nothing happened. Normally, this would require making the rocket electrically 'safe' and draining off the liquid oxygen; but this would have meant delaying the launch. In overall command was Marshal Mitrofan Nedelin, then head of the Soviet Strategic Rocket Forces. Korolev was also present as Chief Designer.

Someone, perhaps Nedelin, ordered technicians to inspect the vehicle on the pad, when its engines suddenly burst into life and several people were burned to death. Officially, the Soviet authorities have never confirmed these attempted space shots. Without going into details, Tass reported the death in October 1960 of Marshal Nedelin in an aircraft accident.

## Venus and Mars

The years that followed were to see a vigorous contest develop between the Soviet Union and United States in robot exploration of the planets. Many of the early space shots ended in failure. Rockets failed to leave orbit, exploded in space, went off course, or suffered frustrating communications failures – but there were also spectacular successes.

Mariner 2, launched from Cape Canaveral for Venus on 27 August 1962, passed within 34,830 km (21,643 miles) of the planet. Instruments probed its surface and recorded a surface temperature of 428°C. Mariner 4 returned the first close-range pictures of Mars as it swung by within some 8,690 km (5,400 miles) on 14 July 1965.

Then, in a series of highly successful missions to Venus, the Soviet Union scored by parachuting instrument capsules into the corrosive atmosphere of the 'veiled planet'. The first direct atmospheric measurements were obtained by Venera 4 on 18 October 1967, but signals stopped at a considerable altitude above the surface, where the temperature was 280°C and the atmospheric pressure some 22 times that at the surface of the Earth.

The first capsule actually to land on Venus and signal its arrival was Venera 7, on 15 December 1970. Venera 8, which arrived on 22 July 1972, reported that the atmosphere was 97 per cent carbon dioxide with not more than 2 per cent nitrogen, and the pressure a crushing 90 kg/cm$^2$ (1,280 lb/sq in). The temperature, 470°C, was above the melting point of lead!

Into this fearsome environment in June 1975 plunged the twin probes Venera 9 and 10,

which carried television cameras. They had been ejected two days before from mothercraft which themselves went into orbit around Venus to make independent studies. Before separating, internal equipment in the probes had been cooled to −10°C to allow radio and instruments to withstand surface heat long enough to send useful data. The Venera 9 capsule entered the atmosphere at a speed of 10.7 km/sec (6.6 miles/sec) at an angle of 20°, protected against frictional heating by clamshell covers. When speed had been reduced to 250 m/sec (820 ft/sec) the probe discarded the clamshells and deployed a parachute. At 50 km (31 miles) the parachute was discarded, and the descent into thickening atmosphere continued by the action of a metal disc brake attached to the body. The capsule hit the surface at 7–8 m/sec (23–6 ft/sec), the landing shock being absorbed by a ring of crushable metal.

The probe had come down on a plateau roughly 2,500 m (8,200 ft) above what would be sea level, and transmitted for 53 minutes. A

**Above** The 1,031 kg (2,272 lb) spacecraft Mariner 9, which became the first artificial satellite of Mars on 13 November 1971. At the top can be seen the rocket-engine nozzle. The four rectangular panels generated electricity from sunlight. Among the 7,329 photographs it transmitted to Earth were pictures of huge equatorial canyons, volcanoes, and features resembling the dry beds of ancient rivers.

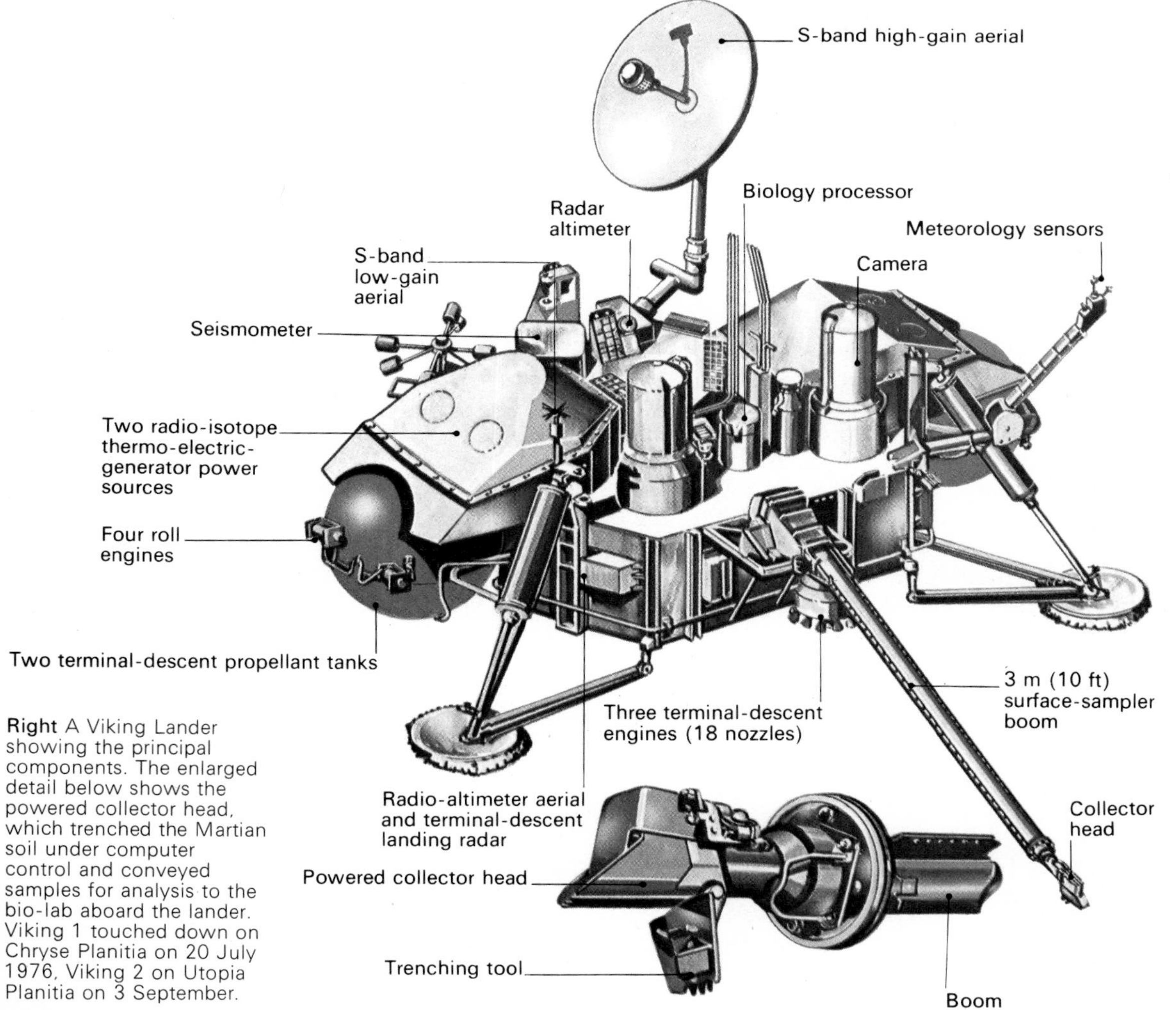


**Right** A Viking Lander showing the principal components. The enlarged detail below shows the powered collector head, which trenched the Martian soil under computer control and conveyed samples for analysis to the bio-lab aboard the lander. Viking 1 touched down on Chryse Planitia on 20 July 1976, Viking 2 on Utopia Planitia on 3 September.

panoramic picture, the first ever seen of the planet's surface, showed level ground strewn with sharp-edged rocks 30 to 40 cm (11.8 to 15.7 in) across. Atmospheric pressure was about 90 atmospheres and temperature 485°C.

The capsule released by Venera 10 came down some 2,200 km (1,367 miles) from Venera 9 and transmitted for 65 minutes. The pictures revealed a vastly different landscape with clearly discernible outcrops of rock among rock debris. The atmospheric pressure was 92 atmospheres (indicating lower terrain), and temperature again 485°C. Scientists concluded that Venus was active, with rocks millions of years younger than those on Earth and possibly still being formed.

The Americans had also been busy. In 1973–4 Mariner 10 made a grand tour of the Inner Planets, photographing the cloud tops of Venus before flying on to take the first pictures of Mercury, the tiny planet nearest the Sun. The photos revealed a Moon-like body heavily pock-marked with craters – and, to the surprise of scientists, there was a weak magnetic field.

It was Mars, however, that captured most attention. As early as 30 July 1969 Mariner 6 had flown within 3,427 km (2,130 miles) of the 'red planet' and returned photos on which could be seen craters. Instruments aboard the craft probed the thin carbon dioxide atmosphere, revealing a surface pressure of only 6–7 millibars – about 1/150th of an atmosphere. More pictures and data came streaming back from Mariner 7, which passed within 3,427 km (2,130 miles) of Mars on 4 August 1969 and photographed the planet's moon Phobos as a tiny speck.

It was now time for Mars to acquire its first artificial satellite. Mariner 9 swung into orbit on 13 November 1971 as a dust storm raged over almost the entire planet. When the dust cleared, to the amazement of scientists the cameras revealed huge equatorial canyons, volcanoes far higher than those on Earth, and features such as the dried-out beds of ancient 'rivers' which seemed to have been carved by water long ago. Altogether the mission returned 7,329 photographs.

It was at the height of the same dust storm that the Soviet Union's Mars 3 attempted to land an instrument capsule. It came down on 2 December 1971, braked by parachute and retro-rocket in the region between Electris and Phaethontis; it began to send television signals, but these ended after only 20 seconds and the resulting picture was blank. Despite further attempts to probe the red planet in 1973–4 with four large spacecraft – Mars 4, 5, 6, and 7 – scientific results were few and landing attempts again ended with vehicle failures.

**Above left** Pioneer 10, which obtained the first close-up pictures of Jupiter in December 1973, will eventually leave the Solar System. It carries Earth's first pictorial greeting to any extra-terrestrial intelligence which may find it millions of years in the future. The launch weight of Pioneer 10 was 258.5 kg (570 lb).

**Above right** The giant planet Jupiter, photographed by Pioneer 10 from a distance of 2,500,000 km (1,550,000 miles). The famous Red Spot, some 48,000 km (30,000 miles) across, could swallow several Earths. Jupiter has at least 13 moons; the dark shadow of one of them, Io, is clearly visible on the planet's surface.

It was left to the United States to bring off the major triumph of landing two Viking spacecraft in 1976. The first touched down in Planitia Chryse on 20 July, the second in Planitia Utopia on 3 September.

While these Mars Landers set about their task of sampling the soil, using programmed instructions fed into an onboard computer from the control centre in California, the two Viking mothercraft photographed the planet comprehensively from orbit and probed its surface with sensitive instruments. Within a year the spacecraft had returned thousands of high-quality photographs in colour, black and white, and stereo. Many samples of soil and rock had been scooped up for robot analysis in their biochemical laboratories. Curiously, despite chemical tests which showed vigorous reactions with the soil, there was no clear evidence of organisms (life), past or present.

The Landers made daily weather reports to Earth, giving air temperature and wind speed and direction, which allowed scientists to predict weather on Mars. One minor Mars-quake was recorded by the Viking 2 Lander. The Vikings confirmed that the exceptionally thin Martian atmosphere (like the thick atmosphere of Venus) was largely carbon dioxide. They discovered small amounts of nitrogen, but dismissed early findings by a Soviet spacecraft that large amounts of argon were present.

Meanwhile, the Viking Orbiters had mapped the distribution of water vapour in the atmosphere and studied surface temperatures. The ice caps were found to contain water ice, with an overmantle of carbon dioxide snow (dry ice) in winter. They photographed in great detail the sinuous channels that resemble river

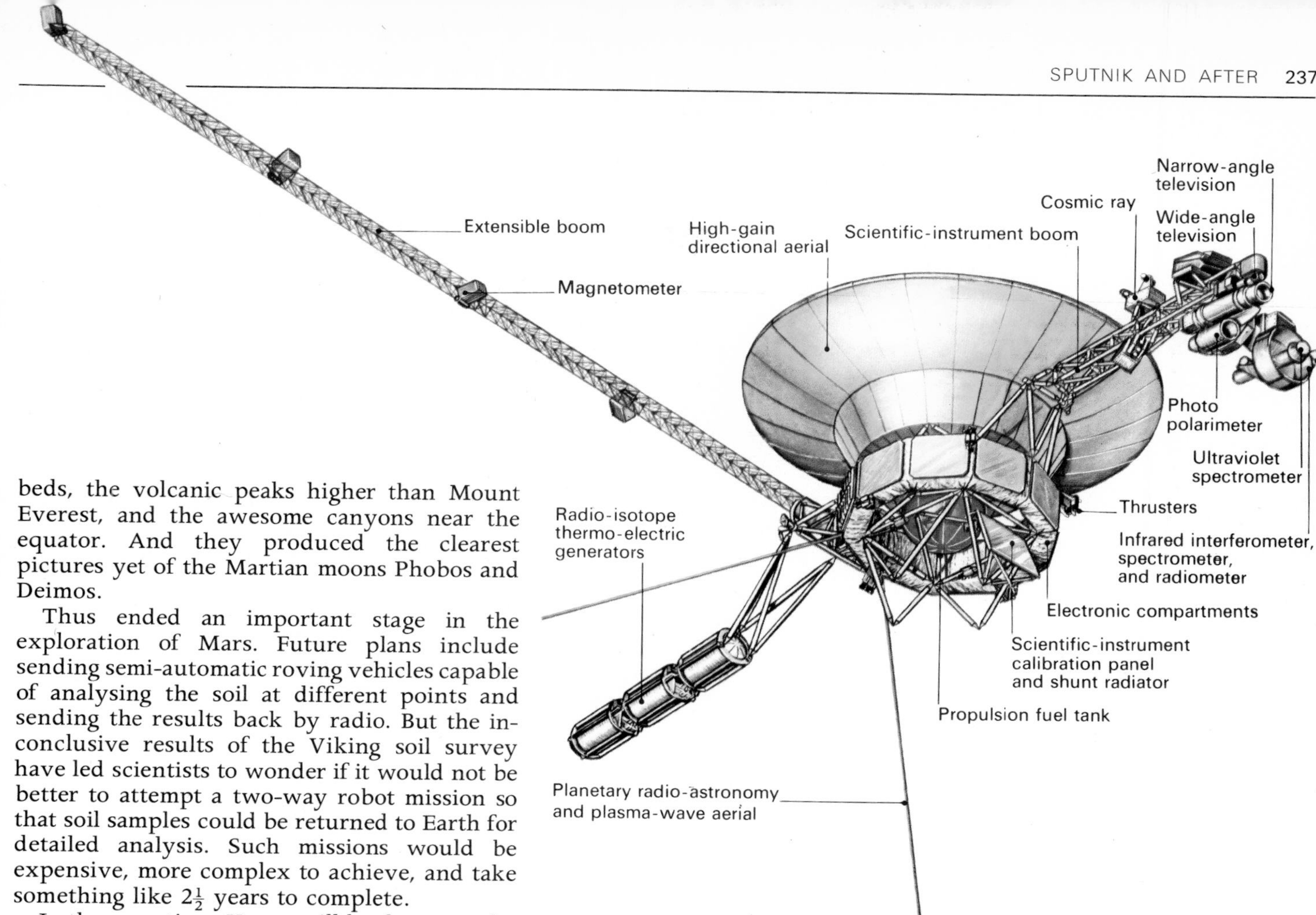


beds, the volcanic peaks higher than Mount Everest, and the awesome canyons near the equator. And they produced the clearest pictures yet of the Martian moons Phobos and Deimos.

Thus ended an important stage in the exploration of Mars. Future plans include sending semi-automatic roving vehicles capable of analysing the soil at different points and sending the results back by radio. But the inconclusive results of the Viking soil survey have led scientists to wonder if it would not be better to attempt a two-way robot mission so that soil samples could be returned to Earth for detailed analysis. Such missions would be expensive, more complex to achieve, and take something like 2½ years to complete.

In the meantime, Venus will be the target for further exploration. In December 1978 Russian and American probes are expected to swing into orbit, and the Pioneer-Venus multiprobe should release cone-shaped bodies to investigate the atmosphere and weather at four different points of entry.

In 1983 the Soviet Union plans to float French-made 8–9 m (26–30 ft) diameter balloons in the corrosive atmosphere of Venus. Instruments suspended from a balloon are designed to pick up data for perhaps four days as it floats some 55 km (34 miles) above the planet's hostile surface.

## Jupiter, Saturn, and Beyond

In the meantime, the giant outer planets have not been neglected. Pioneer 10 took the first close-up photos of Jupiter and its famous 'red spot' in December 1973. A year later a sister craft swung around this extraordinarily puzzling planet on its way to the ringed planet Saturn, which it should reach in September 1979. Eventually, both probes will escape from our Solar System into the infinite spaces beyond. Following them are two more American spacecraft, Voyager 1 and 2, which have the task of investigating Jupiter, Saturn, and their moons in greater detail. If the first Voyager completes its mission to Saturn, the second may be retargeted to fly on to Uranus in January 1986 and Neptune in September 1989.

As both spacecraft will achieve Solar System escape velocity they carry messages in case they are ever found by another civilization. A 30.5 cm (12 in) disc recording includes pictures and sounds depicting the physics of our part of the Galaxy, the characteristics of Earth and its diversity of life, culture, music, and thought, and messages from President Jimmy Carter and Secretary General of the United Nations Dr Kurt Waldheim. The President's message reads (in part):

> This is a present from a small distant world, a token of our sounds, our science, our images, our music, our thoughts, and our feelings. We are attempting to survive our time so we may live into yours. We hope someday, having solved the problems we face, to join a community of galactic civilizations. This record represents our hope and our determination, and our good will in a vast and awesome universe.

In 1981–2 will come another major milestone, the launching from the orbiting Space Shuttle of the Jupiter-Orbiter-with-Probe (JOP) which is intended to make the most detailed study yet of Jupiter, its radiation belts, and its moons. The mission includes an Orbiter which will circle around the planet for at least two years, sending back information, and a Probe which has been designed to plunge deeply into the giant planet's atmosphere.

In 1977 two Voyager spacecraft like this one were launched from Cape Canaveral on 'gravity-assist' missions to explore the giant outer planets of our Solar System. Voyager 1 should reach Jupiter in March 1979 and pass Saturn in November 1980; Voyager 2 should reach Jupiter in July 1979, Saturn in August 1981, and (possibly) Uranus in January 1986. Both spacecraft are expected to leave the Solar System in 1989. On the chance that intelligent life-forms have evolved elsewhere in the Galaxy, the Voyagers carry recorded messages which identify the origin and development of our species. The launch weight of each Voyager was 808 kg (1,782 lb) and the diameter of its high-gain aerial was 3.7 m (12 ft).

# 29 Manned Spaceflight

The combination of Sergei Korolev's brilliance as a rocket designer and Premier Khrushchev's determination to set the pace in world affairs was to create another resounding triumph for the Soviet Union. On 12 April 1961 Major Uri Gagarin of the Soviet Air Force became the first human being to travel in space. Gagarin's 4,725 kg (10,418 lb) spacecraft made a single orbit of the Earth inclined at 65° to the Equator and ranging between 169 and 315 km (105 and 196 miles) altitude. The historic flight was over in just 108 minutes, Gagarin finally parachuting back onto Soviet soil inside his spherical capsule.

Celebrated throughout the Communist world as a great technical and political victory, it was the final act which set the United States on course for a massive assault on the space frontier. On 25 May 1961, before a joint session of Congress, President John F. Kennedy set the scale of the challenge:

> I believe this nation should commit itself to achieving the goal, before the decade is out, of landing a man on the Moon and returning him safely to Earth. No single space project will be more impressive to mankind, or more important for the long-range exploration of space, and none will be so difficult or so expensive to accomplish.

It was some time, however, before NASA could match the capability of Korolev's big rocket. At first the best that could be managed were sub-orbital flights in small Mercury capsules launched from Cape Canaveral by Redstone rockets. On 5 May 1961 astronaut Alan Shepard made a 15-minute ballistic lob in the capsule 'Freedom 7' reaching a maximum height of 185 km (115 miles) and splashing down 478 km (297 miles) downrange. On 21 July Virgil 'Gus' Grissom made a similar flight in 'Liberty Bell'.

On 6 August the Soviet Union responded by launching Gherman Titov in Vostok 2, in which he completed 17 orbits – the first spaceflight to exceed one day.

The first American in orbit, Major John Glenn of the Marine Corps, eventually flew on 20 February 1962 in the nose of a modified Atlas ICBM. His Mercury capsule 'Friendship 7' made three circuits of the Earth in a flight which lasted 4 hr 55 min 23 sec.

Over the next few months the contest was to hot up dramatically. In August the Soviet Union had two manned spacecraft in orbit at the same time. At one stage Andrian Nikolayev in Vostok 3 passed within 5 km (3 miles) of Pavel Popovich in Vostok 4; they completed 64 and 48 orbits respectively. And on 16 June 1963 the first space woman, Valentina Tereshkova, took off in Vostok 6 in a joint flight with cosmonaut Valery Bykovsky in Vostok 5. It appears that she had been introduced into the cosmonaut team on the direct intervention of Premier Khrushchev, who saw in it another propaganda victory for the USSR. Her epic flight of 48 orbits lasted 70 hr 50 min.

In the meantime the last of six Mercury

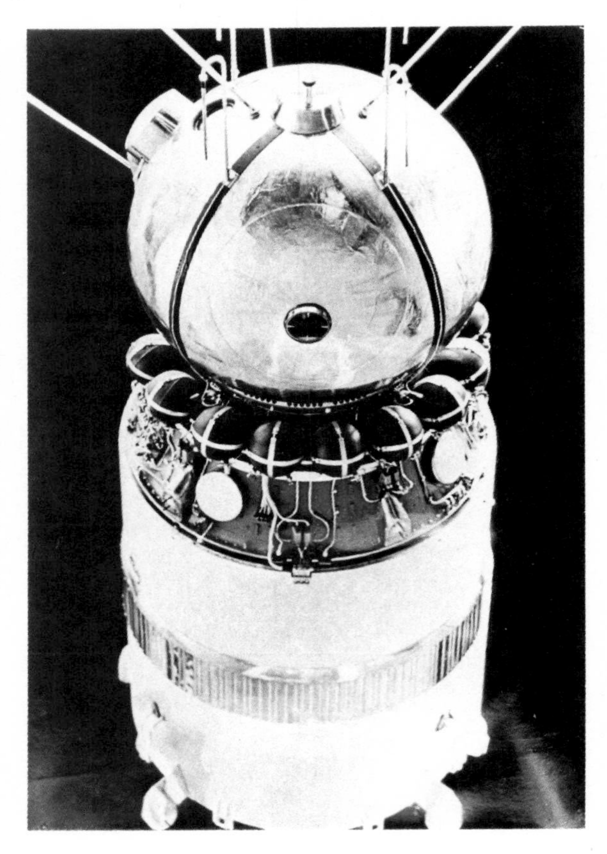

**Left** A Vostok spacecraft and final stage of its carrier rocket. The cosmonaut travelled inside the spherical capsule, which was attached to a service module supplying air and electricity. The service module was fitted with a retro-rocket, which brought the craft out of orbit at the end of the mission and was jettisoned before the capsule re-entered the atmosphere. The spherical re-entry capsule was 230 cm (90.5 in) in diameter, and Vostok's total weight, including capsule, was 4,725 kg (10,418 lb).

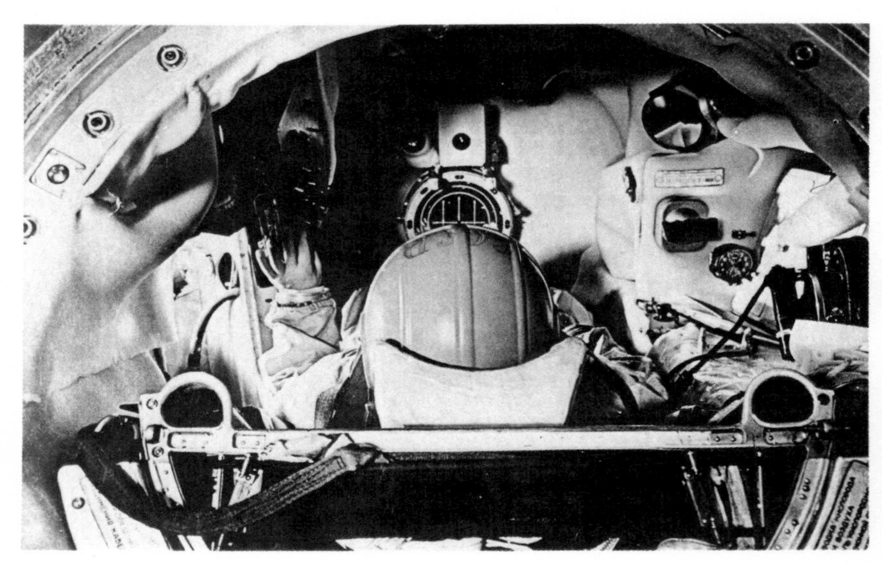

**Below** Interior of Vostok's spherical capsule, viewed through the ejection-seat hatch. The cramped quarters included an instrument panel with revolving Earth-globe, a switch panel, television screen, television camera, and a porthole incorporating an optical-orientation device.

Above The Mercury capsule, designed to carry one man, was launched into orbit by modified Atlas ICBMs. The overall length, including retro-rocket pack and aerodynamic 'spike', was about 7.9 m (26 ft).

Above Astronaut Ed White floats in space on 3 June 1965, secured to the Gemini 4 spacecraft by a 7.6 m (25 ft) umbilical through which he received oxygen. The first Russian spacewalk had occurred 11 weeks previously.

Below The Agena docking target only 13.7 m (45 ft) from the nose of the Gemini 8 spacecraft flown by Neil A. Armstrong and David R. Scott in March 1966. Such experiments gave Apollo astronauts valuable docking practice.

astronauts, Gordon Cooper, had completed a flight lasting 34 hr 20 min on 15–16 May 1963.

Russia now modified the Vostok into Voskhod, squeezing into its tiny capsule a three-man crew who completed 16 orbits on 12–13 October 1964. On 18 March 1965 came the first spacewalk, by Alexei Leonov from Voskhod 2.

## Orbital Rendezvous

Now began a series of spectacular American flights by two-man spacecraft launched by the Titan II. From Gemini 4 Ed White became the first American to 'walk' in space on 3 June 1965; and in Gemini 7 Frank Borman and James Lovell remained aloft for nearly two weeks during which, in a remarkable rendezvous manoeuvre, they approached within one foot of Gemini 6 flown by Walter Schirra and Tom Stafford.

Gemini 8 caused the first upset in space after Neil Armstrong and Dave Scott had docked in orbit with an unmanned Agena target satellite. Barely had the spacecraft come together than a faulty thruster caused the spacecraft to spin and tumble. Armstrong succeeded in extracting his ship and brought it down under manual control.

Later Gemini spacecraft were able to rendezvous and dock with Agena targets as a routine, and the Americans perfected the art of spacewalking. During the last Gemini mission, flown by James Lovell Jr and Edwin Aldrin on 11–15 November 1966, Aldrin was outside his craft for 129 minutes.

In the meantime, working closely with US industry, Wernher von Braun and his team at the Marshall Space Flight Center were already making headway with the huge Saturn rockets of the Apollo programme.

First to appear was Saturn 1 which had a cluster of eight H-1 liquid oxygen/kerosene engines in its first stage. Each engine delivered a thrust of some 85,277 kg (188,000 lb). The second stage had six liquid oxygen/liquid hydrogen RL-10 engines, each of which gave 6,804 kg (15,000 lb) thrust. The first of these vehicles was launched successfully in October 1961 with only the first stage live. The fifth flight, the first with a live second stage, in January 1964 put a 17,191 kg (37,900 lb) payload into Earth orbit.

By the eleventh launching von Braun had moved on to an uprated version, the Saturn 1B with lift-off thrust increased to 725,700 kg (1,600,000 lb). A new S-IVB top stage with a single liquid oxygen/liquid hydrogen J-2 engine produced a thrust of 90,720 kg (200,000 lb).

At the end of 1966 preparations were being made to send the first Apollo astronauts into Earth orbit to make a test of the Command and

**Apollo Project** The main picture shows the three-stage Saturn 5 launch vehicle with Lunar, Service, and Command modules. The total height from the top of the launch escape system to the base of the launch vehicle was 111 m (363 ft). Saturn 5's first stage, powered by five Rocketdyne F-1 engines, developed some 3.4 mn kg (7.5 mn lb) thrust to lift the spacecraft to an altitude of 61 km (38 miles) and 80.4 km (50 miles) downrange within 2½ minutes. Its speed at that point was 9,978 km/h (6,200 mph). The second stage (five Rocketdyne J-2s) burned for 6½ minutes, reaching an altitude of 185 km (115 miles) 1,504 km (935 miles) downrange and a speed of 24,940 km/h (15,500 mph). The third stage (one Rocketdyne J-2) placed the spacecraft into orbit at a speed of about 28,160 km/h (17,500 mph), later increasing to the lunar-transfer velocity of 40,225 km/h (25,000 mph). The first lunar-landing mission lasted about eight days.

Service Modules. The first spacecraft disaster struck on 27 January 1967 when astronauts Virgil Grissom, Ed White, and Roger Chaffee were sealed in an Apollo capsule carrying out tests on launch complex 34 at Cape Canaveral. They died in a fire which broke out in the capsule before they could be rescued. On 22 April 1967 it was the Soviet Union's turn. Vladimir Komarov blasted off from Tyuratam in the first manned test of a new spacecraft, Soyuz 1. After making 18 orbits he died when the parachute lines of his capsule became entangled and he plummeted to Earth.

Soyuz consisted of a three-man re-entry capsule, a service module with an engine for orbital manoeuvres, and an orbital module which was used mainly for scientific experiments but could also serve as a sleeping and dining compartment and as an airlock from which space-suited cosmonauts could emerge into space. Large solar panels extended like wings, generating electricity from sunlight.

While the Russians waited to re-test their Soyuz, the United States launched the first manned flight of the Apollo programme on 11 October 1968. Walter Schirra Jr, Donn F. Eisele, and R. Walter Cunningham completed 163 Earth orbits in a mission lasting 260 hr 9 min. A few days later at Tyuratam came the launch of the unmanned Soyuz 2, followed on 26 October by Soyuz 3 flown by Georgi Beregovoy, who manoeuvred near the unmanned craft during a mission which lasted 94 hr 51 min.

The 'Moon race' was now reaching its climax. In September the Russians had used one of their Proton rockets – three times as powerful as the Vostok launcher – to send an unmanned Zond 5 on a looping trajectory around the Moon and back to Earth. The re-entry capsule, big enough to carry a man, contained turtles and other biological subjects, and Western tracking stations picked up taped voice transmissions from the craft. At the end of the seven-day flight the capsule splashed down in the Indian Ocean. It was picked up by a Soviet research ship, taken to Bombay, and airlifted back to the

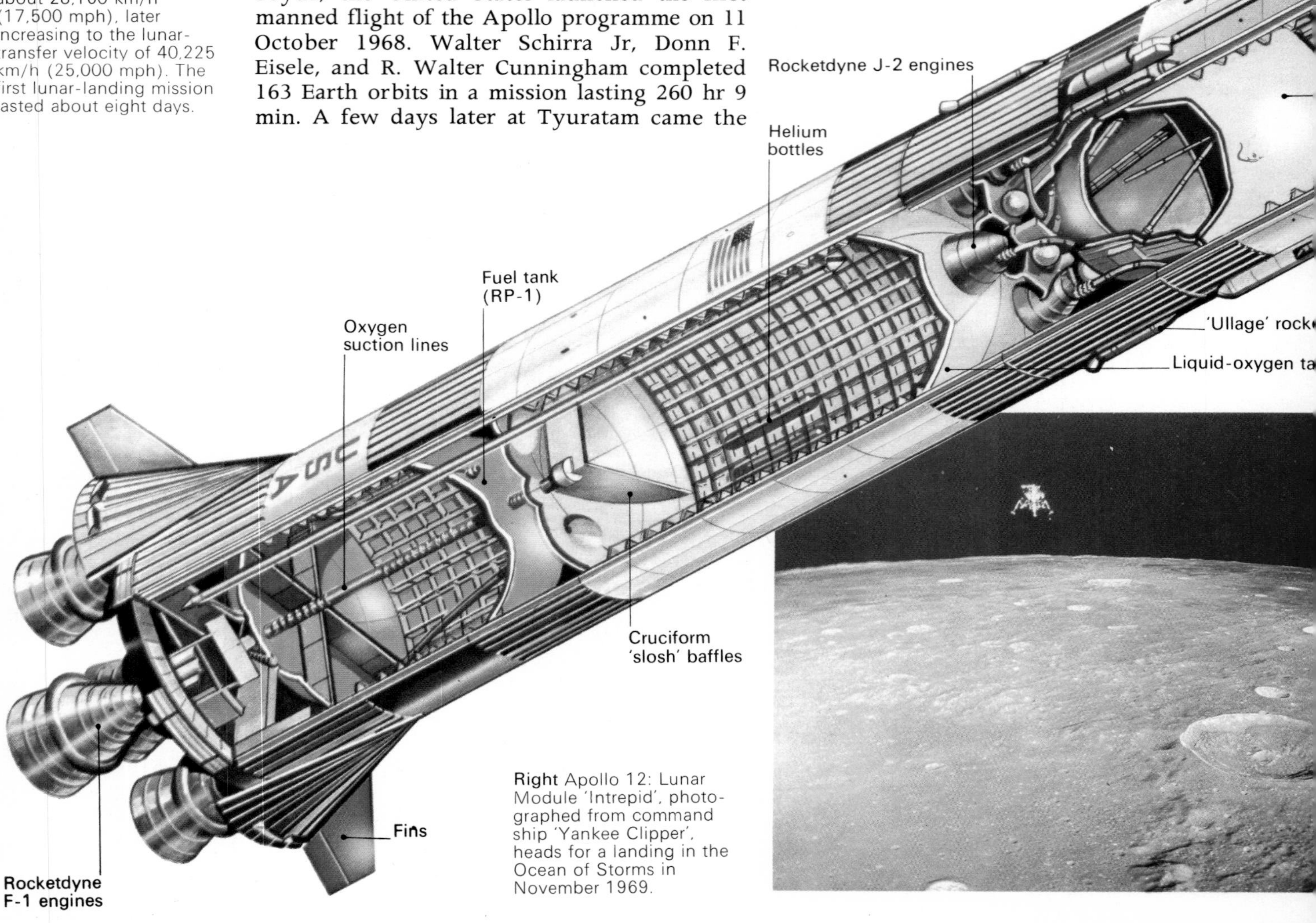


**Right** Apollo 12: Lunar Module 'Intrepid', photographed from command ship 'Yankee Clipper', heads for a landing in the Ocean of Storms in November 1969.

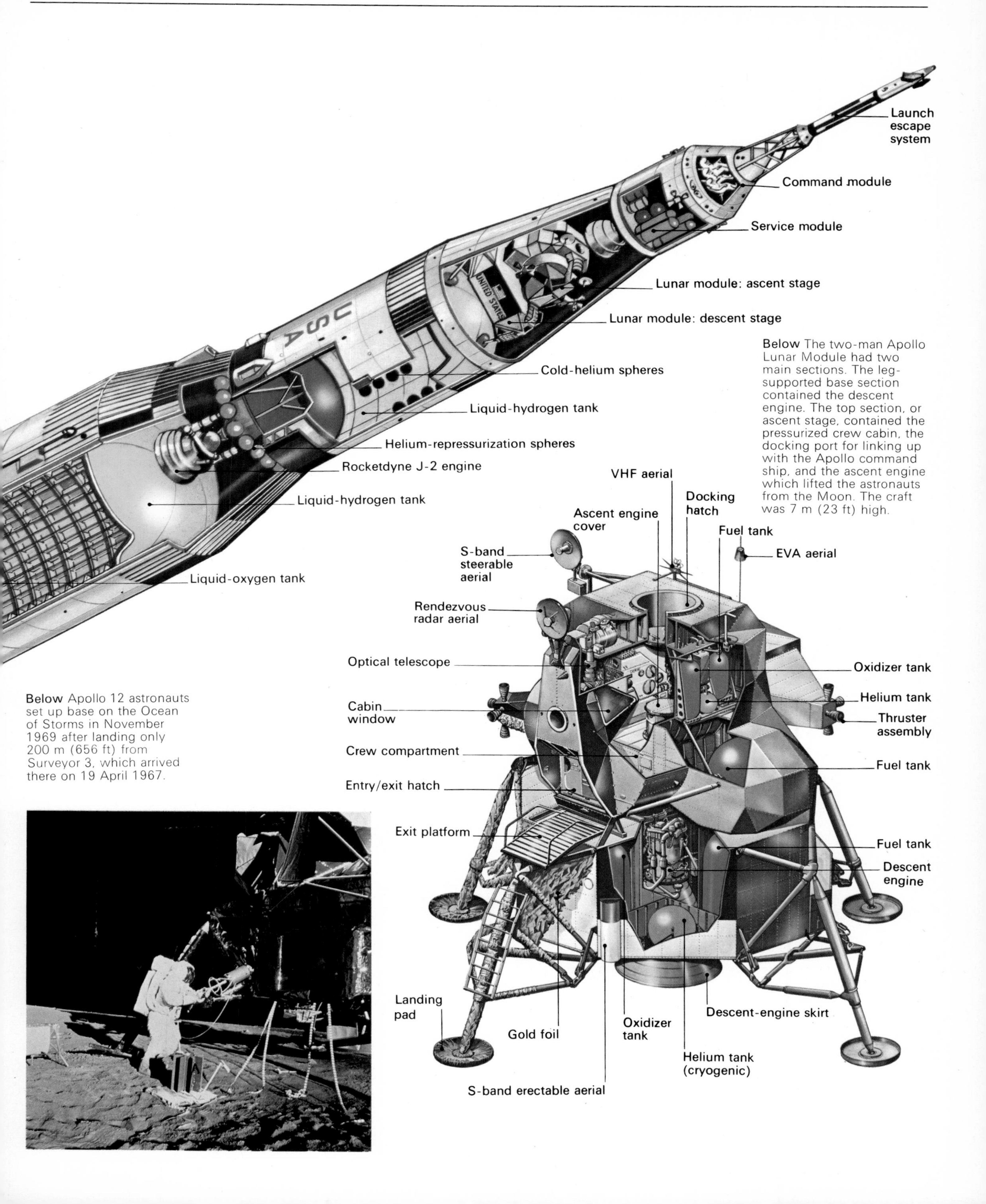


**Below** The two-man Apollo Lunar Module had two main sections. The leg-supported base section contained the descent engine. The top section, or ascent stage, contained the pressurized crew cabin, the docking port for linking up with the Apollo command ship, and the ascent engine which lifted the astronauts from the Moon. The craft was 7 m (23 ft) high.

**Below** Apollo 12 astronauts set up base on the Ocean of Storms in November 1969 after landing only 200 m (656 ft) from Surveyor 3, which arrived there on 19 April 1967.

Soviet Union. Two months later an improved craft, Zond 6, made a similar flight but returned directly to the Soviet Union by making an aerodynamic 'skip' as it re-entered the atmosphere.

The Zonds' successes in rounding the Moon undoubtedly played a part in speeding the decision to launch Apollo 8, fully manned, into orbit around the Moon. On 21 December 1968 Frank Borman, James A. Lovell, and William Anders blasted into space from Cape Canaveral in the nose of a Saturn 5 to complete a copybook flight which included 10 lunar orbits. They succeeded in televising spectacular views of the lunar surface to Earth, and returned in triumph.

In the meantime the Russians were still trying desperately to perfect the techniques by which spacecraft could be docked together in Earth orbit. The first experiment of this kind had been made in 1967 between two unmanned spacecraft, Cosmos 186 and 188. In January 1969 two similar manned craft, Soyuz 4 and 5, were joined for 4 hr 35 min, creating a rudimentary space station. Vladimir Shatalov, the active docking partner, flew alone in Soyuz 4; Soyuz 5 was occupied by Boris Volynov, Alexei Yeliseyev, and Evgeny Khrunov. After the ships were joined, Khrunov and Yeliseyev climbed out from the orbital module of Soyuz 5 wearing space suits. After spending nearly an hour outside doing unspecified 'engineering experiments' they transferred to Soyuz 4 and returned to Earth with Shatalov.

One last test flight remained before Apollo astronauts could land on the Moon. In May 1969 three Apollo 10 astronauts, Tom Stafford, John Young, and Eugene Cernan, achieved lunar orbit and two of them separated in the Lunar Module to swing within 15,240 m (50,000 ft) of the surface.

## Men on the Moon

The day of the lunar-landing mission dawned bright and clear at Cape Canaveral on 16 July 1969, when Neil Armstrong, Edwin Aldrin, and Michael Collins blasted off in their Apollo 11 spacecraft. Three days previously, in a last defiant attempt to steal the United States' thunder, the Soviet Union had launched a robot lander. The spacecraft, Luna 15, was being manoeuvred in lunar orbit as the Americans were attempting to land. It finally crashed on the Sea of Crises some two hours before the Apollo astronauts blasted off from their landing site in the Sea of Tranquillity.

The American landing had been a triumph in every way. While Astronaut Collins waited 'upstairs' in the mother ship making observations, Armstrong and Aldrin flew their Lunar Module 'Eagle' to a safe landing after negotiating a boulder field. At 4.18 p.m. EDT Armstrong was heard calmly reporting: 'Houston, Tranquillity Base here – the Eagle has landed.'

Armstrong placed his left foot on the Moon at 10.56 p.m. saying: 'That's one small step for a man, one giant leap for mankind.' Eighteen minutes later Aldrin joined him on the dusty lunar surface. The two men tested the soil, planted the 'Stars and Stripes', set out the first Apollo scientific station, and collected 20.8 kg (46 lb) of Moon samples. Their stay lasted 21 hr 36 min.

Left behind on the Moon was a metal plaque affixed to the forward leg of the descent stage of the Lunar Module. The inscription reads: 'Here Men from the Planet Earth First Set Foot Upon the Moon, July 1969 AD. We came in Peace for All Mankind.'

A month later the Russians made their third circumlunar flight with an unmanned craft, Zond 7. In November Charles Conrad Jr, Alan Bean, and Richard Gordon Jr blasted off from Cape Canaveral, and had their rocket struck by lightning – happily without disastrous consequences. Conrad and Bean went on to make a precise landing in the Ocean of Storms only 198 m (650 ft) from where the Surveyor 3 robot had come down $2\frac{3}{4}$ years before. They set up their ALSEP – Apollo Lunar Surface Experiment Package – research station, the first to be powered by a nuclear generator, and walked over to inspect the Surveyor, removing from it the television camera and other small items for examination on Earth. This second Apollo landing produced 34 kg (75 lb) of Moon samples.

So far the Apollo landings had gone with remarkable ease. On 13 April 1970 NASA was celebrating the smooth beginning of the Apollo 13 mission, flown by James Lovell Jr, John Swigert, and Fred Haise, when potential disaster struck some 329,845 km (205,000 miles) from Earth. What seemed at first to be a meteoroid impact turned out to be an explosion of one of the fuel-cell tanks in the Service Module, which deprived the ship of electrical power. The mission, which had expected to achieve a landing in the Fra Mauro region of the Moon, was immediately abandoned. Houston and the Apollo contractors worked night and day to devise a rescue scheme.

The solution which brought the astronauts safely home after they had rounded the Moon was to use the attached Lunar Module as a 'lifeboat', drawing upon its oxygen and electrical supplies and using its engine for necessary power manoeuvres. Just before the crew made their return to the Earth's atmosphere the damaged Service Module and the Lunar Module were jettisoned. To everyone's relief the Apollo 13 capsule made a safe re-entry, splashing down in the Pacific Ocean.

## Exploring the Lunar Surface

While the United States worked to improve the safety of Apollo mooncraft, the Soviet Union brought off the triumph that had eluded them during the Apollo 11 mission. In September 1970 the automatic spacecraft Luna 16 went into lunar orbit, and then descended onto the Sea of Fertility. A drilling device extracted a 100 g (3.5 ounce) core sample and automatically loaded it into a return capsule. The capsule was fired back and landed by parachute on Soviet soil.

Another spectacular advance came on 17 November 1970, when the Russians landed Luna 17 in the Sea of Rains. This particular robot extended ramps from which rolled something that seemed to come straight from the pages of Jules Verne. It was the electric powered eight-wheeled rover Lunokhod 1, steered by a group of 'drivers' seated at a console in the Soviet Union using a two-way radio/television link. The robot mooncar made a remarkable journey of 10,542 m (6.5 miles), during which it tested the properties of the lunar soil at various stopping points.

It was now the turn of Alan B. Shepard Jr, Stuart A. Roosa, and Edgar D. Mitchell to take Apollo 14 to the Moon. The launching took place at Cape Canaveral on 31 January 1971. Shepard and Mitchell made the first exploration of the lunar highlands in the area of Frau Mauro and used a two-wheeled pullcart to transport their tools and equipment. It was on this trip that Shepard tried his famous golf shot, driving a ball into a moondust 'bunker'. The astronauts were on the Moon for nearly 34 hours, during which they gathered 41.5 kg (91.5 lb) of rock and soil.

Next to leave, on 26 July 1971, were Apollo 15 astronauts David R. Scott, Alfred M. Worden, and James B. Irwin. This time the destination was the Hadley-Apennines, and the Lunar Module was equipped for a longer stay. After touching down, Scott and Irwin unpacked from the base of the Lunar Module a battery-powered car called a Lunar Rover in which they made long excursions from the landing site, collecting 76.5 kg (168.6 lb) of samples. They were on the Moon for a record time of 66 hr 55 min. Worden in the orbiting command ship made observations with a battery of scientific instruments in the newly installed scientific instrument module, and when the landing party were back on board a small sub-satellite was released which continued to transmit scientific data from lunar orbit for more than a year.

The Russian eight-wheeled, electrically driven robot Lunokhod 1 landed on the Sea of Rains on 17 November 1970. Clearly visible at left are the television 'eyes' which enabled controllers nearly a quarter of a million miles away in the Soviet Union to steer the vehicle by radio. The vertical cone is an omnidirectional aerial, and to its right, set at an angle, is a narrow-beam aerial. The lid of Lunokhod's body opened to expose solar cells, which re-charged the robot's chemical batteries. The vehicle tested the lunar soil at various points and sent data to Earth by radio, continuing its research programme for more than 10 months.

Apollo 16 blasted off on 16 April 1972 with John W. Young, Thomas K. Mattingly II, and Charles M. Duke Jr. This was the first mission to visit a really elevated part of the Moon, the Descartes region. Once again a landing party, Young and Duke, had the use of a Lunar Rover. The astronauts lived on the Moon for 71 hr 6 min, during which three separate excursions were made in the Rover totalling 20 hr 15 min, the longest period men have remained outside their Lunar Module.

Apart from setting up another ALSEP station, they erected the first astronomical observatory on another celestial body, a Far Ultra-Violet camera/spectrograph, and collected 94.5 kg (208.3 lb) of soil and rocks. Before leaving lunar orbit the astronauts released another sub-satellite.

Apollo 17, the last in the series, flown by Eugene A. Cernan, Ronald E. Evans, and Harrison H. Schmitt, was fittingly spectacular. Launched during the night of 7 December 1972, it lit the Florida sky.

The landing was made in the Taurus Littrow region of the Moon, which Cernan and Schmitt – the latter a qualified geologist – explored with the aid of a third Lunar Rover. The men set out the fifth ALSEP station, made three separate excursions from the landing craft, and collected a record 110 kg (243 lb) of lunar samples. They were on the Moon for nearly 75 hours.

The Soviet Union continued to exploit robot techniques, insisting that this was far cheaper than landing men. However, Luna 18 crashed in a highland area on the boundary of the Sea of

Engineering model of the 19,000 kg (18.7 tons) Salyut space station at the Yuri Gagarin Cosmonauts' Training Centre. Above the technician on the ground can be seen the forward docking port, the airlock hatch on the transfer compartment, and docking aerials. Projecting from the body are three large extensible solar panels, which generate electricity from sunlight.

Fertility in September 1971 after making 54 revolutions of the Moon. The following February a sister craft, Luna 20, touched down in a similar mountain region of the same Sea, drilled into the surface, and took off with a small core sample. Luna 21, which made a soft landing in January 1973, discharged a second Lunokhod rover inside the rim of the 55 km (34 miles) diameter Le Monnier crater in the Sea of Serenity. The rover analysed the soil in different places and tested its properties. Another soil sampler, Luna 23, was damaged on landing in November 1974 and could not return a sample. Then, on 18 August 1976, Luna 24 dropped down onto the Sea of Crises to retrieve a core sample from a depth of some 2 m (6.5 ft).

What were the principal findings? Although the moon rocks contain the same chemical elements as those found on Earth, there are important differences. All the rocks are igneous, having been formed by the cooling of molten lava. The biggest difference is that they contain no water at all, whereas almost all terrestrial rocks have at least one or two per cent of water.

In general the light-coloured highlands are richer in calcium and aluminium than the dark-coloured *maria* – so-called 'seas' – which contain more iron, titanium, and magnesium. A random analysis of soil from the Apollo 11 landing site in the Sea of Tranquillity showed the following percentages: oxygen 40, silicon 19.2, iron 14.3, calcium 8.0, titanium 5.9, aluminium 5.6, and magnesium 4.5. In some of the samples were small glass spherules probably caused by the impact of meteoroids.

A measure of the success of the Apollo missions was that instruments in the ALSEP stations were still measuring meteoroid impacts, 'moonquakes', magnetism, and the heat flowing from the Moon until they were switched off in 1977.

## Orbiting Space Stations

With the Moon race ended, the Russians began to consolidate their programme of operations in Earth orbit. Soyuz 6, 7, and 8 made a group flight in which seven cosmonauts participated. In Soyuz 6 experiments were made into the welding of metals. Soyuz 9, flown by Andrian G. Nikolayev and Vitaly Sevastyanov in June 1970, remained in orbit for 424 hr 59 min, exceeding the record set by Gemini 7.

On 19 April 1971 the Salyut 1 space laboratory was blasted into orbit, without occupants, on the nose of a Proton D-1-H from Tyuratam. Five days later three cosmonauts in Soyuz 10 docked with the station for $5\frac{1}{2}$ hours but did not go aboard. The following June, however, a successful boarding was carried out by cosmonauts Georgi I. Dobrovolsky, Vladimir Volkov, and Viktor I. Patseyev, who joined the station in Soyuz 11 to conduct medical, biological, and engineering experiments. After spending 22 days in Salyut 1, the crew were returning to Earth when the cabin of their Soyuz ferry was accidentally depressurized. The capsule soft-landed automatically, but when its hatch was opened the men were found to be dead.

This unhappy accident caused the Russians to reduce the number of cosmonauts aboard Soyuz craft from three to two and to introduce extra safety equipment.

Despite sporadic problems with the docking systems of their spacecraft, by the mid-1970s they had set up a regular ferry service to a series of Salyut stations which can operate both manned and unmanned under command from Russian mission control. Salyut 6 in 1977 was the first to have two docking ports so that two ferries could dock at the same time for replacing crews, carrying out rescue operations, and delivering fuel, air, foodstuffs and equipment.

The United States, too, had turned to space-station activities in their Skylab programme. This had involved converting an S-IVB rocket stage into an orbiting workshop with docking facilities for Apollo spacecraft and an Apollo Telescope Mount (ATM) for making detailed studies of the Sun. Also aboard were cameras and other instruments for studying Earth's natural resources, and a number of professional and student experiments related to human physiology, biology, chemistry, and the processing of materials under conditions of microgravity.

When Skylab was launched by a two-stage Saturn 5 on 14 May 1973 it was badly damaged as it left the atmosphere. The meteoroid shield was torn away, carrying with it one of the station's extensible solar 'wings'. The remaining solar wing failed to open in orbit. Deprived of much electrical power and unprotected from the heat of the Sun because of loss of the shield, the station was overheating badly.

When the first boarding party of astronauts reached the station they brought with them

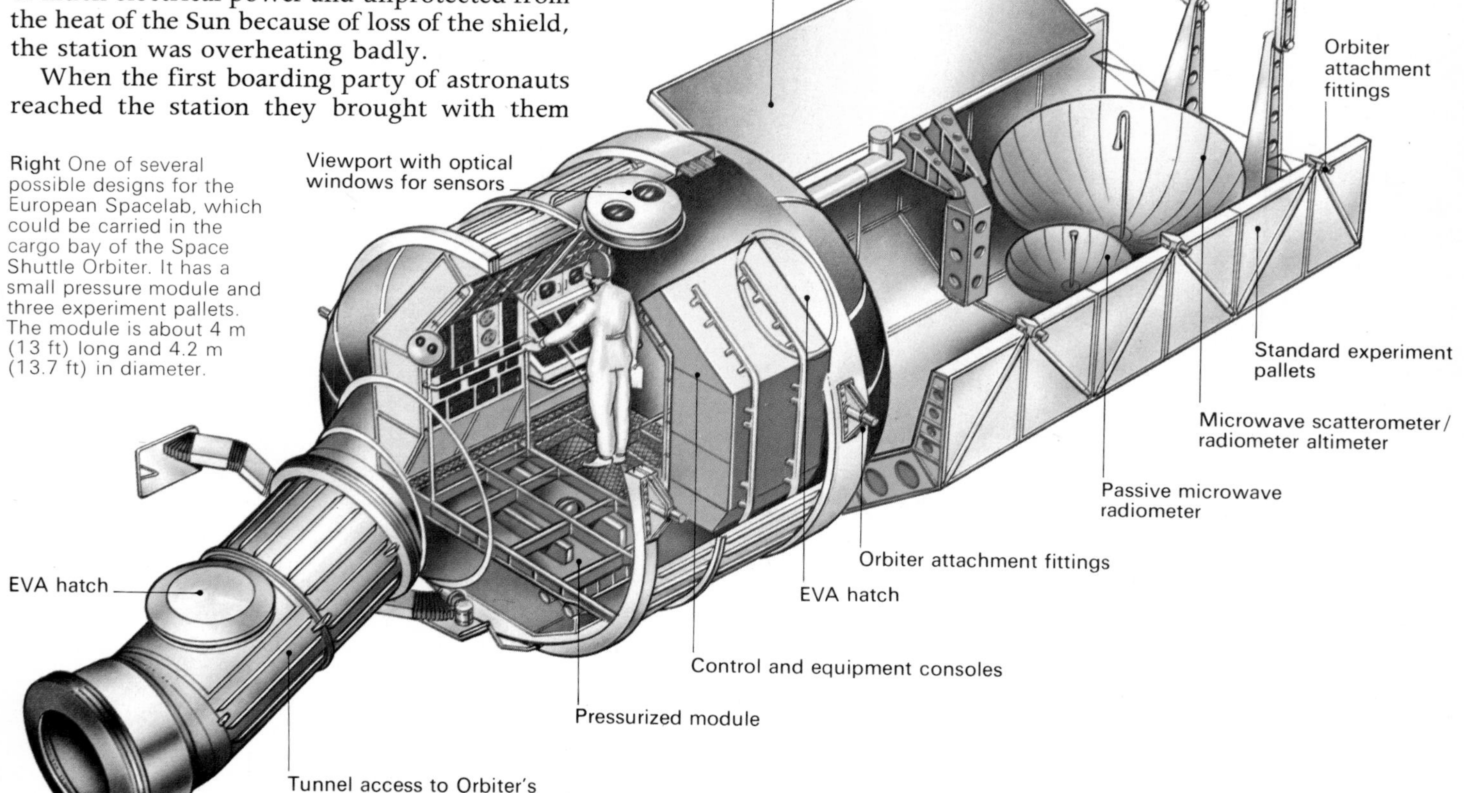


**Right** One of several possible designs for the European Spacelab, which could be carried in the cargo bay of the Space Shuttle Orbiter. It has a small pressure module and three experiment pallets. The module is about 4 m (13 ft) long and 4.2 m (13.7 ft) in diameter.

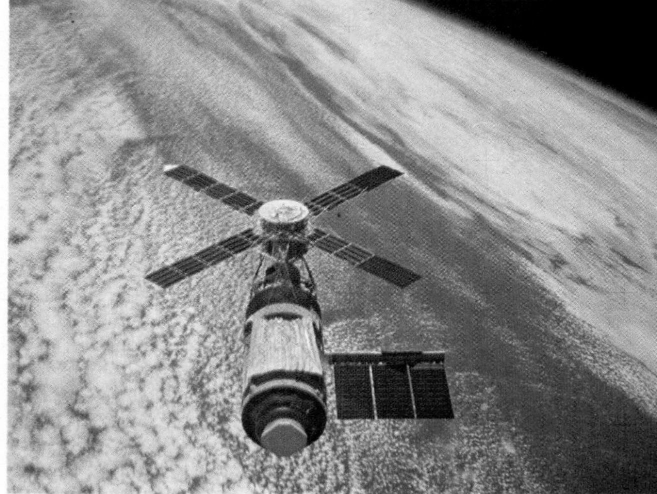

**Near right** Handshake in orbit: Soyuz 19 spacecraft seen from the window of Apollo 18 in July 1975. After docking, the crews visited each other's spacecraft.

**Far right** The 78,400 kg (77 ton) Skylab space station was damaged when it was launched on 14 May 1973. This picture shows Skylab after boarding parties had repaired and opened its one remaining solar wing and erected sunshades in place of the missing meteoroid shield.

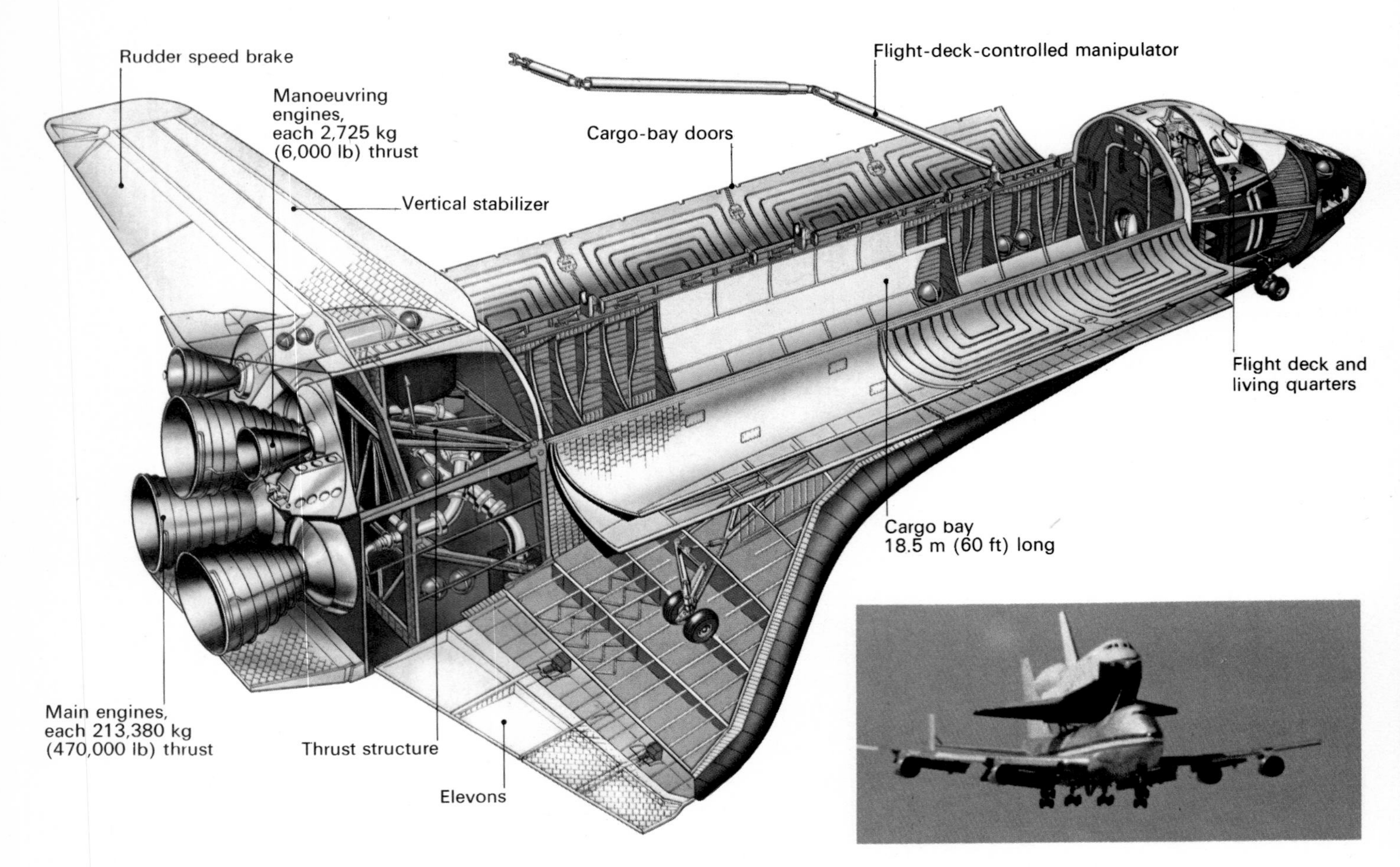


**Above** NASA's Space Shuttle Orbiters can each carry about 29,500 kg (65,000 lb) of cargo and up to seven crew members and scientists into orbit, and return 14,515 kg (32,000 lb) of cargo to Earth. Each is intended to fly at least 100 missions. The Orbiter is 37.1 m (122 ft) long, with a wing span of 23.8 m (78 ft); the landing weight is 84,800 kg (187,000 lb). At lift-off the Orbiter is assisted by two large solid-rocket boosters which each develop a thrust of 1.2 mn kg (2.6 mn lb). They are jettisoned at an altitude of some 46 km (28 miles) and are recovered by parachute. A large external fuel tank is jettisoned just before the Orbiter goes into orbit at a height of 185 km (115 miles).

**Right** The Orbiter was first air-launched from a Boeing 747 'mother' to test its handling at low speeds. This picture shows the combined vehicles shortly after take-off. The 747 is also used to ferry the Orbiter to launching sites.

various tools which had been hastily improvized for making repairs. After entering Skylab they deployed a 'parasol' from the inside through a scientific airlock to cover part of the exposed area, which helped to cool the station. Later, using special cutting tools, they manually deployed the stuck solar wing during a spacewalk.

During two further missions, lasting 59 and 84 days respectively, astronaut teams carried out more repairs and completed a full programme of experiments, proving beyond doubt the value of man in space.

One last flight was made by an Apollo spacecraft. That was in July 1975 during the celebrated link-up with a Russian Soyuz spacecraft 225 km (140 miles) above the Earth. Not only did the two nations collaborate in developing an international docking module, so that crews could be exchanged in space, but spacemen visited each other's countries for training.

Technically and politically this was a major breakthrough and showed that suitably designed spacecraft could carry men and cargo to the space stations of another country. It also showed that spacemen could help each other in an emergency. The Soviet Union and United States have since discussed the possibility of a Space Shuttle docking with a Salyut space lab.

## The Space Shuttle

The Shuttle has been described as a 'maid-of-all-work' space freighter. Not only will it fly scientists into orbit for research purposes but it will launch satellites, space probes, or any other cargo weighing up to 29,484 kg (65,000 lb) that will fit into an 18.3 m (60 ft) long by 4.6 m (15 ft) wide cargo bay. It can be used to retrieve or repair satellites which have failed in orbit, or take into orbit sections of space stations or spaceships for assembly.

An early plan was to prevent the large Skylab space station from falling to destruction in 1979–80 as it spirals closer to Earth under the effects of air friction. A space robot called a Teleoperator Retrieval System (TRS) fitted with a television camera and a docking probe would be released from the Shuttle's cargo bay by an astronaut on the flight deck. Working in conjunction with a television monitor screen, the astronaut then would proceed to steer the TRS into the axial docking port of Skylab so that its low-thrust engine could push the station into a higher orbit, or send it spinning to destruction over the open sea.

When satellites or other cargo are not carried, the Shuttle can fly into orbit with

Spacelab, a large and fully-equipped laboratory built in Europe, in which scientists can work in shirt-sleeve comfort. Unlike early space labs, such as Skylab and Salyut, which had to be abandoned after use, Spacelab returns to Earth each time it is used, for it never leaves the Shuttle's cargo bay. When the spaceplane arrives in orbit the mission specialist on the flight deck opens the cargo-bay doors and Spacelab and its research instruments are exposed to the vacuum of space. To commute to their lab bench, the scientists aboard the Shuttle enter an interconnecting tunnel and emerge directly into the environmentally controlled laboratory. When their work is done, the scientists return to their accommodation in the Shuttle, the cargo-bay doors are closed, and the spaceplane is ready to fly home. After landing, the Lab will be lifted out and made ready for another flight with different experiments.

Although factories in space may seem fantastic, it is likely that we shall see important developments in this direction well before the end of the century. Small electric furnaces aboard Skylab, and the ASTP Apollo spacecraft which docked with Soyuz, showed the promise of making entirely new alloys and very pure crystals under micro-gravity conditions. The manufacture of ultra-pure vaccines and serums holds out equal promise.

New alloys impossible to produce on Earth, because materials of different properties, such as metal and glass, will not mix under normal gravity conditions, could be made in space. They may include new superconducting materials which, when cooled, lose virtually all resistance to electric current. Indeed, there seems to be no practical limit to the opportunities of the space-age industrial revolution over the next few decades.

## Habitats in Space

The idea that one day mankind might expand its Earthly domain by adding space colonies in perpetual orbit has also aroused wide interest. It began with a teasingly simple question which Professor Gerard K. O'Neill put to a session of Princeton University's first-year physics course in 1969. 'Is a planetary surface the right place for an expanding technological civilization?' After long debate – and much to the surprise of students and tutor – the conclusion was: No!

The argument that emerged was broadly as follows. The Moon should be mined for resources and materials, which would be launched into space by electromagnetic catapult and used in the construction of self-contained habitats established at semi-stable points in the Moon's orbit. Thousands of people would live in these artificial worlds, which would be fashioned internally to resemble an Earth's landscape of hills, meadows, lakes and rivers – with trees, flowers, agricultural crops, farm animals, and birds. Natural sunshine would be brought inside by external mirrors. Wastes, air, and water would be recycled. The central structure would spin on its axis, so that artificial gravity would enable the colonists to keep their feet firmly planted on the ground.

Professor O'Neill looks upon the space habitats as a means of exploiting industrial techniques on a very large scale while avoiding terrestrial pollution. One of their most important roles, in our world of dwindling fuel resources, could be to tap the energy of the Sun. This would involve construction, in geostationary orbit, of city-sized solar-power stations capable of tapping the radiant energy emitted by the Sun and beaming it to Earth in the form of microwaves.

It would be very difficult to build such vast power stations by launching them, piece by piece, from the Earth's surface. Moreover, this method would involve using enormous quantities of our most valuable terrestrial resources, such as titanium and aluminium. Whence O'Neill's idea of using raw materials known to exist on the Moon and processing them in the orbiting habitats.

In this and other ways planet Earth, which is steadily exhausting its fund of resources, may be launched into a bold new age. If such visions appear to owe more to science fiction than to hard reality, it should be remembered that, less than three decades ago, the idea of men walking on the Moon seemed to most people little more than a fanciful dream.

**Below** A study at NASA's Ames Research Center, Santa Clara (Cal.), has inspired this artist's conception of a 21st-century space habitat for 10,000 people. The colonists – members of a space-engineering task force – would live on the landscaped inner surface of the central sphere, which would be almost a mile in circumference and would rotate to provide Earth-like gravity conditions. The tyre-shaped toruses on each side of the sphere would be devoted to agriculture. Between the toruses and the sphere are solar mirrors to provide heat and light; the white and orange discs encircling the sphere are also solar mirrors. At the extreme ends of the habitat are micro-gravity industrial areas and docking points for space vehicles. The large rectangular panels nearby are radiators for the disposal of waste heat.

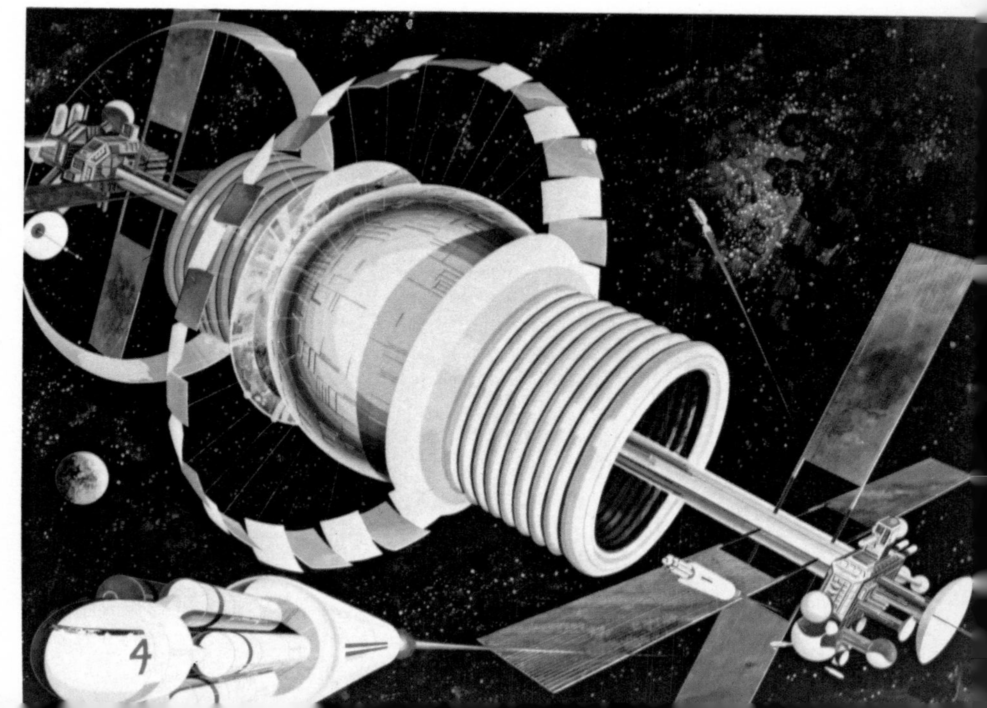

# Index

Page numbers in *italics* refer to captions.

# Acknowledgments

The publishers thank the following organizations and individuals for their kind permission to reproduce the photographs in this book:

Aerofilms Limited 211; Aerophoto-Schiphol BV 213 below; Air Portraits 141; Basil Arkell 191, 192, 193, 194, 195 above and below, 197 below, 198 centre, 205 below, 215; Associated Press 4–5, 48–9, 61 below, 103 above, 246 below; Bayer Armeemuseum 114; Bell Helicopters (USA) 196 above; Best-Devereux & Co., 182 below; Blitz Publications Limited 99 left, 148–9 above; Boeing Airplane Company USA 59, 73, 94–5 below, 102 below, 153 below, 169 below, 198 above; British Aerospace (Aircraft Group) 54, 75 above left, 75 below, 88, 125 above, (Filton Division) 56–7, 94–5 above, (Guided Weapons Division) 123 above, (Manchester Division) 139 centre; British Airways 151 below right; Kenneth J.A. Brookes 18 below, 55, 158 above; David Calkin 18 above; Canadair 207 below; CEGB-BBC (Peter Cook) 31 above, 32–3, 34 above, 35 above and below; Michel Colomban 182 centre; Dallas Airport Authority 213 above; de Havilland Canada 203; Deutsches Museum 227 right; Mary Evans Picture Library 10, 17, 20 above, 22, 25, 41 below, 42, 46, 47, 98; Hugh Field 172 above and below, 176 below; *Flight International* 82, 109 above, 173, 174, 175 above and below, 177, 179 above left, 185 above and below, 189, 218 above, 219 above; Fokker-VFW International (Netherlands) 85; Kenneth W. Gatland 247; General Dynamics – Fort Worth (USA) 110; James Gilbert 60, 78, 104, 139 below, 181; Goodyear Tyre & Rubber Company (USA) 75 above right; Robert Harding Associates 50 left, 180; Professor Clive Hart 11, 13, 15, 16, 19 below, 29 above, centre and below, 37, 41 above; Hawker Siddeley Aviation 80, 139 above; Paul Hilder 28; Michael Holford Library 224; Michael Hooks 66–7, 83, 118, 123 below, 124 below right, 130–1, 136, 162, 205 above, 207 above, 212, 220; Angelo Hornak 226; Imperial War Museum (London) 101 above, 102 above, 112, 113 above, 117 below, 120, 121 above, 122 above and below, 126 below, 132, 134 below; Litton Systems Limited (USA) 219 centre; Lufthansa Airlines 148 below, 152; Luftschiffbau Zeppelin GmbH 146; Marconi Radar Limited 221 below; Martin Marietta Corporation (USA) 107, 118 above, 121 below, 135 above; McDonnell Douglas (UK) 86–7, 159 below; Messerschmitt-Bölkow-Blohm 196 below, 231 above left; Ministry of Defence-RAF (Crown Copyright) 103 centre, 129 below; J.G. Moore Collection 12, 19 above, 96–7, 109 below, 143 below, 206 below; NASA 62, 222–3, 235 above, 236 above left and right, 239 above right and below, 240 below, 241 below left, 245 below left and right; Novosti Press Agency 127, 230 above, 233 above left and right, 238 above and below, 243; Photri 228 above and below, 230 below, 231 above right, 234 below left, 239 above left; Ken Pilsbury 50 right, 51; Plessey Radar Limited 217 above, 218 below, 221 above; *Radio Times* Hulton Picture Library 147 above and below; Rawlings Photography 179 below; Redifon Simulation Limited 143 above right; John Rigby 171 above; Herbert Rittmeyer 118 below; Bruce Robertson 135 below; Rolls-Royce Limited 89; Royal Aeronautical Society 31 below; Royal Aircraft Establishment, Farnborough 57; The Royal Artillery Institution, Woolwich 225; Science Museum (London) 8–9, 14, 23, 24, 30, 38, 39, 40, 43, 44–5, 45, 204; Snark International front and back end-papers, 21; Spectrum Colour Library 20 below, 34 below, 179 above right; John Stroud 155, 156 above, 158 below, 161, 165, 167 below, 170, 171 below; Süddeutscher Verlag 61 above, 99 right, 134 above, 190, 202; Tass 232 below, 234 below right, 244; John W.R. Taylor 6–7, 63 above, 68–9, 69, 84, 101 below, 105, 115, 119 above, 129 above, 142 above left, 159 above, 163, 164 above, 168, 169 above left, 169 above right, 182 above, 186–7, 197 above; Teledyne Ryan Aeronautical 103 below; United Airlines Incorporated (USA) 153 above; US Air Force 119 centre, 198 below; Vickers Limited 63, 133, 157 below, 164 below, 183; Michael Young 117 above; Zefa 208–9.

ILLUSTRATORS

Mike Badrocke, Art Bowbeer, Tim Hall, Eric Jewel, Frank Kennard, John Marsden, and John W. Wood.